POWDER TECHNOLOGY AND MULTIPHASE SYSTEMS

Gas Permeametry and Surface Area Measurement

Ellis Horwood Series in
CHEMICAL ENGINEERING
Series Editor: Dr DAVID SHARP, OBE, former General Secretary, Institution of Chemical Engineers, London

THERMODYNAMICS FOR CHEMISTS AND CHEMICAL ENGINEERS
M.H. CARDEW
ENVIRONMENTAL BIOTECHNOLOGY
C. F. FORSTER and D.A.J. WASE
POWDER TECHNOLOGY AND MULTIPHASE SYSTEMS:
Gas Permeametry and Surface Area Measurement
G. J. I. IGWE
MAJOR CHEMICAL HAZARDS
V. MARSHALL
BIOLOGY AND BIOCHEMISTRY FOR CHEMISTS AND CHEMICAL ENGINEERS
W. J. MITCHELL and J. C. SLAUGHTER
INSTRUMENTATION AND AUTOMATION IN PROCESS CONTROL
M. J. PITT and P. E. PREECE
MASS TRANSFER
J. A. WESSELINGH and R. KRISHNA
BIOLOGICAL TREATMENT OF WASTE-WATER
M. WINKLER

Ellis Horwood Limited also publishes related titles in
APPLIED SCIENCE AND INDUSTRIAL TECHNOLOGY
INDUSTRIAL CHEMISTRY
BIOLOGICAL SCIENCES

POWDER TECHNOLOGY AND MULTIPHASE SYSTEMS

Gas Permeametry and Surface Area Measurement

GODWIN J. I. IGWE Dipl.Ing., M.Phil., Ph.D.
Department of Chemical Engineering
College of Engineering, Prairie View A&M University, Texas, USA

ELLIS HORWOOD
NEW YORK LONDON TORONTO SYDNEY TOKYO SINGAPORE

First published in 1991 by
ELLIS HORWOOD LIMITED
Market Cross House, Cooper Street,
Chichester, West Sussex, PO19 1EB, England

A division of
Simon & Schuster International Group
A Paramount Communications Company

Typeset in Times by Ellis Horwood Limited
Printed and bound in Great Britain
by Bookcraft (Bath) Limited, Midsomer Norton, Avon

British Library Cataloguing in Publication Data

Igwe, G. J. I. (Godwin Joseph Ifeanyichukwu) *1946–*
Powder technology and multiphase systems.
1. Powders. Surfaces. Structure & properties. Determination
I. Title
620.43

ISBN 0–13–680885–9

Library of Congress Cataloging-in-Publication Data

Igwe, Godwin J. I. (Godwin Joseph Ifeanyichukwa), 1946–
Powder technology and multiphase systems: gas permeametry and surface area measurement / Godwin J. I. Igwe.
p. cm. — (Ellis Horwood series in chemical engineering)
Includes bibliographical references and index.
ISBN 0–13–680885–9
1. Powders — Surfaces. 2. Surfaces — Areas and volumes. 3. Gas flow. I. Title. II. Series.
TA418.7.I39 1991
620′.43–dc20

90–22506
CIP

Table of contents

Foreword

The rapid growth of interest in powders and their surface properties in many diverse industries now requires improved apparatus for meeting the increasing demand for rapidity, accuracy, and reproducibility of results and quality control of manufacturing and sizing processes. The originality of this improved design of a gas permeability apparatus for surface area measurement by an outstanding scholar has greatly provided insight and increased our knowledge of the impedance offered to fluid flow by a packed bed of powder particles.

A. Gadalla, PhD
Professor of Chemical Engineering
Texas A & M University
Department of Chemical Engineering
College Station, Texas, USA

Preface

This book gives an account of an original attempt to design an improved apparatus for the measurement of surface area of powders.

Our understanding of the relationship between particulate properties and structure, and the way in which particles become rearranged, must be directly observed under carefully controlled experimental conditions if we are to master the complexities of the motion of particle sets in dynamic environments.

Our industrial and commercial progress is becoming more and more dependent upon our understanding and mastery of the complex and broad science and technology of fine particles. The problem then is how to design suitable equipment to determine the behaviour, characteristics, and properties of the particles so that their role in environmental processes can be evaluated.

For many years the scientist has accepted that matter may exist as solids, liquids, or gases, although the dividing line between the states may often be rather blurred. This classification has been upset by powders, which at rest are solids, when aerated may behave as liquids, and when suspended in gases take on some of the properties of gases.

It is probably not an exaggeration to say that our whole industrial civilization is based upon the processing of particulate matter, yet until recently, it has been largely uninvestigated by scientists and engineers.

In powder technology, we rely on scientific understanding and processing techniques to take economic advantage of the properties of the particles, and to shape and engineer materials into useful products. For instance, the size of fine particles determines the setting time of cement. The particle size is important in determining the activity of chemical catalysts, the hiding power of pigments. The size of fine particles has crucial effects on the potency of drugs, and even the taste of food. The sintering shrinkage of metallurgical powders is also strongly affected by the size of the particles of which the powder is made up.

Powders of natural or synthetic materials are required for the manufacture of, *inter alia*, cosmetics, ceramics, paints, paper, pesticides, rubber, plastics, and a wide range of building and pharmaceutical products. Pigments and fillers (extenders) are used in many of these applications to provide uniformity in colour and texture, and also to enhance the strength of the manufactured material. How about wheat flour and gunpowder, catalyst beds to speed chemical reactions, and corn starch pudding; many intricate metallic shapes whose surfaces are formed as accurately as if they were machined to close tolerances; the substances which give whiteness or colour to most surface coatings, the components giving solidity and strength to dams, concrete highways; the dirt which the wash cloth removes from the skin of the face or body and then is suspended in the wash water along with the particulate grime of the city, coming from many causes; the black smoke from a furnace of diesel, and the dust coming from the wear, tear, and attrition of almost everything to adhere to clothes and then to be removed in the laundry?

Extension of the scenario beyond the above leads us to the crude oil refineries where the surface area and permeability of compacted catalysts of various shapes and sizes are of high economic importance in the production of various refinery products like gasoline, kerosene, diesel, liquified petroleum gas, and so on, where they carry out specific reactions.

Most importantly, consider the crucial health areas of pharmaceutical tablets where permeability of compressed powders or granules in the form of porous pharmaceutical tablets is a common method of delivering the drug substances as medicines. The physical properties of the tablets may influence the manner and means by which the drug is released *in vivo* so that factors such as the rate in which a liquid, e.g. water or gastrointestinal fluid, penetrates the compacted mass may be critical to the biological behaviour of the delivery system.

The goal of this book, therefore, is an attempt to provide a methodical description of gas permeametry and instrument design. It in no way pretends to be exhaustive, because we still wonder why matter in finely divided form behaves as it does.

This book is intended for individuals with limited experience in the field of instrument design and analysis. It therefore devotes a reasonable share of space to discussions of the principles and materials associated with the method. From this section of the book, the reader should begin to see how design relates to measurement methods and the limitations that exist. Over-all, the objective of this book is to give readers information in such a way that they can sift back and forth between design methods and measurements in such a way that a well-defined research or study plan can be developed before the purchase of any equipment or the commencement of any design. In this way the quality and reliability of particle data that are collected can be improved, giving a clearer picture of the problem and importance of how instrument design and measurements are linked and related to the goals and objectives of a particular type of study or investigation.

The book begins with an introduction of the relationship between gas flow rate and surface area, which forms the foundation for understanding the theoretical background and behaviour of gas flow and pressure drop. Chapter 2 examines the various air permeability techniques for the determination of surface areas. It explores extensively viscous, slip, and molecular flow interactions through a packed bed. Chapter 3 describes the options available in instrumentation in surface area evaluation. It amplifies procedure and the design basis of each instrumentation, pointing out both similarities and differences among groups. In Chapter 4, the design of an improved gas apparatus is comprehensively elucidated. It describes the flow diagram of the apparatus, and how the pressure range of the transducers, including the electronic pressure sensors, were controlled and utilized in a practical manner, with emphasis on operating procedure for measuring permeametry through the packed bed.

Chapters 5 through 9 present the experimental results obtained from the new electronic permeameter designed to measure gas flow rate and pressure drops at sub-atmospheric pressures. The results obtained from this new flow apparatus are reproducible. Regression analysis carried out produced high correlations valid at a significance level of 0.1%. Chapter 6 examined dicalcium phosphate dihydrate powder and determined the surface area and the Carman–Arnell constant. A good agreement was found between the surface values obtained by permeametry and microscopy. Similar results were found in Chapters 7 and 8, using spherical bronze, cyclone tin, and copper powders. Based on the linear relationship obtained from this analysis, a statistical correlation is presented, linking the empirical slip coefficient of Carman–Arnell and the Klinkenberg factor, to the mean particle size. Chapter 9 summarizes the results. It is hoped that the research data shown in the above chapters will help the reader to make the best choice of methods.

Chapter 10 expands other available surface area measuring techniques, including B.E.T. Isotherms, Dead Space determination with Helium, Mercury Porosimetry, Pore Size Distribution, and Gas Adsorption. Structural characterization during gas–solid reactions and estimation of porosity is experimentally discussed in Chapter 11.

A more detailed coverage of the general concept of powder characterization and compaction is introduced in Chapter 12 in an attempt to keep pace with a rapidly developing field. The behaviour of powder particles to each other and the effects of the environment are still areas of which very little is known. Or is it the voids between the powders, the particle shape, the friction conditions in a powder mass; how about mixing of powders which depends on the mobility of the particles, where mobility itself depends on friction? We still wonder why matter in finely divided form behaves as it does.

Chapter 13 gives an account of fine particles in the plastic industrial process, their instrumentation and maintenance. It specifically highlights the various polymeric and plastic processing techniques. Finally, Chapter 14 highlights the concept of dust explosion and some attendant health hazards. The book concludes with a glossary of frequently used terms in powder characterization.

It must be stressed that any errors, both of omission and commission, in view of the sheer immensity of the subject, are mine and not those of the great number of my friends both in the Universities and Industries, to whom I have repeatedly turned for information and advice, and to whom I owe a great debt of gratitude for their assistance and forebearance.

I also wish to acknowledge and thank holders of copyright for permission to publish.

It is expected that this book will be of interest to a wider audience, especially students who almost surely have had sufficient exposure to chemistry, physics, and mathematics of gas–solid systems, and engineering designers working in instrumentation and particulate industries. It should also prove useful to workers in Universities and Research Institutes where problems and quick reference materials in industrial, chemical engineering and process analysis are encountered. An instructor, by appropriate selection of topics, can emphasize, for example, instrumentation, provide an overview of particulate materials, concentrate on physical behaviour, or emphasize on physical properties.

Godwin-Joseph I. Igwe

Acknowledgements

The author wishes to express his most sincere gratitude to Dr Terry Allen, Postgraduate School of Studies in Powder Technology, Department of Chemical Engineering, University of Bradford, England (now Senior Consultant, Du Pont, Wilmington, USA), for his genuine interest, invaluable advice, assistance, and constructive criticisms throughout the preparation of this book.

I am also highly indebted to Dr Nayland G. Stanley-Wood, Postgraduate School of Studies in Powder Technology, Department of Chemical Engineering, University of Bradford, England, for his advice and helpful criticisms. It must be emphasized that this book could not have been completed without their always being ready and willing to offer help and encouragement.

The author also wishes to thank Professor Raymond W. Flumerfelt, Head of Department of Chemical Engineering, Texas A & M University, College Station, Texas, USA, firstly, for appointing me as Visiting Professor of Chemical Engineering, and secondly, for the opportunity to use the University facilities.

I am indebted to the many people who have provided the assistance, encouragement, and constructive criticism leading to the preparation of this book. My colleagues at Prairie View A & M University, Dr K. Walter, and especially Dr Adu-Wusu, who has used the text on instrumentation, powder characterization, and compaction in our introductory courses in Unit Operations and Gas–Solid systems, and has provided invaluable suggestions.

I wish to acknowledge the thoughtful reviews of the manuscript offered by Professor A. Gadalla, Texas A & M University, College Station, and Dr A. Kumar, Prairie View A & M University, Prairie View, Texas.

My special thanks to the World Bank in Washington and the Robert S. McNamara Fellowship program, for the 'outstanding scholar' fellowship and providing the financial support. I deeply appreciate the honour.

I thank Professor Ray Anthony, Department of Chemical Engineering, and Professor Walter Bradley, Department of Mechanical Engineering, Texas A & M University, College Station, Texas, for arranging all the plant visits. I am grateful to the Presidents of the following Corporations for making it possible for me to visit their industries and analytical laboratories and to have useful discussions with their Maintenance and Machine Tools Engineers, as well as management and members of staff.

1. Mr Arthur McFerrin, Jr, President, KMCO Inc., Crosby, Texas. (Batch producers of small-scale chemicals).
2. Mr David Drummond, Plant Manager, AKZO Chemie America (Ketjen Catalysts) Pasadena, Texas. (Producers of fluid catalytic-cracking catalysts/chemicals).
3. Mr Granville Hahn, President, Permian Research Corporation, Big Spring, Midland Odessa, Texas. (Plastic extrusion equipments and instruments).
4. Mr Roy Serpa, President, Instamelt Systems, Inc., Big Spring, Midland Odessa, Texas. (Sheet extrusion machines and instruments).
5. Co.-Ex. Pipe Company, Big Spring, Midland Odessa, Texas. (Polyethylene pipes for crude oil and gas).
6. Mr Ken Dunwoody, Plant Manager, Western Container Corporation, Big Spring, Midland Odessa, Texas. (Producers of Coca-cola plastic bottles).
7. Mr Jim Reaves, Vice President, EGC Micromet Inc., Houston, Texas. (Producers of custom injection moldings).
8. Mr H. B. Bub Scott, President, Hub Scott & Associates Inc., Houston, Texas. (Plastic processing machineries and equipments).
9. Mr F. H. Schneider, Technical Manager, Mobil Chemical Company, Petrochemicals Division, Houston, Texas. (Petrochemicals headquarters).
10. Dr David Warner, Plant Manager, Mobil Polyethylene Plant, Beaumont, Texas. (Producers of polyethylene resins/polymers).
11. The Plant Manager, Mobil Refinery, Beaumont, Texas. (Producers of all kinds of petroleum refinery products).
12. Mr W. J. Kerns, Mobil Olefins/Aromatics Plant, Beaumont, Texas. (Producers of olefins/aromatic products).
13. Mr Victor S. Soares, Mobil Chemical Speciality Plant, Beaumont, Texas. (Producers of chemical speciality products).

Finally, and most importantly, I am deeply indebted to my wife, Rose, and children, Maureen, Chuck, and Emeka, for their encouragement, patience, endurance, and self-deprivation, over the past five years of compiling the manuscripts.

1

Flow through packed beds

1.1 INTRODUCTION

The unique property of a powdered material, the one that distinguishes it from other solid bodies on the one hand, and from simple conduits on the other, is its finely divided form, its permeability, and its complicated pore structure.

Fluid flow and pressure drop in packed beds take place within extremely complicated microscopic boundaries that make any rigorous solution of the equations of change in a packed bed practically impossible. This is one reason why some of the brilliant and successful practitioners in the field of flow through packed beds have tried as much as possible to use the continuum approach in which no attention is paid to the importance of molecular flow. The other reason is that the continuum approach is often adequate for the phenomenological description of transport processes in packed beds.

The desire of scientists and engineers to be able to understand and then explain their observations has always been a powerful driving force for progress. There have, therefore, been numerous attempts over the past forty years of so to explain the flow phenomenon in terms of pressure drop across packed beds, where bed resistance will depend on bed porosity, bed structure, and uniformity of packing, particle size distribution, as well as specific surface.

Various equipment has been designed to measure flow rate and pressure drop, but the results have seldom been entirely satisfactory, although with every step a further penetration into an immensely complicated territory has been achieved.

1.2 THE RELATIONSHIP BETWEEN GAS FLOW RATE AND SURFACE AREA

The equation relating the transport of a gas through a bed of powder under non-turbulent flow conditions, to the surface area of the powder is made up of two terms:

$$\frac{\mathrm{d}V}{\mathrm{d}t}=\frac{A\Delta P}{k\mu LS_V^2}\frac{\varepsilon^3}{(1-\varepsilon)^2}+\frac{8}{3}\left(\frac{\delta k_0}{k}\right)\frac{A\Delta P}{L\overline{P}S_V}\left[\frac{2RT}{\pi M}\right]^{\frac{1}{2}}\frac{\varepsilon^2}{(1-\varepsilon)} \tag{1.1}$$

where A = Cross-sectional area of permeameter cell, m^2

$\frac{\mathrm{d}V}{\mathrm{d}t}$ = Volume flow rate of gas through the powder bed, m^3/s

k = Aspect factor (Kozeny constant = 5)

L = Length of powder bed, m

M = Molecular weight of the gas (29 kg/k mol for air)†

$\overline{P}$ = Mean gas pressure in the bed, N/M^2

R = Gas constant, 8.314 41 $J.Mol^{-1}.K^{-1}$

S_V = Volume specific surface, m^2/m^3 (using viscous flow term)

S_V^2 = Powder surface area as given by the Carman–Arnell equation at a range of pressures, m^2/m^3

ε = Powder bed porosity

δk_0 = Variable in Carman–Arnell equation (dimensionless)

μ = Viscosity of air, 186×10^{-7} $N.s.m^{-2}$

ΔP = Pressure drop across the packed bed, N/m^2

The first term applies to viscous (or laminar) flow and relates to the surface of the particles. This equation was first developed by Kozeny [1] and later modified by Carmen [2] who adopted a value of 5 for the constant of proportionality. The Kozeny and Carman equation could be used as a basis for a simple routine permeability method for the determination of surface area.

They [1,2] assumed the bed to be made up of capillaries embedded in a solid matrix, thus extending the Poiseulle's equation [3] by equating the surface area of these capillaries with the surface area of the powder. The assumptions made in the derivation of their equation are so sweeping that the derived results only indicate whether a particular powder is coarser or finer than a similar powder. Powders with the same external specific surface which differ in particle shape or particle size distribution may well give very different indicated surfaces by this method.

The second term applies to molecular (or Knudsen) flow and relates to a surface which includes surface cracks and fissures as well as the voids between the particles of which the bed is composed.

The combined equation was proposed by Carman & Arnell [4] who introduced a 'slip' coefficient in the second term. This coefficient corrects for the nature of the molecule–surface interaction. A fraction of the molecules striking the surface will be specularly reflected, i.e. if the capillary walls are smooth, molecules striking them at any angle rebound at the same angle with the same average velocity and the component of velocity perpendicular to the wall reversed. The surface of packed beds of powder are not smooth, and molecules striking them rebound in any direction, i.e. diffuse reflection.

For any powder, as the inlet pressure falls, the contribution of the viscous flow term to measured surface area will decrease and the contribution of the molecular flow term increase [5]. Further, the finer the powder, the greater the contribution of the slip flow term in the combined equation [6].

† Some common and frequently used conversion tables are given in Appendix A3.1.

Carman & Arnell assumed a constant value for the slip coefficient, and this has been adopted by others [5–10].

In measuring the porosity of sintered plates, however, a related constant, the Klinkenberg factor, is assumed to increase with increasing fineness of the interparticle voids. It was, therefore, decided to investigate these three propositions:

(i) The slip coefficient is constant;
(ii) The slip coefficient depends on the nature of the solid surface;
(iii) The slip coefficient depends on the size of the interparticle voids.

1.3 FLOW RATE

1.3.1 Mass and volumetric flow rate

Continuous processes involve the movement of material from one point to another — sometimes between process units, sometimes from a production facility to a transportation depot, or vice versa. The rate at which a material is transported through a process line is the flow rate of that material [11–14].

The flow rate of a process stream may be expressed as a mass flow rate (mass/time) or as a volumetric flow rate (volume/time). Suppose a fluid (gas or liquid) flows in the cylindrical pipe shown in Fig. 1.1, where the shaded area represents a section

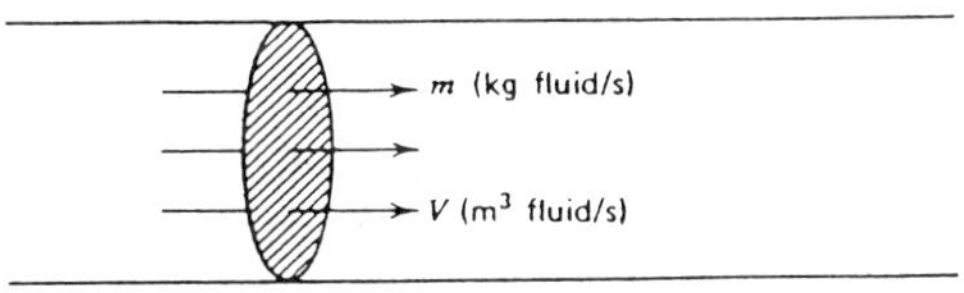

Fig. 1.1 — Flow through a cylindrical pipe.

perpendicular to the direction of flow. If the mass flow rate of the fluid is m(kg/s), then every second m kilograms of the fluid pass through the cross-section. If the volumetric flow rate of the fluid at the given cross-section is V(m^3/s), then every second V cubic metres of the fluid pass through the cross-section. However, the mass m and the volume V of a fluid — in this case, the fluid that passes through the cross-section each second — are not dependent quantities but are related through the fluid density ρ.

$$\rho = \frac{m}{V} \tag{1.2}$$

Thus: The density of a fluid can be used to convert a known volumetric flow rate of a process stream to the mass flow rate of that stream or vice versa.

The mass flow rate of process streams must be known for many process calculations, but it is frequently more convenient to measure volumetric flow rates. A common procedure is, therefore, to measure V and calculate m from V and the density of the stream fluid.

1.3.2 Flow rate measurement

A flowmeter is a device mounted in a process line that provides continuous reading of the flow rate in the line. Two commonly used flowmeters — the rotameter and the orifice meter — are shown in Fig. 1.2. Many other flowmeters are described elsewhere [11–14].

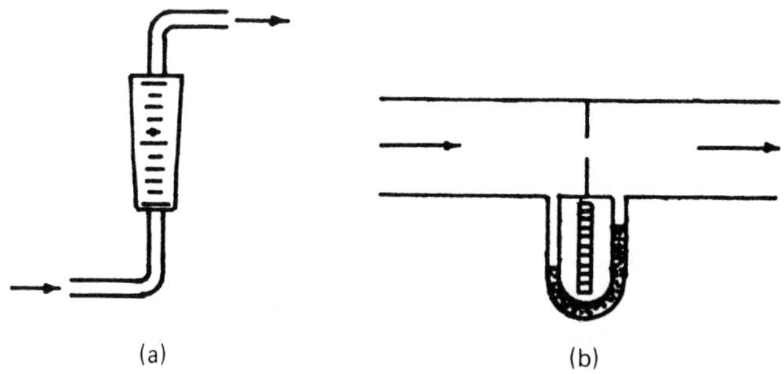

Fig. 1.2 — Flow meter (a) rotameter, (b) orifice meter.

The rotometer is a tapered vertical tube containing a float; the larger the flow rate, the higher the float rises in the tube. The orifice meter is an obstruction in the flow channel with a narrow opening through which the fluid passes. The fluid pressure drops from the upstream side of the orifice to the downstream side; the pressure drop (measured with a differential manometer) varies with the flow rate: the greater the flow rate, the larger the pressure drop.

1.4 PRESSURE

1.4.1 Fluid pressure and hydrostatic head

A pressure is the ratio of a force to the area on which the force acts. Accordingly, pressure units are force units divided by area units (e.g. N/m^2, and lb_f/in^2 or psi). The SI pressure unit N/m^2 is called a pascal (Pa).

Consider a fluid (gas or liquid) contained in a closed vessel or flowing through a pipe, and suppose that a hole of area A is made in the wall of the containing vessel as in Fig. 1.3. The fluid pressure may be defined as the ratio F/A, where F is the minimum force that would have to be exerted on a plug in the hole to keep the fluid from emerging.

Let us introduce an additional definition of fluid pressure to explain the concept of atmospheric pressure and to discuss common methods for measuring pressures of fluids in tanks and pipes. Suppose a vertical column of fluid is h(m) high and has a

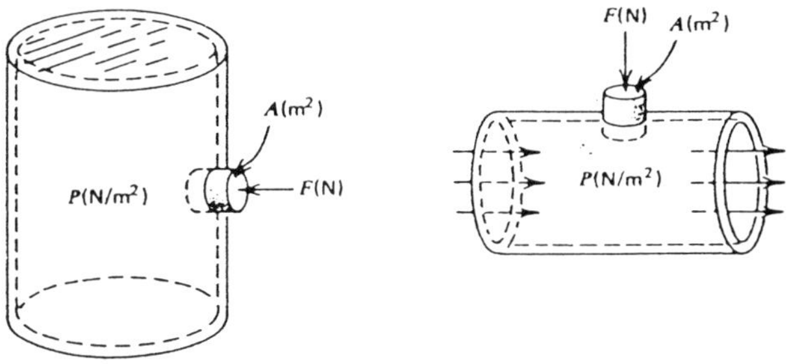

Fig. 1.3 — Fluid pressure in a tank and a pipe.

uniform cross-sectional area $A(\mathrm{m}^2)$. Further suppose that the fluid has a density of $\rho(\mathrm{kg/m}^3)$, and that a pressure $P_0(\mathrm{N/m}^2)$ is exerted on the upper surface of the column as shown in Fig. 1.4. The pressure P of the fluid at the base of the column, called the

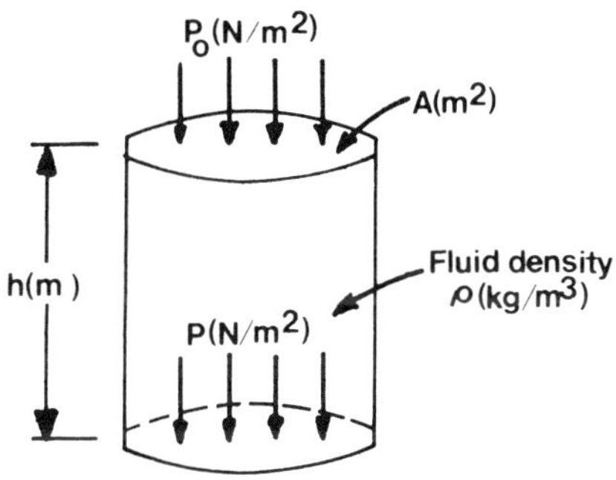

Fig. 1.4 — Pressure at the base of a fluid column.

hydrostatic pressure of the fluid, is, by definition, the force F exerted on the base divided by the base area A. F thus equals the force on the top surface plus the weight of the fluid in the column, or

$$P = P_0 + \rho \frac{g}{g_c} h \tag{1.3}$$

(Note that g is an acceleration and g_c is a factor for converting from one force unit to another. Remember that g varies with position; it is different at different points on the Earth's surface, and different on the Moon than on the Earth, but g_c is a constant and always has the value:

$$g_c \equiv 1 \frac{\mathrm{kg.m/s^2}}{\mathrm{N}} = 1 \frac{\mathrm{g.cm/s^2}}{\mathrm{dyne}} = 32.174 \frac{\mathrm{lb_m.ft/s^2}}{\mathrm{lb_f}} \tag{1.4}$$

$$g = 9.8066\,\text{m/s}^2 \Rightarrow g/g_c = 9.8066\,\text{N/kg})\ . \tag{1.5}$$

Since A does not appear in this equation, the formula is applicable to a column of fluid as thin as a test tube or as wide as the ocean.

In addition to being expressible as a force per unit area, a pressure may be expressed as a head of a particular fluid, that is, as the height of a hypothetical column of this fluid that would exert the given pressure as its base if the pressure at the top were zero. One can thus speak of a pressure of 14.7 psi or equivalently of a pressure (or head) of 33.9 ft of water (33.9 ft H_2O) or 76 cm of mercury (76 cm Hg). The equivalence between a pressure P (force/area) and the corresponding head P_h (height of a fluid) is given by equation (1.3) with $P_0 = 0$:

$$P\left(\frac{\text{force}}{\text{area}}\right) = \rho\,\text{fluid}\,\frac{g}{g_c}\,P_h\,(\text{head of fluid}) \tag{1.6}$$

Example: *Calculation of a pressure as a head of fluid*

Express a pressure of 2.00×10^5 Pa in terms of mm Hg.

Solution:

Solve equation (1.6) for P_h (mm Hg), assuming that $g/g_c = 9.807$ N/kg and noting that the density of mercury is $13.6 \times 1000\,\text{kg/m}^3 = 13\,600\,\text{kg/m}^3$.

$$P_h = P(g_c/g)/\rho\,\text{Hg}$$

$$= \frac{2.00 \times 10^5\,\text{N}}{\text{m}^2}\ \frac{\text{m}^3}{13\,600\,\text{kg}}\frac{\text{kg}}{9.807\,\text{N}}\ \frac{10^3\,\text{mm}}{1\,\text{m}} = 1.50 \times 10^3\,\text{mm Hg}$$

Example: *Pressure below the surface of a fluid*

What is the pressure 30.0 m below the surface of a lake? Atmospheric pressure (the pressure at the surface) is 10.4 m H_2O, and the density of water is $1000.0\,\text{kg/m}^3$. Assume that g/g_c is 9.807 N/kg.

Solution:

First, the hard way, using equation (1.3).

$$P = P_0 + \rho(g/g_c)h$$

$$P(\text{N/m}^2) = \frac{10.4\,\text{m H}_2\text{O}}{}\ \frac{1.013 \times 10^5\,\text{N/m}^2}{10.33\,\text{m H}_2\text{O}}$$

$$+ \frac{1000.0\,\text{kg/m}^3}{}\ \frac{9.807\,\text{N}}{\text{kg}}\ \frac{30.0\,\text{m}}{} = 3.96 \times 10^5\,\text{N/m}^2\,(\text{Pa})$$

Next, the easy way, using: $P(\text{mm Hg}) = P_0(\text{mm Hg}) + h(\text{mm Hg})$ (1.7)

$$P(\text{m H}_2\text{O}) = 10.4\,\text{m H}_2\text{O} + 30.0\,\text{m H}_2\text{O} = 40.4\,\text{m H}_2\text{O}$$

(Verify that the two calculated pressures are equivalent).

1.5 ATMOSPHERIC PRESSURE, ABSOLUTE PRESSURE, AND GAUGE PRESSURE

The pressure of the atmosphere can be thought of as the pressure at the base of a column of fluid (air) located at the point of measurement (at sea level, for example). The pressure P_0 at the top of the column equals 0, and ρ and g are average values of the density of air and the acceleration of gravity between the top of the atmosphere and the measurement point.

A typical value of the atmospheric pressure at sea level, 760.0 mm Hg, has been designated as a standard pressure of 1 atmosphere.

The fluid pressures referred to so far are all absolute pressures, in that a pressure of zero coresponds to a perfect vacuum. Many pressure measuring devices give the gauge pressure of a fluid, or the pressure relative to atmospheric pressure. A guage pressure of zero indicates that the absolute pressure of the fluid is equal to atmospheric pressure. The relationship for converting between absolute and gauge pressure is:

$$P_{\text{absolute}} = P_{\text{gauge}} + P_{\text{atmospheric}} \tag{1.8}$$

The abbreviations psia and psig are commonly used to denote absolute and gauge pressure in $lb_f/in.^2$. Also, it is common to refer to negative gauge pressures (absolute pressures less than atmospheric) as positive amounts of vacuum. For example, a gauge pressure of −1 cm Hg (7.5.0 cm absolute if atmospheric pressure is 76.0 cm Hg) may be called 1 cm of vacuum.

1.6 FLUID PRESSURE MEASUREMENT

Several mechanical devices are used for the measurement of fluid pressures. The most common of these devices is a Bourdon gauge, which is a hollow tube closed at one end and bent into a C configuration. The open end of the tube is exposed to the fluid whose pressure is to be measured. As the pressure increases, the tube tends to straighten, causing a pointer attached to the tube to rotate. The position of the pointer on a calibrated dial gives the gauge pressure of the fluid. A schematic diagram of a Bourdon gauge is shown in Fig. 1.5.

Bourdon gauges are used to measure fluid pressures from nearly perfect vacuums to about 7000 atm. More accurate measurements of pressures below about 3 atm are provided by manometers.

A manometer is a U-shaped tube partly filled with a liquid of known density (the manometer fluid). When the ends of the tube are exposed to different pressures the fluid level drops in the high pressure arm and rises in the low pressure arm. The difference beween the pressures can be calculated from the measured difference between the liquid levels in each arm.

Manometers are used in several different ways, as shown in Fig. 1.6. In each diagram, pressure P_1 is greater than pressure P_2.

Fig. 1.6a shows an open-end manometer: One end is exposed to a fluid whose

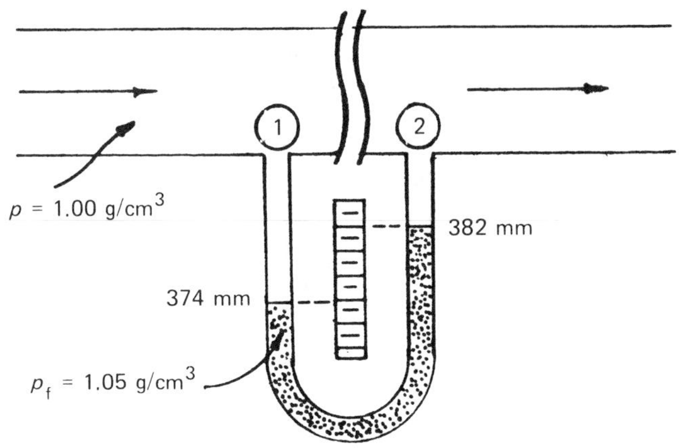

Solution:

$h = (382 - 374)\ \text{mm} = 8\ \text{mm}$. From equation (1.10)

$$P_1 - P_2 = (\rho_f - \rho)\frac{g}{g_c}h$$

$$= \frac{(1.05 - 1.00)\ \text{g}}{\text{cm}^3}\ \frac{980.7\ \text{cm}}{\text{s}^2}\ \frac{1\ \text{dyne}}{1\ \text{g.cm/s}^2}\ 8\ \text{mm}\ \frac{1\ \text{cm}}{10\ \text{mm}}$$

$$= 40\frac{\text{dynes}}{\text{cm}^2}$$

Example:

The pressure of gas being pulled through a line by a vacuum pump is measured with an open-end mercury manometer. A reading of -2 in. is obtained. What is the gas gauge pressure in inches of mercury? What is the absolute pressure $P_{atm} = 30$ in. Hg?

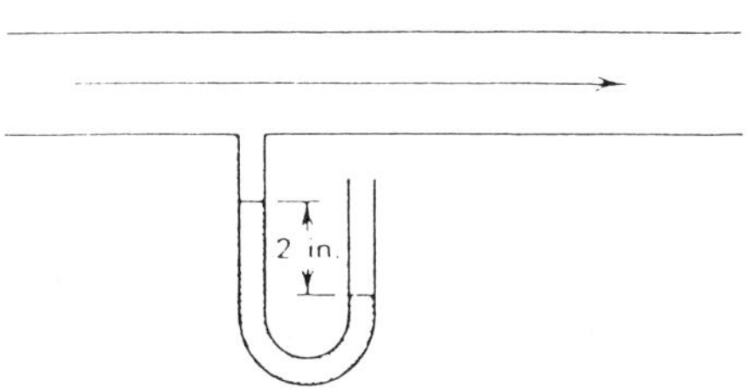

Solution:

From equation (1.12) and the definition of a gauge pressure:

$$P_1 - P_{atm} = P_{gauge} = -2\ \text{in. Hg}$$
$$\downarrow$$
$$P_1 = P_{atm} + P_{gauge} = 30 - 2 = 28\ \text{in. Hg}.$$

REFERENCES

[1] Kozeny, J., *Ber. Akad. Wiss., Wien*, **11a** (136), 271–306, (1927).
[2] Carman, P. C., *J. Soc. Chem. Ind.*, **57**, 225–238 (1938), and **58**, 1–7, (1939).
[3] Poiseuille, J., *Inst. de France Acad. Des. Sci.*, **9**, 433, (1846).
[4] Carman, P. C. & Arnell, J. C., *Can. J. Res.*, **26A**, 128–136, (1948).
[5] Igwe, G. J. I., Allen, T., & Stanleywood, N. G., Surface area measurement and gas permeametry at sub-atmospheric pressures. *Proceedings of the Particle Size Analysis* 1988, *of the Particle Characterization Group of the Analytical Division of the Royal Society of Chemistry* , Surrey UK, p. 243–256, (1988).
[6] Igwe, G. J. I., Gas and particle surface reaction at sub-atmospheric pressures. Paper No. 88–3630. *Proceedings of the First National Fluid Dynamics Congress of the American Institute of Aeronautics and Astronautics, held July* 25–28 *at Hyatt Regency Hotel, Cincinnati, Ohio, USA* (1988).
[7] Allen, T., *Particle size measurement*, 3rd Ed., Chapman and Hall, (1981).
[8] Allen, T., *Powder Technology*, **33**, 195–200, (1982).
[9] Lea, F. M. & Nurse, R. W., *Trans. Inst. Chem. Eng. Symposium*, **25**, 47, (1947).
[10] Lea, F. M. & Nurse, R. W., *J. Soc. Chem. Ind.* (*London*), **58**, 278, (1939).
[11] *Perry's chemical engineers' handbook*, 6th Edition, Chapter 5. McGraw-Hill Book Company (1984).
[12] Felder, M. R. & Rousseau, R. W., *Elementary principles of chemical processes*, 2nd Ed., John Wiley & Sons, New York (1986).
[13] *Process instruments and controls handbook* (Ed. Considine) 2nd Ed., McGraw-Hill, New York (1974).
[14] *Fundamentals of vacuum science & technology* (Ed. Lewin) McGraw-Hill, New York (1965).

2

Air permeability techniques for determination of surface areas

2.1 LITERATURE REVIEW

2.1.1 Introduction

Permeametry is concerned with the study of the flow of fluids whether liquids or gases through granular beds and packed columns. It can further be considered as either an extension of flow around isolated particles or of flow in pipes. The latter is usually used because in packed beds the bed is idealized as a bundle of capillaries.

The permeability can then be related to the size and shape of the particles forming the bed and hence the surface area of the powder. The bed is made up of unconsolidated material and the particles are assumed to be rigid enough not to be deformed by gentle compaction. The permeability coefficient measured in this non-isotropic medium is constant in one direction only and may have a different value in another direction.

The permeability method for measuring the specific surface of a powder has a particular appeal by virtue of its directness and simplicity. The method depends upon a relation between the rate of flow of a fluid through a compacted bed of powder, the voids in the bed, and the surface area of the particles. It is concerned with single-phase slow viscous flow (low Reynolds number flow) [1] through unconsolidated media, specifically non-uniform packings, ideally random, and composed of relatively coarse or fine particles.

Flow of fluids through porous media is also a process of major importance in several well-known areas of technology, including hydrology [2, 3], soil physics (agriculture) [4], soil mechanics (civil engineering) [5–8], petroleum technology [9, 10], chemical engineering [11–14], and mineral [15–17], among others.

As pointed out in Chapter 1, the best-known equation relating the rate of fluid flow through porous media with the parameters of the fluid, medium, and pressure gradient under low Reynolds number conditions is the Kozeny–Carman equation.

2.1.2 Types of flow

The transfer of a gas through a porous medium is more complex than the transfer of a liquid [18] and depends on such factors as the size of the voids formed between the particles, the size of the pores within the pores, the absolute pressure, and the pressure drop across the bed. There are, therefore, three types of flow to consider: viscous flow, in which molecular–molecular interaction predominates; transitional flow in which the mean free path λ of the gas molecules is of the same order as the capillary diameter, the slip term is of the same order as the viscous term, and both have to be evaluated; molecular flow, in which the interaction is predominantly between the gas molecules and the powder surface.

Allen [19] has stressed the importance of the regime in which both these types of flow are present — the transitional region.

2.1.3 Flow of viscous fluid through a packed bed of powder

The first attempt to study experimentally the relationship between flow rate and pressure drop and the parameters of a packed bed was made by Darcy [20], and may be expressed as

$$U = K\frac{\Delta P}{L} \tag{2.1}$$

where U is defined as the apparent velocity of flow, taken as the approach velocity, ΔP is the pressure difference across the powder bed of thickness, L and K is the permeability factor which is constant for a particular system.†

Darcy's law has been found to be valid for a large range of porous media, including colloidal membranes [21], porous plates [22], fine powder [23], and sands [24]. It was, however, found to be invalid for compressible beds using either liquids or gases and all fluids at higher velocities.

The law also breaks down for gas flow at atmospheric pressure through beds of fine particles (i.e. small voids), and this breakdown extends to large voids (i.e. coarse particles) as the pressure is decreased.

Equation (2.1) is closely analogous to Poiseuille's equation

$$U_{\mathrm{m}} = \frac{d^2\Delta P}{32\mu L} \tag{2.2}$$

which was derived for the flow of a viscous fluid through a circular capillary, where

U_{m} = mean velocity of fluid within the capillaries
d = capillary diameter
ΔP = pressure drop across the bed of solid
μ = fluid viscosity
L = bed thickness.

† A full list of symbols appears on page 56.

2.1.3.1 *Poiseuille's equation for flow through a capillary and the definition of coefficient of slip*

Poiseuille's equation [30] is developed below in order that its extension by Carman [25] and Kozeny [26] and its further extension to slip flow may be more readily followed.

Assume a section of a capillary down which a gas is flowing as a result of a pressure drop (ΔP) across its two ends distance L apart.

Consider a small cylindrical element of thickness δr at a distance, r, from the axis of the capillary. The force acting on this element due to the pressure difference is:

$$F_1 = \pi r^2 \Delta P \tag{2.3}$$

When flow reaches steady state, equilibrium exists between this force, F_1, and the force from the viscous drag acting across the boundaries of the surface.

The viscous drag is defined as:

$$F_2 = -\mu S \frac{dU}{dr} \tag{2.4}$$

where S is the area of cylindrical surface, μ, the coefficient of viscosity, and $\frac{dU}{dr}$ is the velocity gradient.

As

$$S = 2\pi r L$$

$$F_2 = -\mu 2\pi r L \frac{dU}{dr}$$

Balancing forces:

$$\pi r^2 \Delta P = -2\pi r L \mu \frac{dU}{dr} \tag{2.5}$$

On integration with $r = R$ and assuming $U = 0$ at the walls of the capillary

$$\int_0^U dU = -\frac{\Delta P}{2\mu L} \int_R^r r dr \tag{2.6}$$

$$U = \frac{\Delta P}{4\mu L}(R^2 - r^2) \tag{2.7}$$

This equation defines a parabolic velocity profile (paraboloid). The volume rate of flow is given by

$$\frac{dV}{dt} = \sum_{r=0}^{r=R} 2\pi r \delta r U = \frac{\Delta P \pi}{2\mu L}\int_0^R (R^2 - r^2) r dr$$

$$= \frac{\Delta P \pi R^4}{8\mu L} \tag{2.8}$$

The average velocity is given by

$$U_m = \frac{1}{A}\frac{dV}{dt}$$

$$= \frac{\Delta P R^2}{8\mu L} \tag{2.9}$$

This is half the maximum velocity as given by putting $r = 0$ in equation (2.7).

Kundt & Warbung [27] found that this equation was not valid when pressures were so low that the mean free path of the gas molecules (λ) was comparable with the radius of the capillary (r); a plot of 'U' against $\Delta PR^2/8\mu L$ giving a minimum marking the transition from one type of flow to another. Poiseuille's equation fails because the velocity at the walls of the capillary is not zero at low pressures as assumed in equation (2.6). It is thus necessary to assume that when $r = R$ there is a frictional force on the surface of the capillary walls and the flowing gas has a velocity v_0 at the walls.

Loeb [28] gave that this frictional force against the wall was, on balancing the forces of viscous drag and pressure difference

$$2\pi RLfv_0 = \pi R^2 \Delta P \tag{2.10}$$

where f is the coefficient of external friction against the wall. The volume flow rate was therefore given as

$$q = \frac{\pi \Delta P}{8\mu L}\left[R^4 + 4\frac{\mu}{f}R^3\right] \tag{2.11}$$

resulting in

$$\frac{q}{A} = U = \frac{\Delta PR^2}{8\mu L}\left[1 + 4\frac{\mu}{fR}\right] \quad (2.12)$$

The ratio μ/f is termed the coefficient of slip, ξ, and gives the ratio of internal friction of the fluid to the coefficient of external friction against the capillary wall. The nature of f and consequently ξ depends on the nature of the impacts between gas molecules and the capillary walls.

2.1.3.2 Derivation of Kozeny and Carman equation

Carman modified Kozeny's equation and assumed that a bed formed from a granular material is equivalent to a group of similar parallel channels such that the total internal surface of pore channels is equal to the total external surface of the powder, and the total internal volume is equal to the pore space in the bed.

The flow through a bed of solid particles takes place in the irregularly shaped channels provided by the voids of the bed. It seems therefore logical that any equation relating to this kind of flow should refer to these channels rather than to the size of the bed forming particle.

The Kozeny hydraulic model assumes that the channels are formed by capillary tubes running parallel to the direction of flow.

Consider fluid flow through a bed of powder represented by N uniform parallel circular capillaries of radius R.

The volume specific surface is defined by

$$S_v = \frac{S}{V_p} \quad (2.13)$$

where

$$\varepsilon = \frac{V_v}{V_v + V_p}$$

$$S_v = \frac{\varepsilon}{1-\varepsilon}\frac{S}{V_v} = \frac{\varepsilon}{1-\varepsilon}\frac{2N\pi RL}{N\pi R^2 L} \quad (2.14)$$

or

$$R = \frac{2\varepsilon}{(1-\varepsilon)S_v} \quad (2.15)$$

where V_p = volume of powder in the bed
S = surface of powder in the bed
V_v = volume of void in the bed
ε = porosity of the bed
R = capillary radius
A = cross-sectional are of the bed

Substituting for (R) in equation (2.9)

$$U_m = \frac{1}{A}\frac{dV}{dt} = \frac{\Delta P}{8\mu L S_v^2}\frac{4\varepsilon^2}{(1-\varepsilon)^2} \tag{2.16}$$

The measured velocity is the approach velocity. That is the volume rate divided by the whole cross-sectional area of the bed, $U = \frac{Q}{A}$ where Q is the rate of flow. Dupuit [29] stated that the actual velocity in the channels will be greater than the approach velocity for two reasons. The volume flow rate approaching the bed is over a cross-sectional area A, whereas in the bed it is over a cross-sectional area εA. Velocity is defined as distance divided by time, and the applicable distance is not L, the length of the bed, but L_e, the average length of the voids within the bed (i.e. the equivalent length).

Correcting for the first effect

$$U_1 = \frac{Q}{\varepsilon A} = \frac{U}{\varepsilon}\ . \tag{2.17}$$

Correcting for the second effect

$$U_m = U_1\left(\frac{L_e}{L}\right)\ . \tag{2.18}$$

Hence, replacing U_m in equation (2.16), remembering that the pressure drop is over L_e and not L,

$$\frac{U}{\varepsilon}\left(\frac{L_e}{L}\right) = \frac{\Delta P}{8\mu L_e S_v^2}\frac{4\varepsilon^2}{(1-\varepsilon)^2}$$

$$U = \frac{\Delta P \varepsilon^3}{\mu L S_v^2\,(1-\varepsilon)^2 2\left(\frac{L_e}{L}\right)^2}$$

If $2\left(\frac{L_e}{L}\right)^2$ is replaced by the aspect factor k.

$$U=\frac{\Delta P\varepsilon^3}{k\mu LS_v^2(1-\varepsilon)^2}$$

resulting in Kozeny's equation:

$$S_v=\left[\frac{\Delta P\varepsilon^3}{k\mu LU(1-\varepsilon)^2}\right]^{\frac{1}{2}} \tag{2.19}$$

where $k=k_0k_1=k_0\left(\frac{L_e}{L}\right)^2$.

2.1.3.3 *The aspect factor, k*

The aspect of factor, k, is the product of two variables: $k=k_0k_1$, where k_0 is a function of capillary shape and size distribution. In the above derivation for uniform circular capillaries k_0 is equal to 2; for other geometrical shapes k_0 varies [30, 31, 32, 78] between 1.2 and 3; for other size distributions k_0 can fall [33] from 2 to 2/3.

The tortuosity factor, k_1, is equal to the square of the ratio of length of tortuous capillary path to length of packed bed. Carman [34] studied the motion of a coloured streamline through a bed of large glass spheres and found that it appeared to adopt an average angle of 45° to the direction of flow, thus making $k_1=2$. Fowler and Hertel [35] stated that the angle to the direction of flow cannot have a constant value, while Hitchcock [36] used a tortuosity factor of $\pi/2$. In all cases tortuosity is considered to be independent of porosity, ε, although it is more probable that a slow variation occurs, since tortuosity approaches unity as ε approaches unity. The effect of ε on surfaces of packed beds will be discussed further in subsequent experimental chapters in this book.

The Kozeny and Carman [37] equation could be used as a basis for a simple routine permeability method for the determination of surface area, and later experimental work confirmed [38] this. The equation can be applied to the flow of both gases and liquids but is valid only for an unconsolidated bed of randomly packed particles larger than about 20 micrometre in size through which flow is largely viscous in character, and applies to the case where the mean free path length of the gas molecules is small compared to the capillary dimensions.

Following Carman's early work on irregularly shaped particles, a value of 5 was quoted for the aspect factor, k, but Carman [25] and other workers [39] have since shown that the value of k is dependent on the composition of the particles, their preferred orientation, and the porosity of the bed.

2.1.3.4 *Effect of porosity on aspect factor, k*

Many workers [35, 40, 41, 78–82] using glass spheres have found that k is constant over a fairly large range of porosities. Schriever [40] found for porosity, $\varepsilon = 0.35 - 0.40$ value of $k = 5.06$; Muskat & Botsel [41] found for $\varepsilon = 0.34$–0.45 a value of $k = 4.5$, and Fowler & Hertel [35] found a value of $k = 5.5$ for $\varepsilon = 0.54$–0.80.

If the porosity exceeds 0.80 the values of k begins to rise sharply [42]. This may be due to the Kozeny–Carman equation being based on channel flow. Thus as porosity tends to higher values, there are fewer channels and the particles become completely independent of each other and the resistance to flow decreases, but for values of porosity below 0.8, k is substantially constant over a fairly wide range. Some workers [43, 44, 45, 78–82] found that the aspect factor k also depends upon particle shape and reported for glass beads, $k = 4.50$, for fibres arranged parallel to flow $k = 3.07$, and for fibres arranged perpendicular to flow, $k = 6.04$.

2.1.3.5 *Other flow equations*

In general it may be argued that the porosity function itself is inadequate or that the aspect factor is a function of porosity. It is also evident that the equations and modifications put forward are valid for only a limited range of porosity which varies from powder to powder.

For a loose bed of powder, increasing specific surface, S_K (S_K is the surface area using the viscous flow term), with decreasing porosity may be attributed to channelling, i.e. flow through the small capillaries is swamped by flow through the larger capillaries, and under these conditions the Carman–Kozeny equation does not hold. For a compacted bed, decreasing S_K with decreasing porosity may be attributed to bed structure. The maximum value of S_K is assumed to be correct since near this point, the rate of change of S_K with ε is small and optimum porosity is available.

If S_K decreases with decreasing porosity at high voidage it is possible that bed fracture is occurring. S_K increasing with decreasing porosity at low porosity may indicate particle fracture.

Conditions vary from powder to powder, but some general observations have been made. Brittle powders fracture at low porosities leading to increasing S_K values; plastic (i.e. soft) powders deform, and S_K rises to a relatively constant value.

Criticism has also been levelled at the porosity function which, if incorrect, can lead to the variations discussed above. The corrections to eliminate this variation as suggested by Carman [38] and later by Keyes [46] has been found to be fully satisfactory [47].

The term 'immobile layer' is used by Harris [48] to refer to gas which does not contribute to the general flow through a packed bed of powder. He stated that the discrepancies usually attributed to inadequacies in the porosity function of non-uniform packing, are in truth due to assumption of incorrect values for porosity and surface area. He assumed that an immobile layer of fluid which does not take part in the flow process surrounds the particles. Therefore the particles have a true and an effective value for density, porosity, volume, and specific surface.

Allen & Maghafarti [49] determined the volume of immobile gas for a range of powders and found that it varied from large positive to equally large negative values. They therefore rejected the hypothesis.

Wasan *et al.* [50, 51] obtained the equation

$$S_K^2 = \frac{A\Delta P}{\mu L(\mathrm{d}V/\mathrm{d}t)} \phi(\varepsilon) \qquad (2.20)$$

where $\phi(\varepsilon)$ is defined as complex function of tortuosity constriction (a factor introduced to compensate for the varying cross-section of the pores in the bed), hydraulic radius, and particle matrix. The model was developed for monosize spheres in a particular packing, but was extended to other regular shapes.

2.1.4 Molecular flow streaming or Knudsen flow

Maxwell [52] developed theoretically the conditions of flow at low pressures, but it was Knudsen [53, 54, 55] who drew attention to the possibility of solving such problems by the use of kinetic theory.

In the relations which have been discussed earlier, it is assumed that the gas or fluid can be regarded as a continuum and that there is no slip between the wall of the capillary and the fluid or gas layers in contact with it. When a gas passes because of a pressure gradient through a system of fine pores, the pressure may be sufficiently low and the pores sufficiently fine that the mean free path of the gas molecules is greater than the pore diameter. In this case, 'molecular flow' is obtained.

In molecular flow the chances of molecules colliding with each other are remote, and viscosity conditions are meaningless. Transport is by a diffusion process from wall to wall where collisions are inelastic and the molecules are diffusely reflected [55].

From Knudsen's law [53], the flow of the gas at low pressure is given as

$$Q = \frac{8}{3} \sqrt{\left(\frac{2RT}{\pi M} \frac{A^2}{\bar{P}}\right)} \frac{\Delta P}{L} \qquad (2.21)$$

where Q is the rate of flow, R the molar gas constant, T the absolute temperature, M the molecular weight of the fluid, and $\bar{P}$ the mean pressure.

2.1.4.1 Gas–wall interaction

Various workers [54, 56, 57, 58] have found that the coefficient of slip is equal to $\xi = 4\lambda/3$ for diffuse reflection on a smooth surface but decreases to $\xi \simeq \lambda$ on a rough surface. This implies that slip is less and the rate of transfer of flow momentum greater on a rough surface than with diffuse reflection on a smooth surface. Gaede [59] intuitively suggested that a projection on a rough wall was struck by more incoming molecules on the upstream side than on the downstream side. Since a surface can only re-emit the molecules received there must be a net re-emission against the direction of flow. A similar effect to this 'back reflection' has been observed for low-pressure flow through a short tube by Davis *et al.* [60]. The internal surfaces of their tubes were machined with various geometrical profiles, and a reduction in gas flow occurred. Fryer [61] in a tentative calculation suggested that a quarter of the flow momentum received from the incident molecules was re-emitted against the gas stream. This may be an over-estimation, but Davis *et al.* have found that gross surface structure can reduce flow by as much as 20%.

The nature of the gas–wall interactions which influences steady state and transition from molecular to streamline flow are not fully understood. Barrer and Nicholson [62] studied the steady state slip flow and specular reflections of rare gases in long single glass capillaries. In the analysis of the specular reflection coefficient they considered the contributions from elastically and inelastically scattered molecules and the effect surface forces and surface roughness would have on the coefficient. No firm conclusions were drawn, but important points were that elastic scattering was apparently important only for light gases, and for heavy gases which are easily condensable, a major factor is the residence time on the surface. Their last observation that a factor $(1-\beta_2)$ which depends on ψ alone (ψ is the angle of inclination of a facet towards the direction of flow), could reduce the flow rate of very light gases below the value for diffuse reflection by severe surface roughening, is in agreement with the work of Davis *et al.* [60].

2.1.4.2 Combination of viscous and molecular flow

Rigden [63] applied the same consideration as used in equation (2.19) to the classical Poiseuille equation corrected for 'slip', and arrived at an equation having the form:

$$Q = \frac{A\varepsilon\Delta P}{k\mu L}\left[\left(\frac{\varepsilon}{S_w\rho_s(1-\varepsilon)}\right)^2 + \frac{2\varepsilon M'\lambda}{S_w\rho_s(1-\varepsilon)}\right] \qquad (2.22)$$

where M' is a constant having the value of 0.874, and λ is the mean free path of the gas molecules. The gas was assumed to behave as though there was a condition of zero velocity at a boundary approximately the length of one mean free path further out than the physical boundary. Rigden did not explain the derivation or the measurement of this slip correction but quoted the work of Millikan [56], Blankenstein [64], States [65], and Traxler & Baum [23].

Klinkenberg [66] had also shown that the discrepancies between the permeability of certain rocks measured with air and liquids could be accounted for by the slip of gas at the walls. Klinkenberg's work was not related, however, to the estimation of surface area, but its extension will be discussed in this chapter.

Carman [67] has since pointed out that the slip correction factor used by Rigden was in the form used for circular capillaries, whereas, for granular beds, the correction should be larger. He recommended the use of a slightly modified equation used by Lea & Nurse [68], which may be written.

$$Q = \frac{A\varepsilon\Delta P}{k\mu L}\left\{\left[\frac{\varepsilon}{S_w\rho_s(1-\varepsilon)}\right]^2 + \left(\frac{2}{f}-1\right)\sqrt{\frac{\pi RT}{2M}}\,\frac{k_0\mu\varepsilon}{S_w\rho_s(1-\varepsilon)\bar{P}}\right\} \qquad (2.23)$$

where f = fraction of molecules evaporating from the surface, i.e. undergoing diffuse reflection, (or $1-f$) = the fraction of molecules undergoing elastic collisions with the surface, hence the transport of fluid through very fine capillaries at low concentrations and reduced gas pressure can be regarded as the molecular flow of gases through capillaries which in turn can be regarded as a random walk phenomenon or diffusional process through a medium. The term $((2-f)/f)$, which takes account of the type of molecular reflections from capillary walls, was a later addition based on the work of Smoluchoski [69, 70] with $\overline{P}=(P_1+P_2)/2=$ mean pressure of gas.

It has been shown experimentally by Knudsen [54] and also by others [23, 56, 65] that when the ratio of mean free path to capillary diameter approaches unity the rate of flow can be expressed in the form:

$$\text{Rate of flow} = \text{Pouiseuille term} + \delta x \text{ Knudsen term} \ . \tag{2.24}$$

The dimensionless factor δ is a complex function of the mean pressure, $\overline{P}$, which has so far not proved susceptible to theoretical treatment. When the ratio of mean free path to capillary diameter is large the Poiseuille term is negligible, and δ approaches the constant value of unity. When the ratio of capillary diameter to mean free path is large $\left(\frac{\lambda}{d} \ll 1\right)$, $\delta \to 0$, the flow is viscous and the second term in equation (2.24) becomes negligible. When the ratio of mean free path to capillary diameter is approximately unity, δ approaches another limiting value (δ_1), which is constant for a given system but which depends both upon the particular gas and upon the material of the capillary wall [71]. Generally, this limiting value lies between 0.7 and 0.9. In brief, then, equation (2.24) covers molecular flow, slip flow, and viscous flow [72].

The application of equation (2.24) to flow in porous media was first carried out by Adzumi [73], but he was unaware of the Kozeny equation for viscous flow in such media. Arnell [74], independently, and later Carman & Arnell [75], by analogy with the Knudsen semi-empirical equation for the flow of gas through capillary tubes at all pressures, arrived at the following equation:

$$Q=\frac{A\varepsilon\Delta P}{\mu L}\left[\frac{\varepsilon^2}{S_w^2 k \rho_s^2(1-\varepsilon)^2}+\frac{8}{3}\sqrt{\frac{2RT}{\pi M}}\,\frac{\delta k_0}{k}\,\frac{\mu\varepsilon}{S_w\rho_s(1-\varepsilon)\overline{P}}\right] \tag{2.25}$$

From a comparison of equations (2.23) and (2.25) it is seen that the first terms on the right-hand sides are equal and identical to the simple Kozeny equation 2.19. The second term on the right-hand side of equation (2.25) is a modified Knudsen term.

Poiseuille's equation was developed by assuming that the velocity at the capillary walls was zero. If it is assumed that the velocity does not reach zero until a distance $M\lambda$ beyond the capillary walls, the modified Poiseuille equation becomes:

$$Q=\frac{Ad^2}{32\mu L}\left[1+\left(\frac{8M}{d}\right)\lambda\right] .$$

Rigden, using the uniform circular capillary model, derived the equation for a powder bed:

$$Q = \frac{A\varepsilon\Delta P}{k\mu L}\left[\left\{\frac{\varepsilon}{S_v(1-\varepsilon)}\right\}^2 + \frac{2\varepsilon M\lambda}{S_v(1-\varepsilon)}\right]$$

[Note: $k = k_0 k_1$ where $k_1 = \left(\frac{L_e}{L}\right)^2$. It seems doubtful that the concept of tortuosity can be applied to the slip flow term, and its omission would modify the term by the ratio (k/k_0)].

Arnell, again using circular capillaries, found that

$$M = \frac{8}{3}\sqrt{\frac{2RT}{\pi M}}\,\frac{k\mu\delta\alpha}{2\varepsilon\lambda\bar{P}} .$$

For granular beds, Carman found that

$$M = \left(\frac{2}{f} - 1\right)\sqrt{\frac{\pi RT}{2M}}\,\frac{k_0\mu}{2\bar{P}\lambda} .$$

Carman compared equations 2.23 and 2.25 and found that:

$$\left(\frac{2}{f} - 1\right) = \frac{16}{3\pi}\,\frac{k\delta\alpha}{k_o\varepsilon} .$$

He next assumed values for δ (0.9) and f (0.79) to show that

$$\frac{\alpha}{\varepsilon} \simeq \frac{k_0}{k} .$$

This extension is considered invalid owing to the different boundary conditions in deriving the two equations.

Within the normal working range of porosities, $\frac{k_0}{k}$ has been found to have a value that lies between 0.4 and 0.6. Allen and Maghafarti [49] found a spread of values for the dimensionless variable $\left(\frac{\delta k_0}{k}\right)$ and recommended its determination by experiment.

2.1.5 The intermediate regime

The equation governing the relationship between the pressure drop across a packed bed of powder and the rate at which a gas will flow through a bed may be written:

$$\frac{1}{A}\frac{dV}{dt} = \frac{\Delta P}{k\mu L S_v^2}\frac{\varepsilon^3}{(1-\varepsilon)^2} + \frac{\Delta P}{k\mu L S_v}\frac{\varepsilon^2}{(1-\varepsilon)} Z\lambda \tag{2.26}$$

where $Z = 2x$ and $x = \left(\frac{2}{f} - 1\right)$.

The coefficient of external friction is defined by Millikan as

$$\xi = \frac{1}{3}\rho_g \bar{u}\left[\frac{f}{2-f}\right] \tag{2.27}$$

where $\bar{u}$ is the mean thermal velocity and ρ_g is the density of the gas.

Equation (2.27) may be transposed to equation (2.26) by using the following relationship derivable from the gas laws:

(1) $$\bar{u} = \left(\frac{8RT}{\pi M}\right)^{\frac{1}{2}}$$

(2) $$\rho_g = \frac{M}{RT}\bar{P} \tag{2.28}$$

(3) $$\mu = \frac{1}{3}\rho_g \bar{u}\lambda\ .$$

Quoted values for the constant in equation (2.28) vary: (10.5, 0.31, 0.35, $\pi/10$, $\frac{1}{3}$, 1.2).

Equation (2.26) is made up of two terms:

The Carman Kozeny term:

$$S_K^2 = \frac{\Delta P}{k\bar{\mu}L}\frac{\varepsilon^3}{(1-\varepsilon)^2} \div \frac{1}{A}\frac{dV}{dt} \tag{2.29}$$

and the Knudsen term

$$S_M = \frac{\Delta P}{k\mu L} \frac{\varepsilon^2}{(1-\varepsilon)} Z\lambda \div \frac{1}{A}\frac{dV}{dt} \tag{2.30}$$

If the Carman–Kozeny term predominates the surface area of the powder may be determined from equation (2.29), taking the aspect factor k to be equal to 5. If the Knudsen term predominates, equation 2.30 may be applied with $Z = 3.4$. The region of overlap may be determined by using the following procedure:

From equations (2.29) and (2.30)

$$S_K^2 = \left(\frac{\varepsilon}{1-\varepsilon}\right) \frac{S_M}{Z\lambda} \ . \tag{2.31}$$

From equations (2.26), (2.29), and (2.30)

$$S_V^2 = S_M S_V - S_K^2 = 0 \ . \tag{2.32}$$

This is quadratic in S_V, having the following solution:

$$S_V = \frac{S_M}{2} + \sqrt{\left(\frac{S_M^2}{4} + S_K^2\right)} \tag{2.33}$$

For $Z = 3.4$ and for air at atmospheric pressure $\lambda = 9.66 \times 10^{-8}$ m, S_K and S_M can be calculated.

2.1.6 The Klinkenberg permeability coefficient

As pointed out in section 2.1.4.2, Klinkenberg measured with air and water the permeability of certain rocks and stated that the enhanced air flow found could be accounted for by the slip of gas at the walls. Most recently, the International Organization for Standardization (ISO) [76] have adopted the permeability coefficient, B, found by Klinkenberg. The value of the Klinkenberg factor B should increase with decreasing pore size, decreasing relative molecular mass, and increasing temperature and viscosity of the gas. It is constant for a given gas and porous material and has the dimensions of pressure.

The expressions and symbols used by ISO may be written and compared in the usual consistent form:

2.1.6.1 *Viscous flow*

The empirical formula for the flow of fluids through porous materials as given by Darcy, assuming that the losses were all due to viscous shear, was written as

$$\frac{\Delta P}{e} = \frac{Q \times \mu}{A \times \psi_v} \tag{2.34}$$

where e = thickness of the test piece.

2.1.6.2 *Viscous and inertia flow*

It is pointed out [76] that experience shows that in the majority of cases concerning the flow of fluids through porous metals, only three mechanisms are usually involved. They are viscous flow, inertia flow, the slip flow. Inertia flow concerns the loss of energy due to the changes in the direction of the fluid in passing through the tortuous porosity and to the onset of local turbulences in the pores, and has been combined with the viscous loss equation of Darcy by Forchheimer to give the equation (slip flow usually absent)

$$\frac{\Delta P}{e} \; \frac{Q \times \mu}{A \times \psi_v} \; \frac{Q^2 \times \rho}{A^2 \times \psi_1} \tag{2.35}$$

At low velocities of flow (Q/A) of viscous fluids, the inertia term ψ_1 of equation (2.35) is usually insignificant compared with the viscous term and can be ignored to give the simpler equation (2.34).

2.1.6.3 *Slip flow*

Equation (2.34) assumes that the pore size is large compared with the mean free path of the molecules of the test fluid. This assumption is most likely to be invalid [76] with a very small pore size and with gases at reduced pressure or high temperature. When the mean free path of the gas molecules approaches the same order of size as the pores of the metal, slip flow occurs. When slip flow is present, the porous metal appears to be more permeable than when slip flow is absent. Also, when slip flow is present, inertia losses are usually absent, so that equation (2.35) may be written in the form

$$\psi_s = \frac{Q \times \mu \times e}{A \times \Delta P}$$

where ψ_s is the permeability coefficient with slip flow present.

The correction for slip flow is given by

$$\psi_s = \psi_v \times \left(1 + \frac{2 \times B}{P_1 + P_2}\right) \tag{2.36}$$

where

ξ_s is the observed viscous permeability with slip flow present;
ψ_v is the true viscous permeabiliuty coefficient;
B is the Klinkenberg factor, which is constant for a given gas and porous material and has the dimensions of a pressure
e is thickness from equation (2.34)
ρ is density of the test fluid at the mean temperature and pressure
μ is absolute dynamic viscosity coefficient as defined by Newton's law.

The relationship between ψ_s and ψ_v may be rewritten in the form

$$\xi_s = B \times \psi_v \times \left(\frac{2}{P_1 + P_2}\right) + \psi_v \ . \tag{2.37}$$

Hence, by measuring ψ_s over a range of different absolute pressures (i.e. P_1 and P_2) and by plotting ψ_s against $2/P_1 + P_2$, a straight line is obtained. The slope of this line is equal to $B \times \psi_v$. The intercept of this line on the ψ_s axis is equal to the viscous permeability ψ_v. As stated earlier, the value of B increases with decreasing pore size, decreasing relative molecular mass, and increasing temperature and viscosity of the gas.

2.1.6.4 Wall and end effects

Equation (2.34), relating to the flow of fluids, assumes that the porosity is uniformly continuous, whereas at the surfaces of a test piece discontinuities occur. In that case, there are two cases [76] to consider:

(a) the wall effect for test pieces edge-sealed into a container;
(b) the end effect at the upstream and downstream surfaces of all test pieces.

In general with granular materials, it is pointed out that if the diameter of a test piece is not less than about 100 times the diameter of the particles comprising the porous metal, the wall effect is usually negligible, and with a test piece diameter of about 40 times the particle diameter, the error is less than about 5%.

End effects are usually negligible when the test piece thickness e is not less than 10 times the diameter of the particles comprising the porous metal. As in the case of the wall effect the end effect also depends upon the difference between the porosity at the surface and the internal porosity.

With the above definitions, we may now systematize the relationships by writing equation (1.1) as:

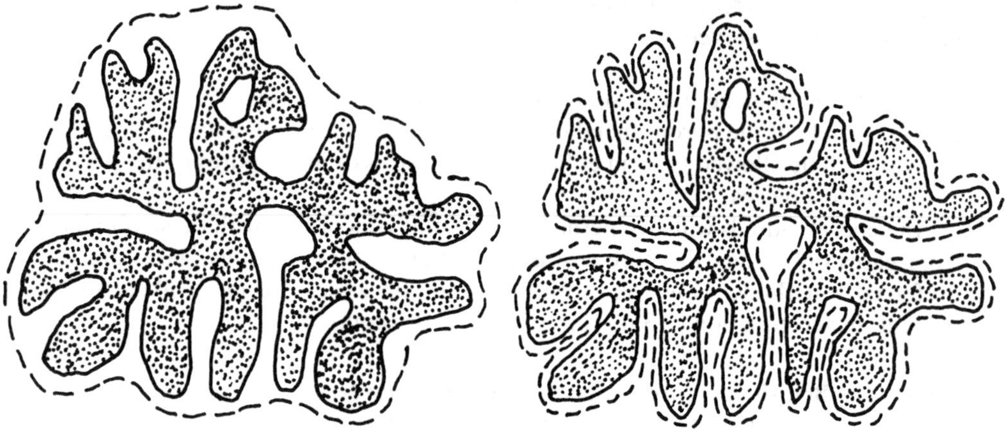

Fig. 2.1 — Open and closed pores. Surface reached by permeametry method. Surface resolved by adsorption method.

changes in the powder solubility and liquid density are negligible. Otherwise the liquid should be saturated with the powder under test and filtered at the test temperature before measuring its density. Frothing may probably occur when wetting agents are used.

2.1.8.2.3 General

The thermostatically controlled water bath shall be maintained at the same operating temperature throughout the following tests. The selected operating temperature shall be 2°C–5°C above room temperature.

2.1.8.2.4 Determination of density of displacement liquid

The procedure shall be as follows.

(a) Clean and dry the density bottle and stopper and weigh to ± 0.0005 g.
(b) Fill the bottle with de-aerated distilled or de-aerated deionized water at room temperature and, without using undue force, insert the stopper with a slight twist, care being taken to avoid trapping air bubbles.
(c) Immerse the bottle nearly to the top of its neck in the water bath and leave for at least 30 min.
(d) After ensuring that air bubbles are absent and that the bottle is filled to the top of the capillary bore, remove with dry filter paper any water on the top of the neck of the bottle where stopper and neck meet, and dry the sides and the top of the stopper, taking care not to withdraw any from the capillary.
(e) Remove the bottle from the water bath and, after allowing it to cool for a few minutes, wipe it free of water with the minimum of handling, and weigh to ± 0.0005 g.
(f) By topping up the bottle with more distilled or deionized water, re-immersing it

in the water bath and proceeding as before, make two additional weighings of the bottle filled with water.

(g) Calculate the mean of these three weighings, and from the difference between this mean weighing and the first weighing of the bottle and stopper alone, calculate the mass (W_1) of water to fill the bottle at the selected operating temperature.

(h) Clean and dry the density bottle and stopper used in the above test and reweigh to ± 0.0005 g. Repeat the above procedure, but filling the bottle with the displacement liquid instead of water, and determine the mass (W_2) (average of three weighings).

Calculate the density of the displacement liquid to the nearest 0.5 kg/m^3 from the formula:

$$\rho_{DL} = \frac{W_2}{W_1} \times \rho_W$$

where

ρ_{DL} is the density (in kg/m^3) of displacement liquid at the selected operating temperature

ρ_W is the density (in kg/m^3) of pure water at the selected operating temperature (see Table 2.1)

Table 2.1 — Density of pure water, ρ_W

Temperature	ρ_W
°C	kg/m^3
18	998.6
19	998.4
20	998.2
21	998.0
22	997.8
23	997.6
24	997.3
25	997.1
26	996.8
27	996.5
28	996.3

W_1 is the mass (in g) of water to fill the bottle at the selected operating temperature

W_2 is the mass (in g) of displacement liquid to fill the bottle at the selected operating temperature.

2.1.8.2.5 Determination of effective solid density of powder

Using the procedure described in section 2.1.8.2.1 calibrate at intervals each density bottle and stopper by determining the mass (W_3) (average of three weighings) of the bottle, stopper, and displacement liquid to fill the bottle at the selected operating temperature.

(a) Clean and dry the calibrated density bottle and stopper and weigh to ± 0.0005 g.
(b) Place a representative sample of the powder in the density bottle by means of a small funnel. The quantity of sample shall be sufficient to approximately half fill the bottle. Reweigh with stopper to ± 0.0005 g and, from the difference between the first and second weighings, calculate the mass (W_4) of the sample in the density bottle.
(c) Place the density bottle in the vacuum desiccator. The lid of the desiccator is fitted with a tap funnel with a long stem which projects into, and touches, the neck of the density bottle so that added liquid runs down the inner surface of the bottle.
(d) Evacuate at a pressure of 0.3 kPa–3.0 kPa and then add displacement liquid, from the tap funnel, to the density bottle until it is approximately $\frac{3}{4}$ full. This addition shall be made with great care to avoid loss of powder by splashing. Leave the suspension to settle, under vacuum, until the liquid above the powder is quite clear.
(e) Admit air to the desiccator and carefully fill the bottle with displacement liquid, ensuring that the liquid above the powder remains clear.
(f) Using exactly the same procedure as described in section 2.1.8.2.4 determine the mass (W_5) (average of three weighings) of the bottle, stopper, powder sample, and displacement liquid to fill the bottle at the selected operating temperature.

NOTE: An alternative method, using a centrifuge, is specified in BS 3483: Part B9.

2.1.8.2.6 Calculation

Calculate the effective solid density of the sample of powder to the nearest 1 kg/m^3 from the formula:

$$\rho_s = \frac{W_4 \times \rho_{DL}}{W_3 + W_4 - W_5}$$

The density of air has been neglected in this equation since its effect is negligible.

2.1.8.2.7 Effective solid volume

These definitions are practical rather than fundamental. The effective solid volume is the mass of the particles divided by the effective solid density. The definitions have to be introduced mainly because the permeability method of measuring specific surface is frequently applied to powders that contain porous particles. The effect of particle porosity or pores within the particles, on the accuracy of measurement and on the interpretation of results, can present difficulties, because the particles that compromise a powder are themselves porous. Permeability measurements indicate only the surface area that contributes to the pressure drop, and this includes little of the area making up the walls of capillary cracks and cul-de-sac pores within the body of a particle. Usually the surface areas obtained by the use of gas adsorption methods are significantly higher since this procedure does include the contributions due to the particle internal pore system.

The possibility that a significant fraction of the particles might contain pores makes it essential that the bed porosity should not be estimated by using a value of solid density extracted from a reference book [77]. The effective solid density determined by the method in section 2.1.8.2.1 shall be used.

2.1.8.2.8 Bed porosity (ε)

The ratio of volume of interparticle voids to the total bulk volume of the compacted bed of powder used for the determination of the effective permeability specific surface. It is calculated from the mass, bulk volume, and effective solid density of the powders as shown below:

2.1.8.2.8.1 Calculation of bed porosity. The bed porosity, ε, required for the calculation of the specific surface from permeability data, is obtained from the relationship:

$$\varepsilon = 1 - \frac{W \times 10^{-3}}{L \times A \times \rho_s} \tag{2.40}$$

where W is the mass powder, in grams, used to prepare the bed;

L is the linear dimension of a bed of powder (in m) parallel to the direction of the flow of air (commonly known as the height of the powder bed)

A is the cross-sectional area of a bed of powder (m^2) perpendicular to the direction of the flow of air.

ρ_s is the effective solid density; ρ_s may be taken as the true density of the material only when it is known, beyond reasonable doubt, that the particles constituting the powder are free from pores. In general, however, this cannot be known with certainty, and the possibility that a significant fraction of the particles may contain pores makes it essential that the bed porosity should not be calculated by using a value for the solid density that has been extracted from a handbook of physical constants [77]. The effective solid density shall be determined by the method [77] given in section 2.1.8.2.1. It should also be noted that the term 'porosity' has wide usage and should not be confused with the internal porosity of individual powder particles, much of which is not accessible to measurement by the methods of air permeability.

2.1.8.2.8.2 Apparent dependence of specific surface area on bed porosity. The Kozeny constant K, is dependent on the pore size distribution within the bed. The pore size distribution will depend not only on the nature of the particles of the powder but also on their packing in the bed; that is, the bed porosity.

With a new material it is advisable to draw a plot of S_v against ε over a range of ε likely to be encountered. Frequently such a plot results in a plateau region where S_v varies little with change in ε. It is desirable to work in this plateau region wherever possible, provided that the bulk density characteristics do not alter appreciably from sample to sample. If the bulk density characteristics do not vary, then for routine use, beds must be packed to a consistent porosity. Where the bulk density of the material does vary, it is preferable to pack the bed at constant pressure.

Where S_v changes significantly with ε, however carefully the bed preparation is carried out, the method is not applicable except for some industrial control purposes where the error introduced is not significant.

2.1.8.2.9 Principles and concepts of air flow

2.1.8.2.9.1 Viscous flow. Air-flow in which adjacent layers of air do not mix except at the molecular level and the velocity of the air at the powder interface is zero.

2.1.8.2.9.2 Carman–Arnell equation. The rate of a flow of fluid moving under the influence of a constant pressure difference through a compacted bed of powder of uniform cross-sectional area is a function of the magnitude of the surface area that the walls of the channels, through the bed, present to the moving fluid. Because there is normally a great variation in, and a lack of precise knowledge of, the shape and dimensions of such channels, rigorous mathematical treatment is impracticable. However, by making a number of simplifying assumptions, a relationship enabling the specific surface area of many powders to be estimated from fluid permeability data has been derived by Carman and Arnell. The equation may be written in the basic form as

$$q = \frac{A\varepsilon\Delta P}{K\mu L}\left(\frac{\varepsilon^2}{S_v^2(1-2)^2} + \frac{8}{3}\sqrt{\frac{2RT}{\pi M}}\,\frac{\delta k_0 \mu\varepsilon}{PS_v(1-\varepsilon)}\right) \tag{2.41}$$

Rather than attempting to solve the equation (a quadratic in S_v) directly, the calculation can be carried out by evaluating two separate terms and combining them to give S_v using equation (2.33).

2.1.8.2.9.3 The Kozeny term (S_k).

$$S_k = \sqrt{\frac{A\varepsilon^3\Delta P}{K(1-\varepsilon)^2 L\mu q}} \tag{2.42}$$

This term is identical to the Kozeny–Carman equation and gives the contribution to the surface area of the powder due to viscous flow. K, the Kozeny constant, is a complex dimensionless number not amenable to theoretical treatment. Its numerical value depends on the pore size distribution within the compacted bed. Where there is sufficient reason to believe that it is advantageous to adopt some other value for K, this is permissible, but the value adopted should be stated when the results are reported. Otherwise K should be taken as 5.0.

2.1.8.2.9.4 The slip flow term (S_m). Enhanced airflow caused by the air velocity at the air–powder interface not being zero as assumed for viscous flow. It becomes significant at low pressures or for very fine powders.

$$S_m = \frac{A\Delta P}{KLq}\frac{8}{3}\sqrt{\frac{2RT}{\pi M}}\frac{\delta k_0 \varepsilon^2}{P(1-\varepsilon)} \tag{2.43}$$

δ is a complex dimensionless function of the mean pressure and k_0 is a dimensionless number in the range 2–3. Neither of these factors has proved amenable to theoretical treatment. In the British Standard, the generally accepted value for the product δk_0 of 2.25 is assumed.

The constants in equation (2.42) can be combined to give a direct relationship between S_m and S_k. When air ($M = 0.029$) is used

$$S_m = S_k^2\, 81.0\, \frac{(1-\varepsilon)}{\varepsilon}\sqrt{T}\,\frac{\mu}{P} \tag{2.44}$$

2.1.8.2.9.5 The combined term (S_v). The combined term, S_v, is shown as equation 2.33 in this book.

2.1.8.2.9.6 Molecular (Knudsen) flow. At very low pressures the mean free path of the gas molecules, λ, increases and intermolecular collisions become increasingly rare until all collisions are between the gas molecules and the void (capillary) walls. The concept of viscosity thus breaks down and molecular transfer is governed by Ficks Law of Diffusion. This regime is called the molecular flow, Knudsen flow, or diffusional flow regime.

2.1.8.2.9.7 Range of measurements. The range of measurements given in the British Standard are applicable to powders with specific surface areas between the limits 2×10^4 and 5×10^7 m^2/m^3. For air at atmospheric pressure (10^5 Pa) and 20°C (293 K) and with a bed porosity of 0.5, the surface area due to viscous flow only, S_k, will be more than 90% of S_v when S_v is less than 1×10^6 m^2/m^3. For the same conditions, the surface area due to slip flow only, S_m, will be more than 90% of S_v when S_v is more than 36×10^6 m^2/m^3.

2.1.8.2.10 Effective permeability volume surface mean diameter (d_{sv})

The diameter of an imaginary solid sphere that has the same volume specific surface as the powder to which the term is applied. If a hypothetical powder is assumed, composed entirely of spheres of this diameter, the effective volume specific surface of the whole will be equivalent to that of the powder under test.

Having reviewed briefly some of the available air permeability principles and concepts used in the determination of surface areas of powders, we shall now proceed to see some of the advances [83–88] made in areas of instrument characterization and technique.

SYMBOLS AND UNITS

A = Cross-sectional area of a bed of powder, m^2

B = Klinkenberg factor, which is constant for a given gas and porous material and has the dimensions of pressure, as used by ISO (N/m^2).

d_{sv} = Surface volume mean diameter, m^2/m$_3$

$\frac{dV}{dt}$ = Volume flow rate of gas through the powder bed, m^3/s

g = Acceleration due to gravity, 9.806 65 m.s^{-2}

ISO = International Organization for Standardization

k, K = Aspect factor (Kozeny constant, equal to 5.0)

k_0 = Shape factor

k_1 = Tortuosity factor

L = Length of powder bed, m

L_e = Length of tortuous path in packed bed of powder, m

M = Molecular weight of the gas (29 kg/kmol for air)

P = The atmospheric pressure at the time of the determination, N/m^2 (Pa)

$\bar{P}$ = Mean gas pressure in the bed, N/m^2

P_1 = Pressure at bed inlet, N/m^2

P_2 = Pressure at bed outlet, N/m^2

q, Q = The rate of flow of an incompressible fluid through a bed of powder, m^3/s

R = Gas Constant, 8.314 41 J. Mol^{-1}.K^{-1}

S_K = Effective volume specific surface (assuming only viscous flow occurs) m^2/m^3

S_V = Volume specific surface, m^2/m^3

S_w = The effective mass specific surface of a powder, m^2/kg

S_M = Surface area using molecular term, m^2/m^3 (assuming only slip-flow occurs)

S_v^2 = Powder surface area as given by the Carman–Arnell equation at a range of pressures, m^2/m^3

T= Absolute temperature (taken as constant), 298 K
t= Time, seconds
u_1= Actual velocity of gas molecules, m^3/s
v, v= Approach velocity, m^3/s; the effective solid volume of a powder sample, m^3
W= Mass of powder in the bed, kg
W_1= The mass of pure water to fill a density bottle, g
W_2= The mass of displacement liquid to fill a density bottle, g
W_3= W_2 plus the mass of the empty density bottle, g
W_4= The mass of sample used to determine the effective solid density of a powder, g
W_5= W_4 plus the mass of the empty density bottle and the displacement liquid to fill the bottle, g
Z= Constant ($z = 2x$).
ΔP= Pressure drop across the packed bed or difference between the pressures on the upstream and downstream surfaces of the porous test material, N/m^2
δk_0= Variable in Carman–Arnell equation (dimensionless)
ε= Powder bed porosity
λ= Gas molecular free path
μ= Viscosity of air, 186×10^{-7} $N.s.m^{-2}$
ρ_s= The effective solid density of a powder, kg/m^3; particle density, kg
ρ_{DL}= The density of the displacement liquid used in the measurement of effective solid density, kg/m^3
ρ_L= The density of a manometer liquid during the determination, kg/m^3
ρ_w= The density of pure water during the calibration of a density bottle, kg/m^3
ψ_I= Inertia permeability coefficient, m
ψ_v= Viscous permeability coefficient, m^2
ψ_s= Permeability coefficient with slip flow present, m^2
ξ= coefficient of external friction

REFERENCES

[1] Happel, J. & Brenner, H., *Low Reynolds number hydrodynamics with special applications to particulate media*, Prentice-Hall, Englewood Cliffs, NJ (1965).
[2] Meinzer, O. E., Hydrology, *Physics of the Earth*, IX, Dover, New York (1949).
[3] Bear, J., Zaslavsky, D., & Irmay, S., Physical principles of water percolation and seepage. Series: *Arid Zone Research XXIX*, UNESCO, Paris (1968).
[4] Baver, L. D., *Soil physics*, Wiley, New York, 3rd Ed. (1956).
[5] Terzaghi, K., *Theoretical soil mechanics*, Wiley, New York (1943).
[6] Lambe, T. W., *Soil testing for engineers*, Wiley, New York (1951).
[7] Terzaghi, K. & Peck, R. B., *Soil mechanics in engineering practice*, Wiley, New York, 2nd Ed. (1968).
[8] Am. Soc. Test. Mater. *Spec. Tech. Publ. 163*, (1954).
[9] Muscat, M., *Physical principles of oil production*, McGraw-Hill, New York (1949).

[10] Trask, P. D., *Applied sedimentation*, Wiley, New York (1950).
[11] Dickey, G. D., *Filtration*, Reinhold, New York (1961).
[12] Brownell, L. E., Filtration, *Kirk-Othmer encyclopedia of chemical engineering technology*, Vol. 9, Interscience, New York, 2nd Ed., pp. 264–286 (1966).
[13] Coulson, J. M. & Richardson, J. F., *Chemical engineering*, Vol. 2, Unit Operations, Pergamon Press, London, 2nd Ed. (1968).
[14] Perry, J. H., *Chemical engineers' handbook*, McGraw-Hill, New York, 4th Ed. (1969).
[15] Harris, C. C. & Smith, H. G., The moisture retention properties of fine coal: A study by permeability and suction potential methods, *2nd Symp. on Coal Preparation, Leeds Univ.*, pp. 57–98, 191–210 (Oct. 1957).
[16] Pryor, E. J., *Mineral processing*, Elsevier, Amsterdam, 3rd Ed. (1965).
[17] Evans, D. J. I. & Shoemaker, R. S., *Intern. Symp. on Hydrometallurgy, Chicago, Feb. 25–Mar. 1, 1973*; American Institute of Mining, Metallurgical and Petroleum Engineers, New York (1973).
[18] Barrer, R. M., *Diss. Faraday Society*, **3**, 61 (1948).
[19] Allen, T., *Particle size measurements*, 2nd Ed. Chapman and Hall (1974).
[20] Darcy, H. P. G., Les Fontaines Publiques dse la Ville de Dijon, *Exposition et Application à Suivre et des Formules à Employer dans les Questions de Distribution d'Eau*, Victor Dalamont, Paris (1856).
[21] Bigelow, N., *J. Am. Chem. Soc.*, **29**, 1675 (1907).
[22] Brunhes, M. & Kolloid, Z., **34**, 1 (1924).
[23] Traxler, N. G. & Baum, S., *Physics*, **7**, 9 (1936).
[24] Fancher, S. & Lewis, T., *Ind. Eng. Chem.*, **25**, 1139 (1933).
[25] Carman, P. G., Determination of specific surface of powders, *J. Soc. Chem. Ind. (Trans.)*, **57**, 225 (1938).
[26] Kozeny, J., Über kapillare Leitung des Wassers im Boden, *Bericht Akad. Wiss., Wien,* **IIa** (136), 271–306 (1927).
[27] Kundt, A. & Warbung, E., *Ann. der Physik,* **155**, 337 (1875).
[28] Loeb, L. B., *Kinetic theory of gases*, 2nd Ed., McGraw-Hill, New York (1934).
[29] Dupuit, A. J. E. J., *Études Théoriques et pratiques sur le mouvement des eaux dans les canaux découverts et à travers les terrains permeables*, Dunod, Paris, 2nd Ed. (1863).
[30] Poiseuille, J. M. L., Recherches expérimentales sur le mouvement des liquides dans les tubes de très petit diamètre, *Mem. Savants Etrange (Paris),* **9**, 433–545 (1846); Engl. transl. *Rheol. Mem.,* **1** (Jan. 1940).
[31] Carman, P. C., *Flow of gases through porous media*, Butterworths, London (1956).
[32] Fair, G. M. & Hatch, L. P., *J. Amer. Water Works Assn.*, **25**, 1551 (1933).
[33] Essenhigh, R. H., *Safety in Mines Res. Est. Report No. 120* (1955).
[34] Carman, P. C., Fluid Flow through granular beds, *Trans. Inst. Chem. Eng.*, **15**, 150–166 (1937).
[35] Fowler, J. L. & Hertel, K. L., Flow of gas through porous media, *J. Appl. Phys.*, **11**, 496–502 (1940).
[36] Hitchcock, R., *J. Gen. Physiol.*, **9**, 755 (1929) (in Carman 1937).
[37] Carman, P. C., Determination of specific surface of powders, *Soc. Chem. Ind., London (Transactions),* **58**, 1–7 (1939).

[38] Carman, P. C., *Symposium on particle size analysis*, p. 118–120 (1947) Suppl. to *Trans. Inst. Chem. Eng.*, **25** (1947).
[39] Sullivan, R. R. & Hertel, K. L., The permeability method for determining specific surface of fibres and powders; *Adv. Colloid. Sci.*, **1**, 38–80 (1940); and *J. Appl. Phys.*, **11**, 761 (1940).
[40] Shriever, W., *Trans. Amer. Inst. Min. Eng.*, **86**, 329 (1930).
[41] Muskat, M. & Botsel, H. G., *Physics*, **1**, 27 (1931).
[42] Sullivan, R. R., *J. Appl. Phys.*, **12**, 504 (1941).
[43] Rose, H. E., *The measurement of particle size in very fine powders*, Constable, London, (1953); and *Int. Mech. Eng., (London), Proc.* **160**, 493 (1949).
[44] Cornell, D. & Katz, D. L., *Ind. Eng. Chem.* **45**, 2145 (1953).
[45] Wyllie, R. J. & Gregory, A. R., *Ind. Eng. Chem.*, **47**, 1379, (1955).
[46] Keyes, W. F., Determination of specific surface by permeability measurements, *Ind. Eng. Chem. Anal. Ed.* **18**, 33–34 (1946).
[47] Allen, T., *Particle size measurement*, 3rd Ed. Chapman & Hall (1981).
[48] Harris, C. C., *Powder Technology*, **14**, 209 (1977).
[49] Allen, T. & Mahafarti, R. P., *Particle Size analysis Conf. Loughborough University* (1981).
[50] Wasan, D. T., Wnek, W., Davies, R., Jackson, M., & Kaye, B. H., Analysis and evaluation of permeability techniques for characterizing fine particles. Part 1, Diffusion and flow through porous media, *Powder Technol.*, **14**, 209–228 (1976).
[51] Wasan, D. T., Wnek, W., Davies, R., Jackson, M., & Kaye, B. H., Part II, Measurement of specific surface area, *Powder Technol.*, **14**, 229–244 (1976).
[52] Maxwell, J. C., *Phil. Trans.*, **170**, 231 (1879) (in Loeb (1934)).
[53] Knudsen, M., Die Gesetze der Molekularströmung und der inneren Reibungsströmung der Gase durch Röhren, *Ann. der Physik, Lpz.*, **4**. Folge, Band 28, 75–130a (1909).
[54] Knudsen, M., *Ann. der Physik, Lpz.*, **4**, Folge, Band 28, 999–1016b (1909).
[55] Knudsen, M., *Ann. der Physik, Lpz.*, **4**. Folge, Band 34, 593–656 (1911).
[56] Millikan, R. A., Coefficients of slip in Gases and the law of reflection of molecules from the surfaces of solids and liquids, *Phys. Review,* **21**, No. 3, 217–238 (1923).
[57] Stacy, L. J., Determination by the constant deflection method of the value of the coefficient of slip for rough and for smooth surfaces in air. *Phys. Review,* **21**, No. 3, 239 (1923).
[58] Van Dyke, K. S., *Phys. Review,* **21**, No. 3, 250 (1923).
[59] Gaede, W., *Ann. der Physik,* **41**, 289 (1913).
[60] Davis, D. H., Levenson, L. L. & Millerton, N., *J. Appl. Phys.*, **35**, 529 (1964).
[61] Fryer, G. M., *Proc. Roy. Soc.*, **293A**, 329–341 (1966).
[62] Barrer, R. M. & Nicholson, D., *Can. J. Chem.*, **43**, 896, 913 (1965).
[63] Rigden, P. J., Discussion of report of air-permeability method for determining fineness of cement, *Am. Soc. Test. Mater. Bull,* **123**, 49 (1943) and *J. Soc. Chem. Ind.*, **66**, 130, (1947).
[64] Blankenstein, E., *Phys. Review,* **21**, 703 (1923).
[65] States, M. N., *Phys. Review,* **21**, 662 (1923).

[66] Klinkenberg, L. J., The permeability of porous media to liquids and gases, *Am. Petrol. Inst. Drilling and Production Practice*, 200–213 (1941).
[67] Carman, P. C., Slip flow in granular beds, *Nature*, **160**, 301 (1947).
[68] Lea, F. M. & Nurse, R. W., Permeability methods of fineness measurement, *Symp. on Particle Size Analysis* (Feb. 1947) p. 47–56. Suppl. to *Trans Inst. Chem. Eng.*, **25**, (1947).
[69] Smoluchowski, M. von., *Ann. der Physik*, **33**, 1559 (1910).
[70] Smoluchowski, M. von., *Ann. der Physik*, **35**, 1000 (1911).
[71] Adzumi, H., Studies on the flow of gaseous mixtures through capillaries: II: The molecular flow of gaseous mixtures, *Bull. Chem. Soc. Japan*, **12**, 285–91 (1937); III: The flow of gaseous mixtures at medium pressures, *ibid.* **12**, 292–303 (1937). On the flow of gases through a porous wall, *ibid.* **12**, 304–12 (1937).
[72] Carman, P. C. & Malherbe, P. le R., Routine measurement of surface of paint pigments and other fine powders. *J. Soc. Chem. Ind.*, **69**, 134–143 (1950).
[73] Adzumi, H., On the flow of gases through a porous wall, *Bull. Chem. Soc. Japan*, **12**, 304–312 (1937).
[74] Arnell, J. C., Permeability studies: surface area measurements, *Can. J. Research* **A24**, 103–16 (1946), A25, 191–209 (1947) and A26, 29–38 (1948).
[75] Carman, P. C. & Arnell, J. C., Surface Area measurements of fine powders using modified permeability equations, *Can. J. Res.*, **126(A)**, 128–136 (1948).
[76] International Organization for Standardization (ISO) Ref. No. *ISO.4022-1977(E) Document No. 82/80603*.
[77] BSI Draft Standard *Methods for the determination of specific surface of powders*. Part I: Air permeability methods (revision of BS.4359: Part 2: 1971). Document 81/70580 (March 1981).
[78] Davies, W. & Rees, W. J., The specific surface and grain shape of silica sands used for glass making. *Trans. Soc. Glass Technol.*, **29**, 279–288 (1945).
[79] Orr, C., *Particulate technology*, Macmillan, New York (1966).
[80] Scheidegger, A. E., *The physics of flow through porous media*, Macmillan, New York, 2nd Ed. (1960).
[81] Harris, C. C., Flow through porous media. Examination of the Immobile Fluid Model. *Powder Techn*, **17**, 235–252 (1977).
[82] Rose, W. D., Permeability and gas-slippage phenomena. *Amer. Petrol. Institute Drilling and Production Practice*, 209–217 (1948).
[83] Igwe, G. J. I. & Allen, T., Advances in gas permeability measurements. *J. Chem. Eng. Techn.*, **10**, No. 2, 86–92 (1987).
[84] Igwe, G. J. I., Improved apparatus for measurement of surface area of powders. *J. Ind. and Eng. Chem. Res.*, **26**, 6, 1254–1259 (1987).
[85] Igwe, G. J. I., Electronic amplification of pressure drops through packed beds at sub-atmospheric pressures. *Indian J. of Technology,* **25**, p. 490–497 (1987).
[86] Igwe, G. J. I. Porosity and permeametry at sub-atmospheric pressures. *J. Chem. Eng. Techn.* To be published in No. 6 (1988).
[87] Igwe, G. J. I., Allen, T., & Stanleywood, N. G., Surface area measurement and gas permeametry at sub-atmospheric pressures. *Proceedings of the Particle Size Analysis 1988, of the Analytical Division of the Royal Society of Chemistry*, 243–256 (1988).
[88] Igwe, G. J. I., Gas and particle surface reaction at sub-atmospheric pressures.

Paper No. 88-3630 of the *Proceedings of the First National Fluid Dynamics Congress of the American Institute of Aeronautics and Astronautics, held July 25–28th in Cincinnati, Ohio, USA* (1988).

3

Instrumentation in surface area measurement

3.1 INTRODUCTION

A large number of gas permeameters have been designed for measuring the flow rate and pressure drops across packed beds. They can be classified into two basic types according to the way in which flow resistance is measured.

3.2 COMPARISON OF CONSTANT FLOW RATE AND CONSTANT VOLUME METHODS

With constant flow rate permeameters, air is passed through the bed of powder at a constant volume rate of flow, with the entry or exit of the apparatus set at atmospheric pressure and the pressure drop measured by a manometer. Constant flow rate methods have some advantage in that the effect of flow rate variation on bed permeability and calculated specific surface can readily be studied. They may be easily used as primary standard methods, particularly if used with bubble flow meters, which eliminate the need for frequent calibration when the apparatus is in constant use. In general, the apparatus is more cumbersome [1] than that used in constant volume methods.

With constant volume permeameters, on the other hand, a fixed volume of air is passed through the bed of powder by creating a pressure difference across the bed and allowing the air to flow from the high pressure side until a predetermined decrease in the pressure difference is attained. Constant volume methods do not require a separate flow meter. The apparatus is usually more robust and often portable.

3.3 MEASUREMENT OF FLOW RESISTANCE

3.3.1 Constant flow rate permeameters

Lea & Nurse [2] designed an air permeability apparatus (Fig. 3.1) which was connected to a capillary manometer and a flow meter. Dry air was forced through the sample bed under pressure or by suction, enabling flow rates and pressure drops to be determined.

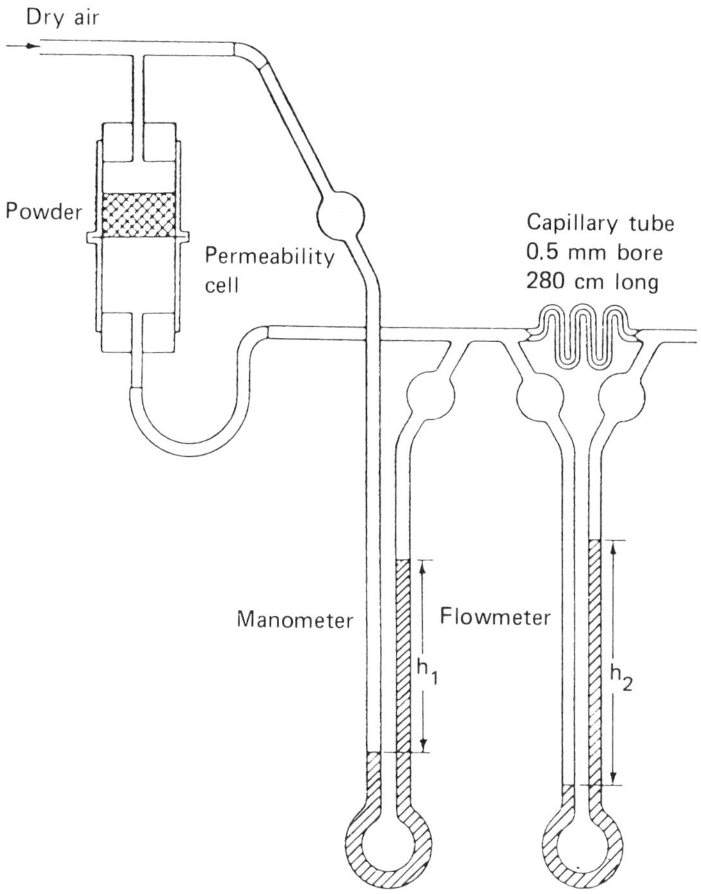

Fig. 3.1 — Lea & Nurse permeability apparatus with manometer and flow meter.

The simplicity of the method is readily apparent and has been found to be quite reliable for regular and irregular particles provided that they have no internal surfaces and are above 10 micrometres in diameter. The specific surface of the sample powder was calculated by using equation 2.19.

Lea & Nurse [3] modified their apparatus with a fixed plunger giving a fixed height of 1 cm . Since the form of correction for molecular flow had then not obtained general acceptance, the results were given in the uncorrected form.

3.3.1.1 Using the Lea and Nurse Apparatus

3.3.1.2 The permeability cell

When the apparatus is first received, the cell diameter should be checked by suitable means as discussed below, and the permeability cell should be examined periodically for wear and corrosion. Corrosion may take place whether or not the cell is in use.

Examples of cell design are shown in Fig. 3.2. The materials indicated in the figures are generally suitable, but may not be so in all instances. For example, metals may be dangerous when used in testing explosive powders.

The cells in general use fall into two categories:

3.3.1.2.1 Variable volume cells

Here, a bed of powder is formed by some suitable compaction technique. In general these cells produce the most satisfactory beds, but the bed height must be measured in order to calculate the porosity of the bed. It is difficult to prepare beds of powder with a precise, predetermined porosity when using this type of cell.

3.3.1.2.2 Fixed volume cells

Here, a bed of constant height is formed, the bed porosity being predetermined by the mass of powder used in the determination. Though not of such general application, this type of cell produces a compaction that is quite satisfactory for some powders, for example, Portland cement.

3.3.1.2.3 Measurement of permeability cell dimensions

One satisfactory type of constant volume permeability cell is illustrated in Fig. 3.2(a). The dimensions marked 'D' and 'L' should be known to within 0.1%. The cell design may be varied, but the principles hereunder should be observed. The bed is formed on a filter paper supported by a perforated plate. The paper disc is clamped between the two parts of the cell, and the way in which the cell is to be positioned is as shown in Fig. 3.1.

The cell has a circular internal cross-sectional area which is accurately known, and it should be easy to connect to the flow resistance measuring apparatus. The compacting device should be capable of producing from a known mass of powder a uniform bed of measurable height. It is important that the effect of the cell wall should be negligible. This will generally be so if the cell diameter is not less than 12 mm, although with some materials smaller cells have been used satisfactorily. The flow resistance of the empty cell should be measured separately, and, if necessary, allowance made for it before calculating the specific surface of any powder from permeability measurements made with the cells.

The internal diameter in the region to be occupied by the powder bed should be measured on at least three diameters, using a telescopic gauge and micrometer to an accuracy of at least 0.1%. 'A' shall be calculated to an accuracy of $\pm 0.20\%$. For fixed volume cells 'L' should be measured by using a test piece of hardened metal of a diameter that will easily fit into the cell and of a length $(L+0.1)$ mm measured to 0.01 mm. The test pieces should be placed in the assembled cell (including the appropriate number of filter paper discs) and the plunger should be inserted. The mean gap between the shoulder of the plunger and the top of the cell should be measured with a cathetometer so that L can be calculated to an accuracy of 0.25%.

For variable volume cells the length between well-defined fixed marks on the cell and on the plunger body shall be measured to 0.01 mm when the plunger is inserted in the assembled cell (include the appropriate number of filter paper discs). In use the length between the marks should be measured to a precision such that the length of the powder bed, L, can be calculated to an accuracy of 0.25% or better.

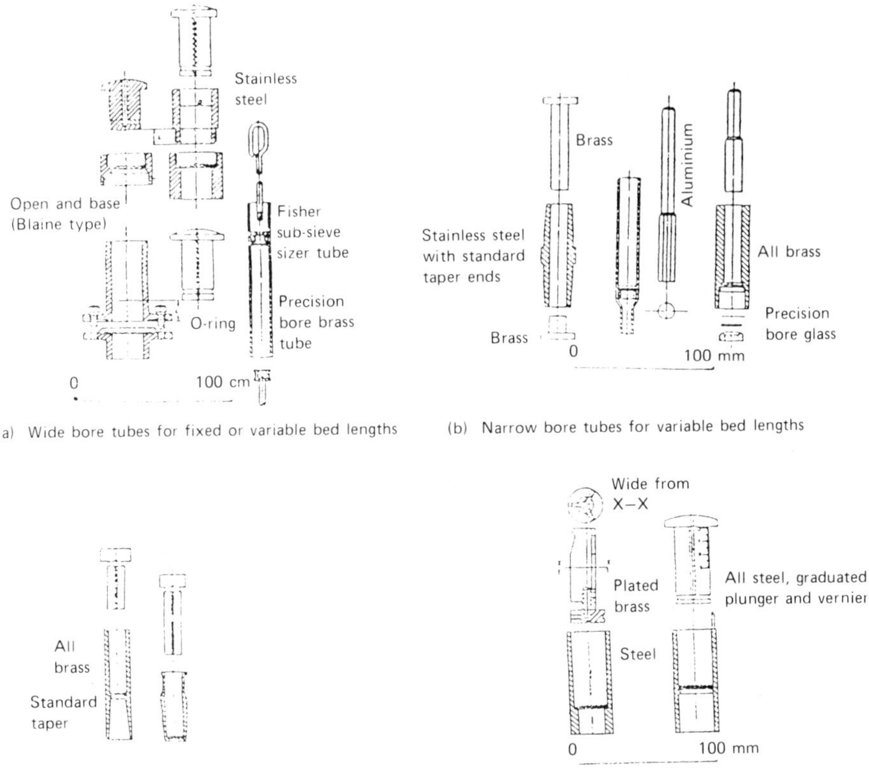

(a) Wide bore tubes for fixed or variable bed lengths

(b) Narrow bore tubes for variable bed lengths

(c) Narrow bore tubes with interchanges plungers for fixed or variable bed lengths

(d) Wide bore tubes for variable bed lengths

Fig. 3.2 — Permeability cells.

3.3.1.2.4 Effect of cell dimensions

The dimensions of the cell not only affect the dimensions of the powder bed but also its porosity. Relatively small errors in A and L (and hence ε) therefore have a larger effect on S, and it can be shown that the magnitude of the errors is as follows:

$$\text{Error in } A = \pm x\% \qquad \text{Resulting error in } S_k = \pm \frac{3\,x}{2\,\varepsilon}\%$$

$$\text{Error in } L = \pm y\% \qquad \text{Resulting error in } S_K = \pm \frac{(3-2\varepsilon)y}{2\,\varepsilon}\%$$

where ε is the true porosity of the powder bed. It is therefore necessary to measure A, and, for fixed volume cells, L, regularly and to use the dimensions found rather than the nominal values.

3.3.1.3 *Manometers*

The manometers used for measuring the pressure difference across the cell and across the flow meter resistance as shown in Fig. 3.1, should have arms about 600 mm in length and should contain kerosine or other non-volatile liquid of low density and viscosity.

3.3.1.4 *Flow meters*

The capillary tube of the flow meter should have a bore of not less than 0.5 mm and a length that gives pressure differences across the flowmeter manometer within the range 150 mm to 550 mm for flow rates that give pressure differences of not less than 150 mm across the bed. For many purposes 3 m will be found to be a convenient length for capillary bores of about 0.8 mm diameter.

3.3.1.5 *Determination of the flow meter constant*

When the apparatus is in regular use, the flow characteristics of the capillary flow meter may change owing to deposition on the capillary wall. The appropriate frequency of calibration should therefore be decided in the light of experience, probably every three months. After passing dry air through the flow meter at a constant rate for a measured time, collect the issuing air over kerosine and measure its volume. Having noted the flow meter reading, calculate the flow meter constant C as follows:

$$C = \frac{V \mu}{t h_1 \rho_L}$$

where†

V = Volume of dry air passed (in m^3)
μ = Viscosity of air at the calibration temperature (in Ns/m^2)
t = time during which air is collected (in s)
h_1 = flowmeter reading (in m)
ρ_L = density of the manometer liquid at the calibration temperature (in kg/m^3)

The viscosity of air in $N.s/m^2$ at temperatures in the range of 15°C to 25°C is given in Table 3.1.

Table 3.1 — Variation of air viscosity with temperature. The viscosity of dry air in $N\ s/m^2$ at temperatures in the range of 15°C to 25°C is shown below: Multiply by 10^6

Temperature °C	15 to 16	17 to 18	19 to 20	21 to 22	23 to 24	25
Viscosity	17.8	17.9	18.0	18.1	18.2	18.3

† A list of symbols and definitions will be found on p. 93.

Repeat this process for five flow meter readings over the range of 0.250 m to 0.550 m. Calculate the average value of C and express the result to the nearest 0.1×10^{-15} metric units.

3.3.1.6 Assembling and testing the apparatus

In re-assembling the permeability cell, it should be ensured that the two parts are firmly bolted together, and the cell and its connections should be tested for leakage. This is best done by disconnecting at the manometer the rubber tube leading from the lower end of the cell, sealing the tube with a screw clip, applying air pressure until the manometer shows a difference in level of at least 500 mm, and then sealing off the air inlet. The reading of the manometer should not change by more than 0.5 mm in a period of 1 min.

After preparing the bed of powder, the porosity should be calculated. The test temperature should be within 5°C of the flow meter manometer calibration temperature. In connecting up the cell, first insert the upper bung, or make the cone and socket joint. Turn on the air slowly and insert the lower bung carefully to avoid forcing the air in the wrong direction through the bed. Adjust the rate of air flow until a reading greater than 150 mm is obtained on each manometer. Record these readings at different air flow rates to ensure that steady conditions have been obtained (this is indicated by the constancy of the ratio of the two manometer readings), and take the mean of three determinations. The filter paper should be changed in the cell after every determination.

Using the appropriate equation and correction factors, depending on the type of flow meter used and whether the pressure difference producing the air flow is created by reducing the pressure at the exit end of the apparatus or by increasing the pressure at the inlet end, calculate both compared to atmospheric. If required, make the full calculation, taking slipflow into account.

In reporting results, show the values that have been used for bed porosity and effective solid density. Report which correction factors (if any) have been applied and whether the results take slip flow into account. The report should be given in terms of effective volume or mass specific surface.

3.3.2 Constant flow rate method

3.3.2.1 The constant flow rate method using the Fisher sub-sieve sizer

Gooden & Smith [4] modified the Lea and Nurse apparatus by incorporating a self-calculating chart which enabled specific surface to be read off directly, using the relationship [5] $S_V = \frac{6}{d_{sv}}$, where d_{sv} is the average particle diameter. This greatly reduces the time required per sample. A commercial instrument has been developed from the Gooden & Smith method, called the Fisher Sub-sieve Sizer [6]. This uses a standardized computation procedure, and the average particle diameter of a powder is read directly from one of a family of curves on a chart incorporated in the instrument, at porosities within the range 0.80–0.40. The resulting figure for this diameter varies with the porosity of the powder, inadequate compression of the sample yielding a high result. The chart gives reproducible results, and its calibration has been discussed by Dubrow [7]. Extension of the porosity range has also been discussed by Edmondson & Toothill [8].

The Permaran specific surface area apparatus [9] measures flow rate and pressure drop across packed beds. Air flow in the measuring circuit is generated by means of an air pump, and the pressure drop through the sample is measured by means of a liquid pressure gauge using vacuum pump oil as the liquid.

Specifically, the Fisher sub-sieve sizer method of testing covers the procedure for the determination of the volume surface mean diameter expressed in micrometres, from which the effective volume specific surface area may be calculated. The values obtained will not normally be absolute, but will be suitable for control purposes [10].

3.3.2.2 The apparatus

The Fisher sub-sieve sizer is an air permeability apparatus in which the variables associated with the Kozeny–Carman equation for the determination of specific surface area are made constant. Fig. 3.3 shows a diagram of the instrument. The

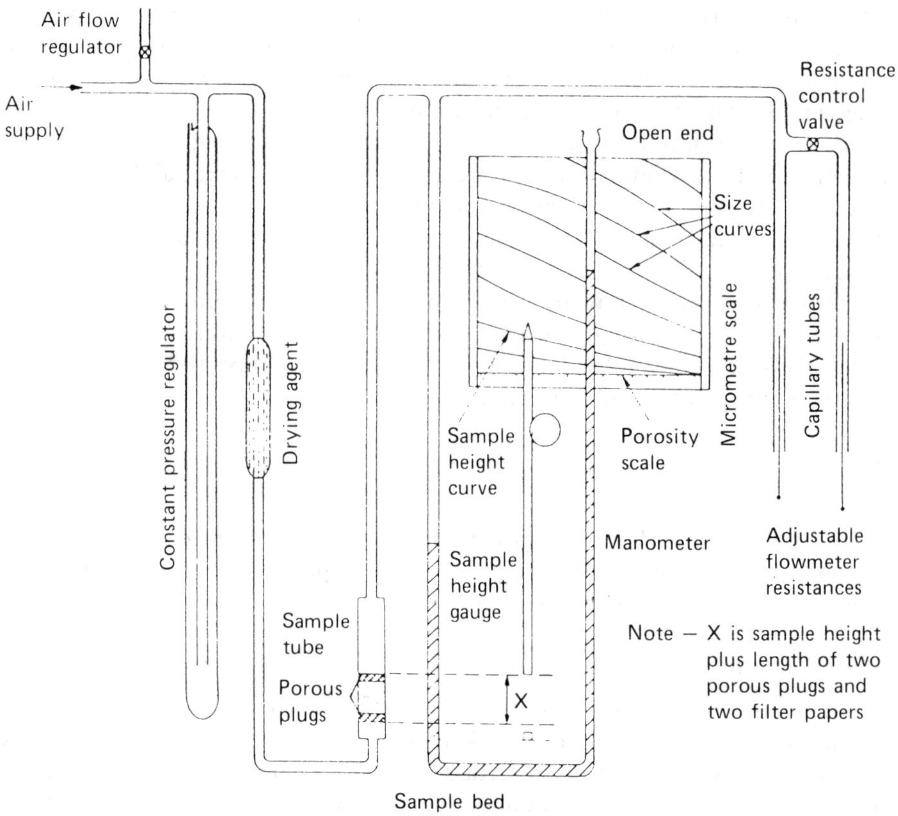

Fig. 3.3 — Fisher sub-sieve sizer.

basic instrument consists of an air pump, an air pressure regulating device, a precision bore sample tube, and an adjustable, standardized, double-range flow meter. Flow meter resistances on earlier instruments consisted of capillary tubes where the resistance of each was varied by a sliding wire in its lumen. Instruments are now fitted with adjustable needle valves. A flow meter and sliding chart permit the mean particle diameter to be read directly.

Ancillary equipment includes two perforated brass plugs, a plug manipulator, filter paper discs, a jewel orifice tube for calibration purposes, a sample pressure calibrator accessory for control of bed compression, a rubber sample tube stand, and a small funnel.

A recommended modification in the British Standards [10] is to include a water manometer having a narrow bore section to damp oscillations caused by the pump, together with two expansion bulbs to accommodate the initial pressure surge.

The Fisher jewel orifice tube, standardized against the US National Bureau of Standards cement sample No. 114, is used as a standard for both ranges of the instrument. As an additional check of calibration at the porosity values normally used, it is useful to set up an internal working standard. This should be a well-blended sample, representative of the material being examined, and determinations should be carried out on several occasions by different operators to establish a mean value and acceptable range for the standard. Alternatively, a suitable reference standard may be purchased [10].

3.3.2.3 Preparation

Before standardization, it is recommended [10] that the following checks must be made:

(a) that the flow meter resistances, air lines, and manometers are clean;
(b) that the calculator chart slides freely without rocking and is correctly aligned vertically with the rack and pinion and horizontally with the indicator pointer;
(c) that there are no leakages at the sample tube rubber seals or elsewhere in the apparatus;
(d) that the sample packing pillar is adjusted to the correct height (using the two brass plugs and two filter discs to determine any adjustment necessary);
(e) that the perforated brass plugs and sample tube are not worn;
(f) that the drying agent is in an effective condition;
(g) that the standpipe water level is at the correct height;
(h) that the manometer level is maintained at the correct height in relation to the chart baseline. It should be checked and adjusted only after the flow meter manometer has been filled and emptied several times, then allowed to drain for at least five minutes.

No adjustments should be made while the instrument is operating. It should also be noted that when replacing the flow meter manometer (for example, after breakage) both arms must be replaced by a matched pair to prevent errors due to differences in level in the two arms. A convenient method of checking this is by fitting millimetre scales to both arms of the manometer so that a rise in one arm may be compared with a corresponding fall in the other.

3.3.2.4 *Standardization*

The instrument may be standardized in the following manner: After items (a) to (h) have been checked, clamp the jewel orifice tube in position between the rubber seal rings and set the instrument range control to the 0.2 μm–20 μm range position.

Two chart readings are marked on the side of the calibration tube. The higher one is applicable to the range control set in the 0.2 μm–20 μm range, and the lower one is used when the range control is set to 20 μm–50 μm.

Slide the calculator along the rails until the porosity reading indicated by the pointer corresponds to that printed on the side of the calibrator tube. Start the air pump by operating the switch on the right-hand side of the instrument. Adjust the air pressure control on the front panel so that bubbles in the standpipe rise at the rate of two to three per second. Assist the flow meter manometer above its maximum reading by closing the standpipe flexible vent with the forefinger. Take care not to eject water from the manometer. Allow the water level in the manometer to settle for at least five minutes. Turn the rack and pinion mechanism until the upper face of the horizontal cross-bar is coincident with the water level in the manometer. Read off the particles diameter as indicated by the tip of the pointer on the chart. If the readings obtained are not within 0.5 μm of the value printed on the side of the calibrator tube, adjust the instrument as follows.

Turn the instrument to bring the right-hand side (as seen from the front) nearest to the operator. Remove the rear panel to expose the two flow meter resistances on the right-hand side. Adjust the nearest flow meter resistance until the meniscus attains the correct reading, allowing time for stabilization between successive adjustments. Set the range switch to the 20 μm–50 μm position and repeat the procedure for the second flow meter resistance.

The values on the calibration tube have been determined at 25°C, and a correction of ±0.1 μm must be applied for each 6°C above or below this temperature. The calibration should be checked at least once per day when the instrument is in regular daily use and adjustments made as necessary. An internal working standard may also be run on a daily basis as a further check of instrument calibration. Follow the procedure for determination of mean particle diameter.

3.3.2.5 *Determination of effective solid density*

Determine the effective solid density of the powder by the method described in section 2.1.8.2.1.

3.3.2.6 *Use of the Fisher sample pressure calibrator*

This device provides a convenient method of reproducibly compressing powder samples to the same degree of compaction. It consists essentially of a torque screwdriver with a dial graduated in 1 inch-pound increments, and a spring loaded tube assembly graduated from 20 psi to 80 psi in 10 psi increments.

Before making particle size measurements, connect the screwdriver to the instrument in place of the rack and pinion control, and place the tube assembly over the sample pillar in place of the sample tube. Establish torque to pressure values by plotting inch-pounds against psi to produce a calibration chart. Take torque readings at 30 psi, 50 psi, and 70 psi. (1 psi – 6898 N/m^2).

Note that commercial equipment calibrated in SI units is not available.

3.3.2.7 Determination of mean particle diameter

Place a brass plug (with the perforated end facing upwards) on the sample pillar and lay a filter paper disc evenly over the top of the plug. Holding the sample tube vertically, carefully push it down over the filter paper disc, plug, and pillar, the filter paper being crimped round the edges of the plug. Transfer the sample tube to the rubber stand, keeping the paper side of the plug uppermost. Weigh out to within 0.005 g a quantity of the sample equal in grams to the effective solid density of the powder, expressed in grams per cubic centimetre. Transfer the material to the sample tube, using a small powder funnel (it may be necessary to use a camel-hair brush to complete the transfer). Tap the side of the tube to settle the powder. Using the plug manipulator or the forefinger push a paper disc and the second plug (perforated end facing downwards) into the top of the sample tube. Carefully remove the sample tube from the rubber stand and place it over the sample pillar on the instrument so that the lower plug is again in contact with the pillar. Lower the rack to make contact with the upper plug and continue turning until the required compaction is attained, using either the rack and pinion control knob or the torque screwdriver. Move the chart laterally until the tip of the pointer is level with a point on the sample height curve. Pressure on the sample must be relaxed before this operation is carried out, and the chart must not be moved again before a particle diameter reading has been taken at the indicated porosity. Raise the rack and transfer the sample tube to the rubber cushion supports, clamping it in position with the upper cap to obtain an airtight seal. Lower the rack until the tip of the pointer is level with the base line of the chart, and record the indicated porosity.

Set the range control to the 0.2 μm–20 μm range, switch on the air pump, and raise the flow meter manometer level to a maximum, assisting it above the rest point by closing the vent pipe with the forefinger as previously described. If the meniscus rises above 20 on the instrument chart, readings may be taken by setting the range control to the 20 μm–50 μm position and doubling the mean particle diameter. The time necessary to reach this maximum can vary from 30 s to several minutes, depending on the fineness of the material under test.

Turn the rack up until the upper edge of the crossbar coincides with the meniscus of the manometer. The particle diameter is indicated by the position of the tip of the pointer in relation to the curves on the chart. When taking readings at successively lower porosities the above procedure should be repeated, compressing the powder to successively lower porosities until the maximum bed compression is reached. Drainage time must be allowed between successive determinations when using the multiple porosity technique. For process control purposes a single compression to constant torque (corresponding to constant sample compaction pressure) will usually suffice. For more accurate estimations it is often preferable to perform two or three multiple porosity determinations, recording the mean specific surface calculated from each.

3.3.2.8 Conversion to specific surface

The effective volume surface mean diameter can also be expressed in terms of the effective volume specific surface area measured in square metres per cubic metre according to the following relationship.

$$S_v = \frac{6 \times 10^6}{d_m}$$

Expressed on a mass basis (square metres per kilogram)

$$S_w = \frac{6 \times 10^3}{d_m \times \rho_s}$$

3.3.2.9 *Reporting results*

In reporting results it is important to state the parameters used in the determination, e.g. porosity or powder bed compaction pressure (in the case of single point readings) and effective solid density. These should be quoted, where applicable, in addition to the values obtained for the effective volume specific surface.

3.3.2.10 *Extension of the instrument range*

The Fisher sub-sieve sizer incorporates a chart covering a working range of porosity from 0.40–0.80. Using the chart supplied with the instrument it is possible to test samples which are not adequately compressed at a porosity of 0.40 by increasing the weight of sample normally used in the test, and applying corrections to the measured porosity and indicated mean particle diameter [11]. The instrument chart is designed to be used with a standard sample volume of 1 cm^3 excluding voids. If the weight of sample taken is increased, then the bed porosity is related to the indicated porosity by the following equation.

$$\varepsilon = 1 - v_1 (1 - \varepsilon_1) \ .$$

Similarly, the mean particle diameter can be calculated if the indicated particle diameter is known.

$$\frac{d_m}{d_1} = v_1 \frac{(\varepsilon_1)^{3/2}}{(\varepsilon)} \ .$$

It has been found useful in many cases to take a weight of sample (in grams) equal to the density $\times 1.25$. Table 3.2 shows correction factors by which the indicated diameter from the chart must be multiplied to give corrected diameter.

3.3.2.11 *Basis of the nomogram*

The instrument uses a chart covering a working range of porosity from 0.40–0.80. This is a nomogram consisting of a series of curves obtained by plotting flow meter manometer heights as ordinates against porosity values as abscissa for a range of values of surface mean diameter in micrometres.

The nomogram is based on work by Edmunson [11] where:

Table 3.2 — Porosity values

Sample volume $v_1 = 1.25$ cm ^{3}

Indicated porosity ε_1	Corrected porosity ε	Diameter factor $v_1 \frac{(\varepsilon_1)^{3/2}}{(\varepsilon)}$
0.800	0.750	1.377
0.760	0.700	1.414
0.720	0.650	1.457
0.700	0.625	1.482
0.680	0.600	1.508
0.660	0.575	1.537
0.640	0.550	1.569
0.620	0.525	1.604
0.600	0.500	1.643
0.580	0.475	1.687
0.560	0.450	1.735
0.540	0.425	1.790
0.520	0.400	1.853
0.500	0.375	1.925
0.480	0.350	2.008
0.460	0.325	2.105
0.440	0.300	2.220
0.420	0.275	2.359
0.400	0.250	2.530

$$d_m = \frac{60000\ (C^1\ n)^{1/2}\ LF^{1/2}}{14\ (AL-1)^{3/2}\ (P-F)^{1/2}}$$

This equation reduces to the following form when the values of instrument constants are inserted:

$$\frac{F}{2} = \frac{25\ d_m^2\ (1.267\ L - 1)^3}{14.444\ L^2 + (1.267\ L - 1)^3\ d_m^2}$$

where $F/2$ is the observed manometer height.

The porosity is related to L by the equation:

$$\varepsilon = \frac{L - 0.789}{L}$$

3.3.3 Constant volume rate method

3.3.3.1 Constant volume rate permeameters

A different geometrical approach to air permeability measurements (Fig. 3.4) was taken by Rigden [12,13]. In this procedure the time taken for the pressure to fall by a given amount is measured rather than the rate of flow under a steady pressure. This is also the basis of the method developed by Blaine [14,15]. The essentials in this method are that the powder is compacted into a bed in a cell, and the two ends of the cell are connected to the two ends of a U-tube containing an oil at vapour pressure. The oil is displaced initially by drawing up on one side; as it returns to its equilibrium level, it forces air through the powder bed. The pressure differential thus diminishes as flow proceeds. The time for the oil to travel between two marks on the U-tube limb is measured and the specific surface is then calculated. The Blaine apparatus is schematically shown in Fig. 3.5.

The Griffen [16] surface area apparatus is a commercial version (Fig. 3.6) of the

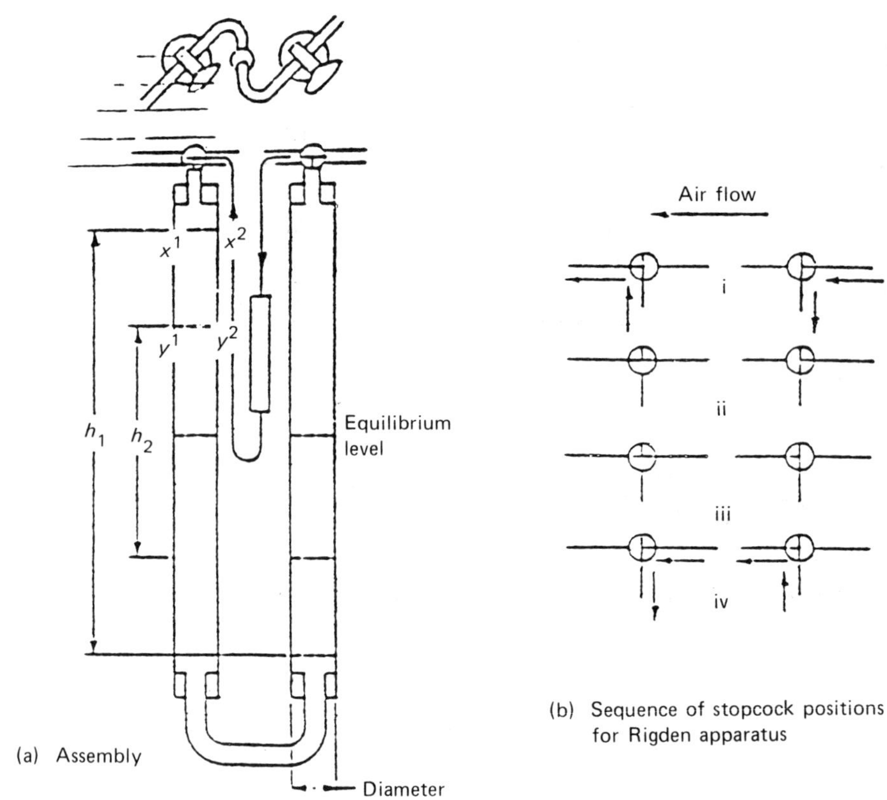

Recommended dimensions

Length mm	900	600	400
Diameter mm	25	35	50
Bore of stopcock	3mm		

Fig. 3.4 — Rigden apparatus assembly.

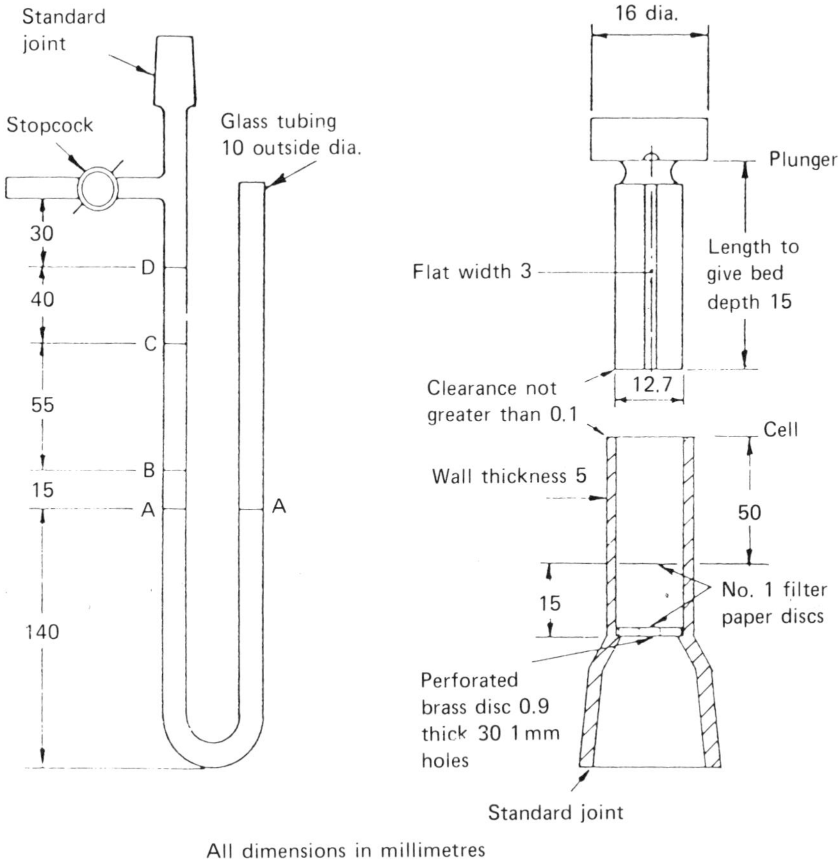

Fig. 3.5 — (a) Blaine apparatus, (b) cell and plunger for Blaine apparatus.

Rigden. Basically, air is caused to flow through a bed of powder by the pressure of oil displaced from equilibrium in two chambers which are connected to the permeability cell and to each other in a U-tube fashion. The oil is brought to the start position by using a rubber hand pump with tap open to the atmosphere. The taps are then rotated so that the oil manometer rebalances by forcing air through the powder bed. Timing is from start to height h_A (for fine powders) or h_B (for coarse powders), the height differences denoting different flow volumes.

More recently, Allen & Maghafarti [17] have modified the George and Griffen apparatus to enable measurements to be carried out at low pressures. Allen & MacSporran [18] showed, after rigorous mathematical proof, that as a gas flows through a bed of powder it expands, and a correction has to be applied to the two term Carman–Arnell equation. This correction has since been adopted by the British Standards Institution [1,10].

3.3.3.2 The constant volume method of Rigden

The Rigden constant volume apparatus consists essentially of a U-tube not quite half filled with a low viscosity, non-volatile liquid. The ends of the tube are connected

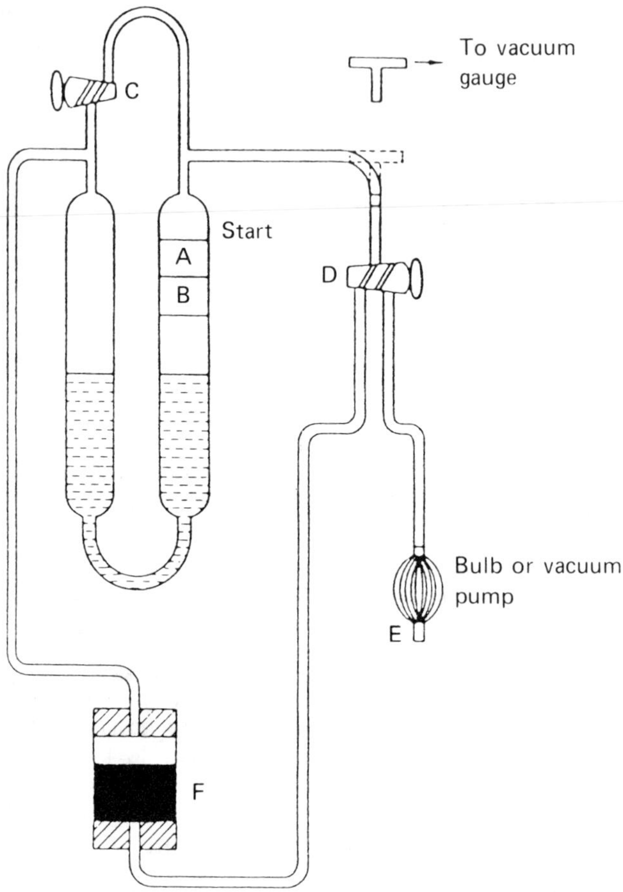

Fig. 3.6 — Griffin surface area of powder apparatus.

across a packed permeability cell. A pressure difference is created across the ends of the U-tube so that the level of the liquid on one side is raised nearly to the top of the limb. The return of the liquid under the action of gravity to its original equilibrium position causes air to flow through the bed of powder.

In the operations that follow, the stopcocks, whose positional sequence is shown in Fig. 3.4(b) should be turned in a clockwise direction only.

(a) Connect the packed permeability cell to the U-tube, the stopcocks being in the positions shown in the assembly diagram.
(b) Turn the stopcocks to the positions shown at (i) in Fig. 3.4(b).
(c) Apply pressure to the right limb to raise the level of the liquid in the other limb above the timing mark x–x′ (Fig. 3.4(b)).
(d) Arrest the flow of air and open the supply line to the atmosphere. Turn the left-hand stopcock to position (ii) simultaneously.

(e) Turn the right-hand stopcock to position (iii). The air in the left-hand limb is now under reduced pressure, whilst that in the rest of the apparatus is at atmospheric.
(f) Turn the left-hand stopcock to position (iv) and start the timer as the meniscus of the liquid passes the mark x–x′.
(g) Stop the timer as the meniscus passes the mark y–y′ and record the time interval.

Repeat operation (b) to (g) twice and record the mean of the three time intervals. In operation (f) the entry of the air at atmospheric pressure from the space between the permeability cell and the U-tube may produce an initial sharp fall in the level of the raised liquid. If the level falls below the timing mark x–x′, turn both stopcocks to position (i) and recommence the determination at operation (c), raising the liquid level to a height that will permit the initial fall to be checked at a point about 1 cm above x–x′.

Another variation of the variable flow technique is the Reynolds and Branson auto-permeameter, in which air is pumped into the inlet side to unbalance a mercury manometer [5]. The tap is then closed and air flows through the packed bed to atmosphere. On rebalancing, the mercury contacts start–stop probes attached to a timing device. The pressure difference (Δp) between these probes and the mean pressure $\overline{P}$ are instrument constants. The flow rate is given by:

$$\frac{\mathrm{d}V}{\mathrm{d}t} = \frac{1}{\overline{P}}\frac{\Delta P}{t}$$

Substituting this in the Carman–Kozeny equation yields a similar equation to the Rigden equation. The automatic timing device on this instrument makes it preferable to the Rigden [5].

With the Ridgen apparatus in general, a U-tube length of approximately 600 mm used in conjunction with a permeability cell of 20 mm to 30mm diameter is suitable for the examination of most powders within the range 5×10^5 to 1.5×10^6 m^2/m^3. The apparatus can also be used for testing powders whose specific surfaces are outside this range. It may be found necessary to extend or reduce the length and diameter of the U-tube in cases where the permeability of the compacted bed of powder is such that a higher or lower hydrostatic pressure is needed in order to achieve a satisfactory flow time. It may also be advantageous to use a different size permeability cell and/or to replace the manometer liquid with one of a different density. Whatever modifications are made to the apparatus components, the flow time should not be less than 2 min in order to minimize errors due to the drainage time of the liquid. Extremely long flow times should also be avoided. To establish stable flow conditions the upper timing mark, x–x′, should be far enough from the top of the limb to allow the liquid to fall at least 10 mm before timing is commenced, and the lower timing mark should be at least 50 mm above the equilibrium level of the liquid. (If the lower timing mark is too near the equilibrium level an error is introduced owing to the compressibility of air.)

The connections between the U-tube and the permeability cell should be kept as short as possible and should offer negligible resistance to flow.

3.3.3.2.1 Calibration of the U-tube

If precision bore tubing is not available, calibration is necessary. This may be carried out as follows.

Attach a pinchcock or tap to one end of a 1.5 m length of clean glass tubing and fasten a temporary scale along the full length. The intervals between the marks on the scale should correspond to approximately one fifth of the working length required. Clamp the tube in a vertical position and fill to the uppermost mark with distilled water. The temperature should remain constant during the succeeding operations.

Run out the water from the first to the second mark into a tapered vessel and record the weight. Repeat this procedure for each marked interval. Calculate the average cross-sectional area of the tubing between each successive mark from the weight of water and the length of the interval. Select and cut two pieces of the required length that do not vary in their cross-sections by more than ±1%. Compensation for any slight taper effect can be made by ensuring that the direction of taper in the U-tube limbs is reversed. Clean and dry all apparatus components before assembly and erect the U-tube in a vertical position. Fill the tube with the manometer liquid to a level slightly below the middle of the limbs. Etch or incise the timing marks x–x′, y–y′, on one limb and the equilibrium position on both limbs. (If a printed scale is attached to the U-tube limb the position of the timing marks can be varied.)

3.3.3.2.2 Operation

A hand bellows is recommended for displacing the liquid to the starting level, but whatever method is used for this purpose it must be possible to arrest the flow of air and allow the pressure on the pump side of the U-tube to return to atmospheric, so that the subsequent manipulation of the stopcocks does not disturb the bed of powder in the permeability cell.

3.3.3.3 The Blaine constant volume method

In the Blaine apparatus, as in the Rigden, air is made to pass through a bed of powder by the flow of liquid in a U-tube system, but with the difference that one side of the cell and one end of the U-tube are open to the atmosphere (Fig. 3.5).

The U-tube, cell and plunger dimensions and design of the components can be modified to suit individual requirements.

The cell is mounted at the top of the narrow bore glass U-tube and the manometer liquid (water or oil at low vapour pressure) is raised from the equilibrium position A–A (Fig. 3.5) to the initial position above D by applying suction through the stopcock. The time necessary for the meniscus to fall between the two timing marks C and B is recorded.

In a larger form of the apparatus the permeability cell is mounted separately, a connection being made from one limb to the top of the cell. When the liquid is drawn up through the opposite limb of the U-tube and released, air is drawn downwards through the bed.

With some types of air permeameter the design permits adequate compensation for instrument variables, enabling a single constant to be used in the calculation. This does not apply to the Blaine instrument since differences in the linear dimensions of the various components between one instrument and another can cause a marked instrument to instrument variation.

Calibration is usually carried out by using a standard sample (or samples), whose specific surface has been previously determined on a primary apparatus. The specific surface of the calibration sample should not differ greatly from that of the sample under test.

For a given instrument and a given material, if the viscosity of the air is assumed constant, then at a fixed porosity

$$S_k = K_b t^{1/2},$$

where K_b is a constant and t is the flow time in seconds. If required make the correction taking slip-flow into account.

The apparatus can be modified to make it semi-automatic in operation, and commercial forms of the instrument are available. One type incorporates a pressurized reservoir to pass air through the bed. A mercury manometer is connected to the reservoir and is fitted with two contacts mounted a suitable distance apart. As the pressure falls the mercury actuates an electric clock. For a given powder and fixed volume cell the clock can be calibrated in units of specific surface. The method is rapid and capable of reproducing reliable results.

3.3.4 Accuracy of measurement

Manometer heights, manometer liquid density, and volume flow rate should each be measured to an accuracy of $\pm 1\%$ or better. The temperature should be known to within ± 2 K, and the appropriate value of air viscosity used. With a pressure drop flow meter the calibration constant, C, should be calculated to an accuracy of $\pm 1\%$ or better.

With constant volume apparatus, the time, t, should be measured to an accuracy of $\pm 0.5\%$ or better, and unless the apparatus is used as a secondary standard method the internal cross-sectional area of the manometer tube should be measured to an accuracy of $\pm 1\%$ or better.

3.3.5 Preparation of the powder bed

3.3.5.1 General

Owing to the wide diversity of the physical properties of powders, in general it is not possible to specify a unique technique for the preparation of the bed of powder. The aim must be to produce a bed that is homogeneous in structure and which possesses a flow resistance appropriate to the permeability apparatus used. The mass of powder required for the test (the test sample) should be representative of the material as a whole and should be obtained from the bulk sample by rotary riffling or other suitable technique as detailed in BS 3406: Part 1. Materials are usually tested in the as-received condition unless they are moist, when they should be dried by an appropriate technique. Whilst most materials are dried at 100°C–105°C or at other temperatures appropriate to the properties of the powder under test it must be realized that raising the temperature during the drying process can affect the specific surface area result obtained. Vacuum drying without heat in the presence of a desiccant is suitable if it can be achieved in a reasonable time. If heat is required it is advisable to use the minimum temperature necessary. Soft agglomerates may be broken down by light pressure, due regard being given to the possible frangibility of individual particles. A suitable method is by lightly rolling the material when it is spread out in a thin layer on glazed paper.

Use of the permeability method is precluded when:

(a) Agglomerates are not easily broken down (unless the specific surface area required is that of the agglomerated form).
(b) The particles fracture or undergo deformation when the powder bed is being compacted.

3.3.5.2 Preparation

The bed cell should be prepared by using a bed support of minimum resistance to flow especially with coarse powders and when no correction is made for resistance due to the bed support. A hardened coarse filter paper circle supported by a perforated disc is suitable. Other filter papers are traditionally used with some apparatus but are not recommended.

Structural homogeneity within the bed is often obtained, especially if long (>10 mm) beds are to be formed, by transferring the test sample to the cell in increments, though with some powders and some apparatus reproducible results may be obtained with the use of a single increment.

Each increment should be consolidated by tamping followed by application of the cell plunger by hand, by machine press (for reproducibility purposes or with fine or ultrafine powders), or centrifugally. In centrifugal compaction place a freely moving plunger on top of the powder to provide the necessary compaction force and to ensure a correctly formed top surface to the bed. In all cases the objective is to produce a homogeneous bed of suitable porosity with a length capable of being measured to better than 0.25% accuracy (i.e. using a cathetometer or calibrated plunger with vernier scale). In preparing the bed care should be taken to avoid:

(a) crushing the powder (do not use excessive compaction pressure);
(b) disturbing the surface of the bed when removing the plunger;

    Although the plunger is designed to avoid the creation of a partial vacuum when it is withdrawn, this is effective only if the plunger is withdrawn carefully, slowly, and with a very slight rotary action. Rapid or careless removal of the plunger may lead to the bed being lifted from its support so that the bottom of the bed is disturbed. This disturbance is not necessarily apparent on visual examination of the bed, but is a common cause of low results.
(c) glazing the surface (do not rotate the plunger or drift whilst it is in contact with the bed);
(d) segregation of different size fractions owing to excessive tamping.

Where the packing characteristics of the powder are unknown tests should be made to ascertain the amount of sample required, the air flow rate being checked at different bed heights or different cell diameters, if necessary. Care should be taken in the use of fixed volume cells as slight changes in the particle characteristics of the product under test can mean that in compacting to a fixed bed length excessive voids may occur in the bed space or particles may be fractured during compaction. It would appear advantageous to prepare beds by using fixed compaction pressures rather than fixed bed lengths.

3.3.6 Measurement of permeability

3.3.6.1 Constant flow rate method

3.3.6.1.1 Procedure

It is assumed that the apparatus constants have previously been measured. Prepare the powder bed, and for variable volume cells measure height L and calculate the bed porosity. Connect the cell to the apparatus so that there are no leaks, taking care to avoid disturbing the powder bed. This can be achieved by ensuring that there is an air flow in the correct direction through the bed whilst making the connections. Adjust the flow of air to give a pressure drop across the bed of 1.5 kPa to 4.0 kPa (manometer height for liquid of relative density 1.0 of 150 mm–400 mm). Record the manometer height, h, and measure the equivalent flow rate of air. When a bubble flow meter is used measure the time for a given volume of air; for a pressure drop flow meter measure the manometer height. Make several measurements of the flow rate and the corresponding pressure drop.

3.3.6.1.2 Calculation

Equations (2.41) to (2.46) are true for SI units. The equations following are for SI units, and the appropriate factors should be used if practical units are substituted.

When a bubble flow meter is used to measure the rate of flow of air

$$S_{\mathrm{k}} = \sqrt{\frac{A_{\varepsilon}^{3} h \rho_{\mathrm{L}} g}{5(1-\varepsilon)^{2} \mu L\, q}} \tag{3.1}$$

$$S_{\mathrm{m}} = S_{\mathrm{k}}^{2} \cdot 81.0 \frac{(1-\varepsilon)}{\varepsilon} \cdot \sqrt{T} \cdot \frac{\mu}{P} \tag{3.2}$$

S_{v} is calculated from equation (2.33).

When a pressure drop flow meter is used, the liquid in each of the two manometers being of the same relative density

$$S_{\mathrm{k}} = \sqrt{\frac{A\, \varepsilon^{3}\, hg}{5\,(1-\varepsilon)^{2}\, LCh_{1}}} \tag{3.3}$$

S_{m} and S_{v} are calculated from equations (3.2) and (2.33) respectively.

3.3.6.2 Constant volume method

3.3.6.2.1 Procedure

It is assumed that the apparatus constants have previously been measured. Prepare the powder bed, and for variable volume cells measure height L and calculate the bed porosity. Connect the cell to the apparatus so that there are no leaks, taking care to avoid disturbing the powder bed. Apply pressure or suction, as appropriate, to raise the manometer liquid level to above the upper timing mark, ensuring that no air flows through the powder bed in the 'wrong direction'. Turn the appropriate stopcock so that as the manometer liquid level falls to its equilibrium position it draws air through the powder bed, and measure the time taken for the level to pass between the upper and lower timing marks.

3.3.6.2.2 Calculation

The full equation which allows for the expansion of the air as it passes through the bed is:

$$S_k = \sqrt{\frac{2A\,\varepsilon^3 g\,\rho_L t}{5(1-\varepsilon)^2\,\mu La\left(\log_e\frac{h_1'}{h_2'}\cdot\frac{B-h_2'}{B'-h_1'} - \frac{1}{2}\log_e\frac{B-h_2'}{B-h_1'}\right)}} \tag{3.4}$$

$$S_m = S_k^2\;81.0\frac{(1-\varepsilon)}{\varepsilon}\sqrt{T}\cdot\frac{\mu}{P} \tag{3.5}$$

S_v is calculated from equation (2.33).

When the flow of air is arrested well before equilibrium conditions are reached, the decrease in pressure difference is small, and equation (3.5) approximates to:

$$S_k = \sqrt{\frac{2A\varepsilon^3 g\rho_L\,t}{5(1-\varepsilon)^2\mu La\,\log_e\frac{h_1'}{h_2'}}} \tag{3.6}$$

S_m and S_v are calculated from equations (3.5) and (2.33) respectively.

When the constant volume apparatus is used as a secondary standard method, calculate the apparatus constant (including the cell), K_a, by measuring time, t, for a powder of known S_k packed to a known porosity and depth.

$$K_a = S_k\sqrt{\frac{(1-\varepsilon)^2\,L}{\varepsilon^3\,t}} \tag{3.7}$$

It is not possible to calculate a simple apparatus constant in terms of S_v direct.

For the powder under examination

$$S_k = K_a\sqrt{\frac{\varepsilon^3\, t}{(1-\varepsilon)^2\, L}} \tag{3.8}$$

S_m and S_v are calculated from equations (3.5) and (2.33) respectively.

For fixed volume cells packed to a standard porosity, ε and L may be incorporated into an apparatus constant K_b such that $S_k = K_b\sqrt{t}$

3.3.7 Variables affecting the result

3.3.7.1 Use of a compressible fluid

Since air is compressible and q in equation (2.41) is the volume rate of flow of an incompressible fluid, some modification to the equation may be necessary.

If P_1 is the entry pressure, P_0 the exit pressure, $P_1 - P_0 = \Delta p$, and q_0 the measured volume rate of flow *from* the powder bed, the appropriate value to be used for q in equation (3.1) may be calculated from either of the following equations:

$$q = q_0\frac{P_1 - \Delta p}{P_1 - \dfrac{\Delta p}{2}} \quad (3.9a) \qquad \text{or} \qquad q = q_0\frac{P_0}{P_0 + \dfrac{-\Delta p}{2}} \tag{3.9b}$$

Alternatively, if the measured flow rate *into* the bed is q_1,

$$q = q_1\frac{P_1}{P_1 - \dfrac{\Delta p}{2}} \quad (3.10a) \qquad \text{or} \qquad q = q_1\frac{P_0 + \Delta p}{P_0 + \dfrac{\Delta p}{2}} \tag{3.10b}$$

When a pressure drop flow meter is used, the pressure applied to the bed being atmospheric, multiply h_1 in equation (3.3) by

$$\frac{P - h\,\rho_L g}{P - \dfrac{h\,\rho_L g}{2}}$$

When the pressure applied to the bed is greater than atmospheric multiply h_1 in equation (3.3) by

$$\frac{P + h_1\,\rho_L g}{P + \rho_L g\left(\dfrac{h}{2} + h_1\right)}$$

The magnitude of these correction factors, which are sometimes called 'mass flow corrections', will depend on the pressure drop, Δp, across the bed. For a pressure drop not exceeding 4 kPa, the correction to S_k will be less than $+1\%$, to S_m less than $+2\%$, and S_v less than $+1.2\%$.

A pressure drop of 4 kPa would be indicated by a height of 400 mm in a manometer filled with a liquid of relative density 1.0.

Comparative tests of the air permeability methods by six laboratories is shown in Table 3.3.

3.3.7.2 Temperature

Temperature enters directly into the equations for the calculation of S_m and indirectly into the other equations by its effect on air viscosity (see Table 3.1) and manometer liquid density, and for bubble flow meters by its effect on the water vapour pressure within the meter. In general, temperature variations of ± 5 K cause error in S_V of less than $\pm 1\%$. When pressure drop flow meters are used it is recommended that measurements be made at a temperature within ± 5 K of the calibration temperature.

3.3.7.3 Effect of cell dimensions

The effect of cell dimensions as a variable affecting the results is as discussed in section 3.3.1.2.4.

3.3.7.4 Effect of powder density

Errors in the measurement of powder density can be shown to give the following errors in S_K.

$$\text{Errors in } \rho_s = \pm 2\% \qquad \text{Resulting error in } S_K = \pm\frac{(3-\varepsilon)}{2\varepsilon}\cdot 2\%$$

where ε is the actual porosity of the powder bed. Results of cooperative trials on the measurement of density of Portland cements show [19,20,21] 95% confidence limits of the order of $\pm 2\%$ before rejection of outliers. It can be assumed that most other materials will give similar results. However, fundamental errors in the accuracy of density determination, as opposed to precision, may cause much larger effects for porous materials [22–44], as already discussed in section 2.1.8.2.8.

3.3.7.5 Water vapour pressure in bubble flow meters

The air in a bubble flow meter becomes saturated with water vapour (this does not apply if non-aqueous bubble system is used), and a correction for this should be applied:

$$q = \frac{V}{t}\cdot\frac{P - P_v}{P}$$

[illegible] Comparative tests of air permeability method of six samples by six laboratories

Test laboratory	Permeability cell	Aluminium (a)		Aluminium (b)		Titanium dioxide		Nitroguanidine		Silica		Copper phthalocyanine	
		S_v	ε	S_v	ε	S_v	ε	S_v	ε	S_v	ε	S_v	ε
A	Lea & Nurse	3550	0.490	3880	0.490	69 000	0.690	11 500	0.690	347 000	0.900	131 400	0.690
	25 mm diameter	3580	0.490	3970	0.490	70 000	0.690	11 500	0.690	382 000	0.900		
						71 200	0.690	11 500	0.690				
B	Lea & Nurse	3590	0.512	4080	0.495	54 300	0.686	10 500	0.771	326 000	0.909	111 000	0.757
	25 mm diameter	3650	0.507			87 500	0.659	11 900	0.631	466 000	0.871	143 400	0.701
		3650	0.503			103 800	0.635	12 100	0.576	513 000	0.856	163 300	0.659
		3620	0.502			105 500	0.631	12 100	0.576				
		3620	0.502			106 000	0.628						
C	Lea & Nurse					70 300	0.678	11 300	0.780	291 000	0.906	88 400	0.759
	25 mm diameter					75 200	0.675	11 600	0.744	284 000	0.906	92 500	0.759
						74 800	0.672	11 500	0.741	277 000	0.906	88 000	0.759
C	Rigden					43 400	0.685	12 200	0.667	332 000	0.902	60 400	0.778
	25 mm diameter					81 600	0.683	11 800	0.656	281 000	0.901	94 100	0.766
						62 900	0.678	12 000	0.655	362 000	0.898	105 000	0.747
						63 000	0.666			358 000	0.895	77 400	0.743
						82 300	0.665			391 000	0.894		
D	Lea & Nurse 25 mm diameter	4200	0.510	4350	0.495	100 900	0.645	11 800	0.681	440 000	0.862	140 000	0.630
D	Lea & Nurse 19 mm diameter	4050	0.487	4300	0.497	101 000	0.601	11 900	0.600	470 000	0.871	144 000	0.675
D	Lea & Nurse 13 mm diameter	3600	0.482	3950	0.483	105 500	0.648	11 700	0.595	450 000	0.869	141 300	0.688
A	Fisher	3240	0.475	3800	0.475	84 500	0.690	11 400	0.690	—	—	142 900	0.690
	13 mm diameter	3190	0.473	3620	0.460	90 900	0.685	11 300	0.680	—	—	142 900	0.670
		3160	0.470	3680	0.455	90 900	0.670	11 500	0.660	—	—	146 000	0.670
		3370	0.450	3750	0.455	90 900	0.670						
		3370	0.450	3630	0.450	88 250	0.670						
		3370	0.440	3570	0.440	100 000	0.660						
E	Fisher	3500	0.436	3900	0.439	100 000	0.663	12 000	0.630	—	—	154 000	0.690
	13 mm diameter	3500	0.435	3950	0.437	100 000	0.657	12 100	0.633	—	—	152 000	0.693
F	Fisher	3500	0.499	3770	0.489	63 200	0.700	11 200	0.668	—	—	120 000	0.715
	13 mm diameter	3500	0.498	3770	0.487	65 200	0.695	11 300	0.657	—	—	120 000	0.714
		3500	0.435	3750	0.487	66 700	0.694	11 200	0.654	—	—	120 000	0.712
D	Fisher	3530	0.495	4140	0.490	100 000	0.665	11 550	0.690	—	—	141 000	0.715
	13 mm diameter	3530	0.495	3900	0.483	100 000	0.655	11 580	0.690	—	—	138 000	0.710
		3540	0.482	4100	0.480	103 000	0.650	11 550	0.690	—	—	139 000	0.702
A	Rigden	3750	0.499	4060	0.499	71 100	0.690	11 400	0.690	304 000	0.900	131 100	0.690
	25 mm diameter	3750	0.499	4180	0.499	71 200	0.690	11 400	0.690	356 000	0.900		
		3570	0.490	3860	0.490			11 600	0.690				
		3580	0.490	3930	0.490								
E	Rieden	4060	0.466	4310	0.439	—	—	12 500	0.630	400 800	0.890	—	—
	13 mm diameter	3890	0.434	4570	0.407			12 500	0.630	401 200	0.884		
F	Rigden	3680	0.444	4340	0.451	57 500	0.680	12 000	0.695	382 000	0.882	118 100	0.715
	25 mm diameter	3530	0.435	4080	0.439	64 600	0.637	12 200	0.690	431 000	0.874	120 300	0.699

where P_v is the water vapour pressure at the temperature of test. At normal laboratory temperature and pressure (20°C,/1×10^5 Pa) the effect of ignoring this correction is to overestimate S_K by 1.5%, S_M by 3.0%, and S_V by 1.8%.

3.3.7.6 *Density of manometer liquid and reading manometer heights*

Errors in either of these measurements of $+a\%$ will lead to corresponding errors in S_k of $+1/2a\%$.

3.3.7.7 *Time*

Errors in measuring time, for both constant flow rate methods using bubble flow meters and for constant volume methods, of $+b\%$ will lead to corresponding errors in S_k of $+1/2b\%$.

3.3.7.8 *Effect of varying operator and instrument*

Whilst all the above factors lead to varying results it is considered that the major effect will be caused by unsatisfactory preparation of the powder bed. The following test for bed uniformity is recommended.

3.3.7.9 *Test for bed uniformity*

Measure the specific surface of a powder using two different masses of the material compacted to the same porosity. If the bed is uniform the results should be the same. With a variable volume cell some considerable skill is needed to ensure that the porosity is the same for each test. In the case of a fixed volume cell (in which the volume of the compacted powder is normally controlled by the position of a shoulder on the plunger), it is necessary to use two plungers of different lengths for this purpose. The bore of the cell should be checked at regular intervals for wear and/or distortion, and adequate safety precautions should be taken when compacting materials that are liable to ignite or detonate as a result of the generation of heat or friction.

3.4 MODIFIED PECHUKAS AND GAGE APPARATUS FOR FINE POWDERS

The permeameter for surface area measurement designed by Pechukas & Gage [25] operates within the 0.10–1.0 μm size range. Two fundamental errors were made in deriving their data [5], because firstly, they did not correct for slip, and secondly, although the inlet pressure was near atmospheric and the outlet pressures was low, no correction was made for compressibility. Carman & Malherbe [45] modified and automated their permeameter as shown in Fig. 3.7. The plug of material is formed in the brass sample tube A. Clamp E controls the mercury flow into the graduated cylinder C, the pressure being controlled at atmospheric by the manometer F. The side arm T_1 is used for gases other than air. The plug is formed in a special press by compression between hardened steel plungers. By taking known weights of powder the measurements may be carried out at a known and predetermined porosity, e.g. 0.45. Carman & Malherbe recommended that the final stages of compression be

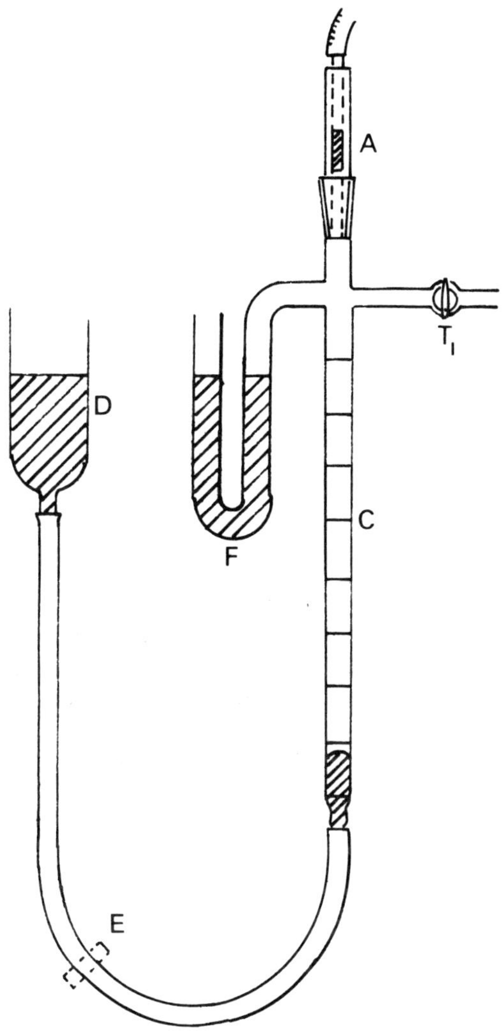

Fig. 3.7 — Modified Pechukas & Gage apparatus for fine powders.

carried out in small increments and that the plungers be removed frequently to prevent jamming.

3.5 NON-STEADY-STATE DIFFUSIONAL FLOW AND TRANSIENT FLOW APPARATUS OF KRAUS AND ROSS

This apparatus [46] consists essentially of two 4-1 reservoirs connected through the cell holding the powder. A mercury manometer was used to measure the pressure on the high-pressure side, and a calibrated thermocouple vacuum gauge was used to measure the pressure on the discharge side (Fig. 3.8). Before the experiment, the

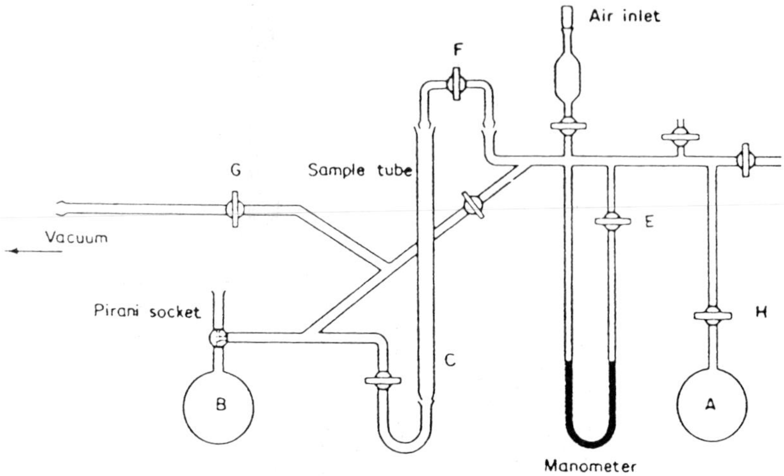

Fig. 3.8 — Transfer flow apparatus (Kraus & Rose).

whole system was evacuated and flushed with the gas being used. The system was then pumped down to 1 or 2 μm of mercury and shut off from the pumps with stopcock G. Stopcocks E and F were closed and the desired inlet pressure established by bleeding gas into reservoir A through H. At zero time, stopcock F was opened and the gas was allowed to diffuse through the cell C into the reservoir. A typical flow rate curve is shown in Fig. 3.9. The time lag t_L is determined by extrapolation of the straight-line, steady-state portion of the curve to the initial pressure in the cell and discharge reservoir.

3.6 THE STEADY-STATE DIFFUSIONAL FLOW APPARATUS

When processing compacted beds of very fine powders and gases near atmospheric pressures or with coarse powders and gases at reduced pressure, the mean free path of the gas molecules is of the same order as the capillary diameter. This results in slippage at the capillary walls so that the rate of flow is higher than that calculated from Poiseuille's law. If the pressure is further reduced until the mean free path is much greater than capillary diameter, viscosity takes no part in flow, since molecules collide only with capillary walls and not with each other. Such free molecular flow is really a process of diffusion and takes place for each constituent of a mixture against its own partial pressure gradient even if the total pressure at each end of the capillary is the same [5].

Allen, *et al.* [5,17,28,31,35], constructed a simple apparatus (Fig. 3.10) which enabled energy flow rate $V\,(\mathrm{d}p_1)/\mathrm{d}t$ to be plotted against P_2, giving a straight line from which S_V can be determined. The graph is shown in Fig. 3.11. These results were found to be in good agreement with BET adsorption values. The mechanism of the instrument works thus: The system is first evacuated with tap 4 closed. Taps 1, 2, and 3 are then closed and gas is allowed in on the inlet side by opening through the

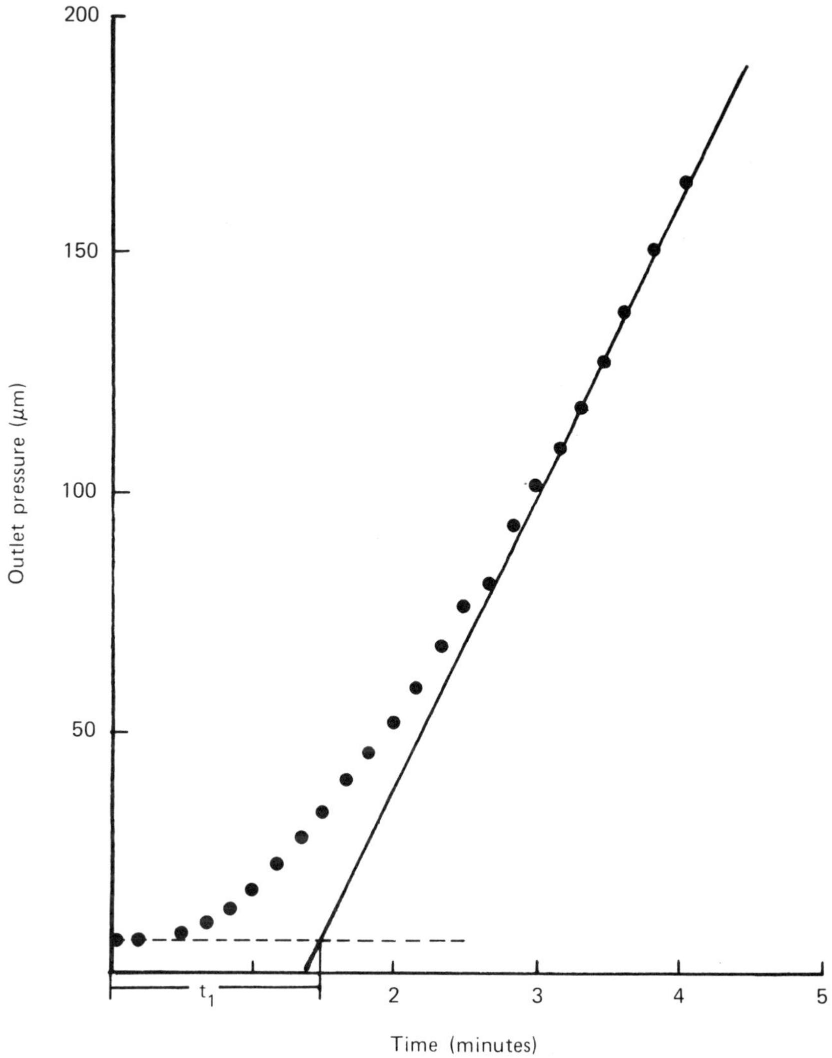

Fig. 3.9 — Flow-rate curve for the transient flow apparatus.

powder plug, the flow rate being monitored by the changing inlet pressure which is recorded as a deflection θ on a pen recorder graph. A timer is incorporated in the commercial version of this instrument (Alstan).

3.7 LIQUID-PHASE PERMEAMETER

In the apparatus [47,48,49] shown in Fig. 3.12, the powder is formed into a bed in a uniform tube A, and rests on the metal gauze B. The gauze is supported horizontally by a loosely wound spiral. The flow of liquid is adjusted to a steady rate with a

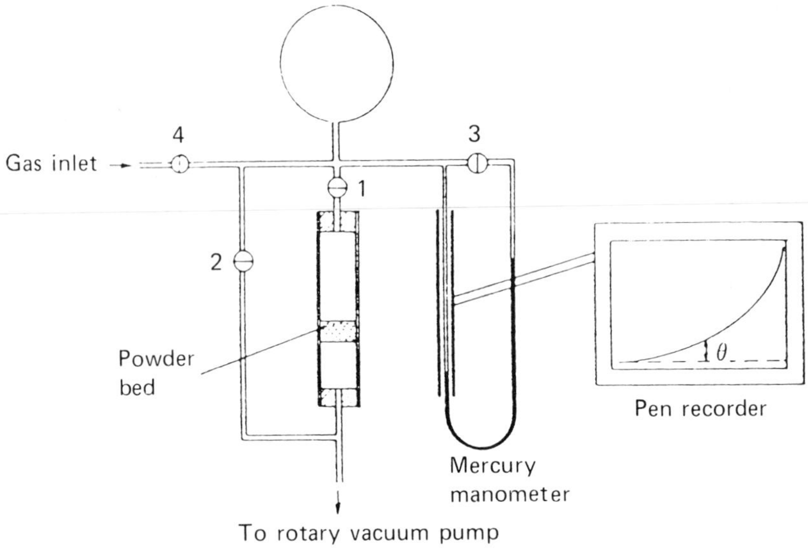

Fig. 3.10 — Simple gas diffusion apparatus.

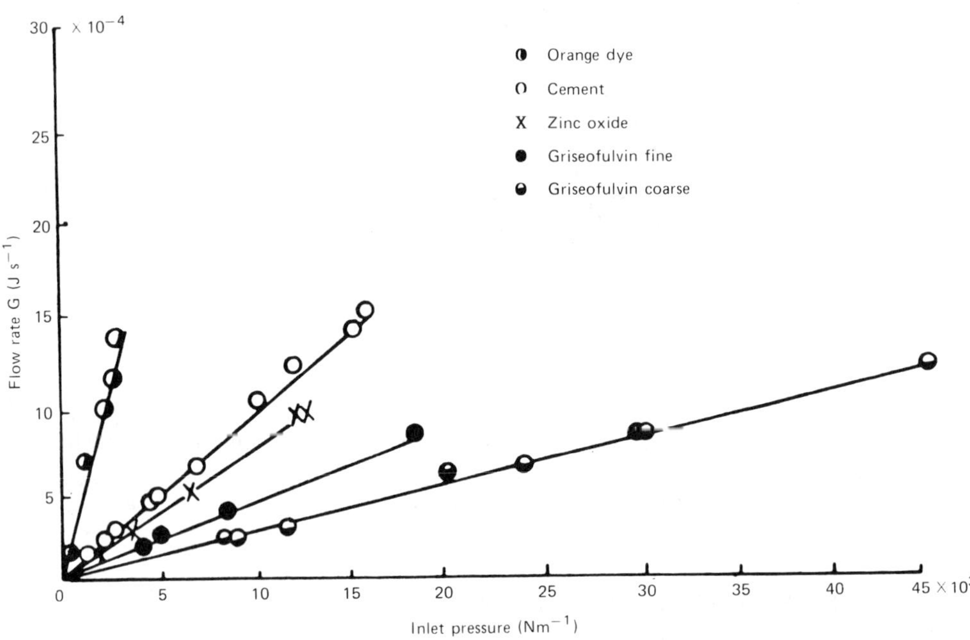

Fig. 3.11 — Gas flow rate against inlet pressure.

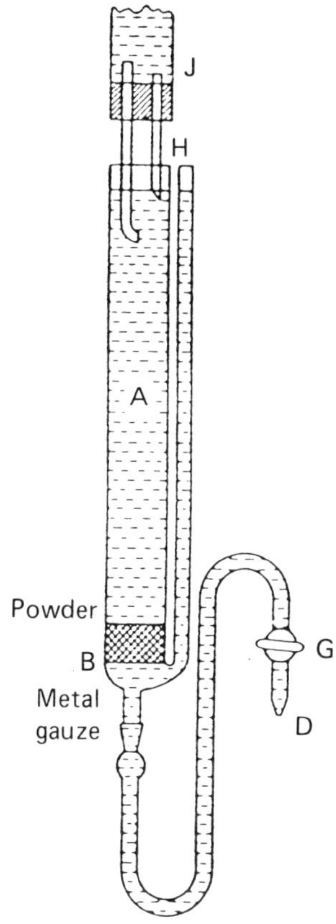

Fig. 3.12 — Liquid phase permeameter.

stopcock G, and the difference in level between D and the constant level in A gives the pressure drop causing flow. Air bubbles enter the tube H, causing a constant level to be maintained in A, and the volume of liquid supplied in known time is given by the graduated reservoir J. The bed is formed by washing a known weight of powder into A, using small quantities at a time and allowing each to settle into place with the assistance of gentle suction. The thickness is measured with a cathetometer.

It is reported [5] that in the early stages of development, liquid permeameters were favoured. As long as there is no appreciable size fraction below 5 μm, this is still the easiest technique. Below 5 μm, the use of liquids becomes unsatisfactory, owing to settling and segregation, the difficulty of removing air bubbles, aggregation, and wetting problems. Gas permeametry was also more attractive owing to the higher permeabilities of air and other gases. However, the surface areas determined by air permeametry were less than those determined by liquid permeametry, and the difference increased with decreasing size. Though gas permeameters were introduced, they were not placed on a satisfactory footing until the difference between liquid and gas permeabilities was shown to be due to slip in gases and corrections to the Kozeny equation were derived.

The difference between Fig. 3.12 and Fig. 3.13 is that in Fig. 3.13, the head decreased during an experiment [50].

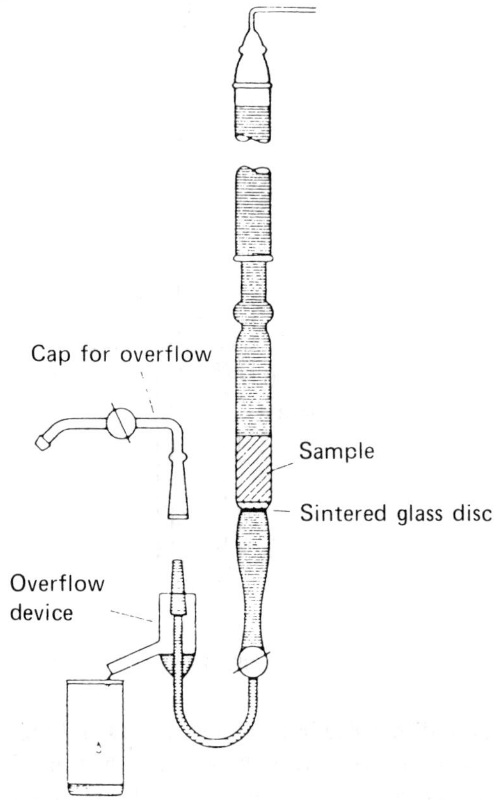

Fig. 3.13 — Variable-head permeameter.

3.8 CONCLUSION

We have attempted to review and elucidate, in the chapter, some of the available fundamental basis and operating standards on surface area measuring techniques using air or gas permeability methods.

There are, however, some other not so rapid but sophisticated surface area measuring methods, and we shall look at these further in Chapters 10 and 12. It is now necessary to demonstrate how to design an improved gas-solid flow apparatus.

SYMBOLS AND DEFINITIONS

Symbol	*Definition*	*Unit*	*SI unit*
A	The cross-sectional area of a bed of powder perpendicular to the direction of the flow of air.	cm^2	10^{-4} m^2
a	The manometric cross-sectional area of a constant volume apparatus	m^2	m^2
B	The atmospheric pressure at the time of the determination as indicated by a height of manometer liquid	m	m
C	The constant of a capillary flow meter	—	—
C^1	The constant of the flow meter resistance used in the Fisher sub-sieve sizer	—	—
d_m	The effective volume surface mean diameter	μm	μm
d_1	The mean particle diameter indicated on the Fisher sub-sieve sizes chart ($d_1 = d_m$ when 1 cm^3 of powder is used in the determination)	μm	μm
F	The pressure difference (water head) across the flow meter resistance in the Fisher sub-sieve sizer	cm	10^{-2} m
g	Acceleration due to gravity	m/s^2	9.81 $m.s^{-2}$
h	The pressure difference across a bed of powder as indicated by a height of manometer liquid	m	m
h_1	The pressure difference across a capillary flow meter resistance as indicated by a height of manometer liquid	m	m
h'_1, h'_2	The manometric heights corresponding to the first and second timing marks in a constant volume method	m	m
K	The Kozeny constant (usually taken as 5)	—	—
K_a	An apparatus constant, representing all the factors in a specific surface/permeability relationship that are derived solely from the apparatus dimensions	—	—
K_b	An apparatus constant for a constant volume apparatus used with a fixed volume cell and a constant bed porosity	—	—
L	The linear dimension of a bed of powder parallel to the direction of the flow of air (commonly known as the height of the powder bed).	cm	10^{-2} m
M	Molecular mass	kg	
P,P	The pressure (water head) applied to a bed of powder in the Fisher sub-sieve sizer; or	cm	10^{-2} m
	The atmospheric pressure at the time of the determination	N/m^2	(Pa)
P_0	The pressure of air leaving the bed of powder	N/m^2	(Pa)
P_1	The pressure of air entering the bed of powder	N/m^2	(Pa)
P_v	Water Vapour pressure	N/m^2	(Pa)
q	The rate of flow of an incompressible fluid through a bed of powder	m^3/s	
q_0	The rate of flow of air from a bed of powder	m^3/s	$m^3.s^{-1}$

q_1	The rate of flow of air into a bed of powder	m^3/s	
R	The Universal gas constant	8.31 J/ k.mol	
R'	An apparatus and material constant in the Rigden method	—	
S_V	The effective volume specific surface of a powder	m^2/m^3	m^2/m^3
S_w	The effective mass specific surface of a powder	m^2/kg	m^2/kg
S_K	The effective volume specific surface of a powder assuming only viscous flow occurs in the determination	m^2/m^3	m^2/m^3
S_M	The effective volume specific surface of a powder assuming only slip flow occurs in the determination	m^2/m^3	m^2/m^3
T	The temperature at the time of the determination	K	
t	Time	s	
V	The volume of air passed through a constant flow rate apparatus during the determination (or during calibration of a capillary flow meter).	m^3	
v	The effective solid volume of a powder sample	m^3	
v_1	The effective solid volume of a powder sample used in the Fisher sub-sieve sizer	cm^3	$10^{-6}\ m^3$
W	The mass of a bed of powder	g	
μ	The viscosity of air at the temperature at the time of the determination	$N\ s/m^2$	$N.s.m^{-2}$
ΔP	The pressure difference across a bed of powder	N/m^2	(Pa)
δk_0	A constant in the Carman-Arnell equation	—	
ε	The porosity of a bed of powder	—	
ε_1	The porosity of a bed of powder indicated on the Fisher sub-sieve sizer	—	
ρ_{DL}	The density of the displacement liquid used in the measurement of effective solid density.	kg/m^3	
ρ_s	The effective solid density of a powder	kg/m^3	
ρ_w	The density of pure water during the calibration of a density bottle	kg/m^3	

REFERENCES

[1] British Standards Institution. Private circulation Document 81/70580: Draft Revision of BS 4359: Part 2: 1971. Air Permeability Methods: (January 1981).

[2] Lea, F. M. & Nurse, R. W., The specific surface of fine powders, *J. Soc. Chem. Ind.*, **58**, 277–283 (1939).

[3] Lea, F. M. & Nurse, R. W., Permeability methods of fineness measurement. *Symp. on Particle Size Analysis*, 47–56 (1947) Discussion—71, 72, 75, 114, 118, 120: Author's Reply: 145, *Suppl. to Trans. Inst. Chem. Eng.*, **25** (1947).

[4] Gooden, E. L. & Smith, C. M., Measuring the average particle diameter of powders. *Ind. Eng. Chem., Anal. Ed.*, **12**, 479–482 (1940).

[5] Allen, T., *Particle size measurement*, 3rd Ed., Chapman and Hall, (1981).
[6] Fisher Scientific Company, Fisher subsieve sizer; Manufacturer's literature, Fisher Scientific Co., 633 Greenwich Street, New York.
[7] Dubrow, B., Calibration of air permeability particle sizers. *Anal. Chem.*, **25**, 1243 (1953). Also, Dubrow, B. & Nieradka. Determination of specific surface of sieve size powders. *Anal. Chem.*, **27**, 302 (1955).
[8] Edmundson, I. C. & Toothill, J. P. R., *Analyst* **88**, No. 1051, 805–808 (1963).
[9] *Permaran specific surface area meters*. Outokumpu Oy Co., Instruments Division, Tapiola, Finland.
[10] British Standards Institution. *Glossary relating to Powders*. BS.2955: (1958).
[11] Edmundson, I. C., *Analyst* **91**, No. 1082, 306–315, (1966).
[12] Rigden, P. J., The specific surface of powders. A modification of the air-permeability method for rapid routine testing. *J. Soc. Chem. Ind.*, **62**, 1–5 (1943) (b) Discussion of report of air-permeability method for determining fineness of cement., *Am. Soc. Terst. Mater. Bull.*, **123**, 49 (1943).
[13] Rigden, P. J., The specific surface of powders: A modification of the theory of the air-permeability method. *J. Soc. Chem. Ind.*, **66**, 130–136 (1947).
[14] Blaine, R. L., A simplified air permeability fineness apparatus. *Am. Soc. Test. Mater. Bull.*, **123**, 51–55 (1943).
[15] Ober, S. S., & Frederick, K. J., A study of the Blaine fineness tester and a determination of surface area from air permeability data. *Symp. on Particle Size Measurement, 61st Annual Meeting of ASTM, Boston, June, 1958,* and *Am. Soc. Testing Materials,* Philadelphia, 279–301 (1958).
[16] Griffen and George Limited, London and Birmingham.
[17] Allen, T. & Maghafarti, R. P., *Particle Size Analysis Conference, Loughborough, Anal. Div. Royal Soc. Chem.* (1981).
[18] Allen, T. & MacSporran, C., Gas flow at reduced pressures. *Powder Technology,* **33**, 195–200 (1982).
[19] Community Bureau of Reference, BCR, Rue de la Loi 200, B1049, Brussels.
[20] National Bureau of Standards, Washington, DC 20234. USA.
[21] National Physical Laboratory, Teddington, Middlesex TW11 0LW.
[22] Keyes, W. F., Determination of specific surface by permeability measurements. *Ind. Eng. Chem. Anal. Ed.,* **18**, 33–34 (1946).
[23] Rose, H. E., *The measurement of particle size in very fine powders*. Constable, London (1953).
[24] Orr, C., Jr. & Dallavalle, J. M., *Fine particle measurement, particle surface and pore volume*. Macmillan, New York (1959).
[25] Pechukas, A. & Gage, F. W., *Ind. Engng. Chem. Analyt. Ed.,* **18**, 370–3 (1946).
[26] Schultz, N. F., Measurement of surface area by permeametry. *Intern. J. Miner. Process.,* **1**, 65–79 (1974).
[27] Wasan, D. T., Wnek, W., Davies, R., Jackson, M., & Kaye, B. H., Analysis and evaluation of permeability techniques for characterizing fine particles. Part I, Diffusion and flow through porous media. Part II, Measurement of specific surface area. *Powder Technol.,* **14**, 209–228, 229–244 (1976).
[28] Allen, T., Stanley-Wood, N. G., & Krishnamoorthy, T. S., *Particle Size Analysis Conf. Loughborough. Soc. Analyt. Chem.* London.

[29] Dallavalle, J. M., *Micromeritics — the technology of fine particles,* Pitmans, New York, 2nd Ed. (1948).
[30] Richardson, J. G., Flow through porous media, In: Streeter, V. L., (Ed.) *Handbook of fluid dynamics,* McGraw-Hill, New York, Sect. 16–1 to 16–112 (1961).
[31] Allen, T., *Silic. Ind.,* **36**, 718, 173–85 (1971).
[32] Collins, R. E., *Flow of fluids through porous media,* Reinhold New York (1961).
[33] Bear, J., *Dynamics of fluids in porous media.* American Elsevier, New York (1972).
[34] Prandtl, L. & Tietjens, O. G., *Applied hydro and aeromechanics.* McGraw-Hill, New York (1934).
[35] Allen, T., *Proc. Conf. Particle Size Analysis, Salford* (ed. Groves, M. J.) *Chem. Soc. Anlyt. Div. Heyden* (1977).
[36] Blake, F. C., The resistance of packing to fluid flow. *Trans. Am. Inst. Chem. Eng.,* **14**, 415 (1922).
[37] *Flow through porous media, sixth State of the Art Symp.,* American Chemical Society Publications, Washington, DC (1970).
[38] Wiest, R. J. M., (Ed.) *Flow through porous media.* Academic Press, New York (1969).
[39] Orr, C., Specific surface area by low-pressure permeametry. *J. Anal. Chemistry,* **39**, No. 7, 834–836, (1967). Also, Orr, C., *Particulate technology,* Macmillan, New York (1966).
[40] Poole, J. B. & Doyle, D., *Solid–liquid separation: a review and bibliography,* HMSO, London (1966).
[41] Kellomaki, A., Kuula-Vaisanen, P., & Nieminen, P., Measuring of surface areas by ethylene glycol monoethyl ether. *Proceedings of the Particle Size Analysis* (1988) (ed. Lloyd, P. J.), page 213–216, John Wiley & Sons, Chichester, New York (1988).
[42] Carrott, P. J. M., Roberts, R. A., & Sing, K. S. W., Choice of adsorptive for determination of surface area and pore size distribution. *Proceedings of the Particle Size Analysis* (1988) (ed. Lloyd, P. J.), 217–227, John Wiley & Sons, Chichester, New York (1988).
[43] Hearn, J., The preparation of and characterization of porosity in macrorecticular resins for use as gas/vapour sorbents. *Proceedings of the Particle Size Analysis 1988* (ed. Lloyd, P. J.), 229–241, John Wiley & Sons, Chichester, New York (1988).
[44] Igwe, G. J. I., Allen, T., & Stanleywood, N. G., Surface area measurement and gas permeametry at sub-atmosphereic pressures. *Proceedings of the Particle Size Analysis* 1988 (ed. Lloyd, P. J.), page 243–256, John Wiley & Sons, Chichester, New York (1988).
[45] Carman, P. C. & Malherbe, P. le R., Routine measurement of surface of paint pigments and other fine powders. *J. Soc. Chem. Ind.,* **69**, 134–143 (1950).
[46] Kraus, G., Ross, R. W., & Girifalco, L. A., *J. Phys. Chem.,* **57**, 330, (1953).
[47] Carman, P. C., Determination of specific surface of powders. *J. Soc. Chem. Ind., (Trans.),* **57**, 225 (1938).
[48] Walther, H., *Kolloid-Z.,* **103**, 233 (1943).

[49] Wiggins, E. J., Campbell, W. B., & Maas, O., *Can. J. Res.*, **26A**, 128, (1948).
[50] Dodd, C. G., Davis, J. W., & Pidgeon, F. D., *Ind. Engng. Chem.*, **55**, 684 (1951).

4

Design of an improved gas flow apparatus

4.1 INTRODUCTION

Viscous flow has been well described by Poiseuille's law [1]. It relates to a gas near ambient pressure and temperature, forced by a small pressure gradient to flow through the channels of a packed bed of powder. At room temperature, a gas molecule can be adsorbed on a solid surface for an extremely short time but not less than the time required for one vibrational cycle or about 10^{-13} seconds [2]. When the adsorbed molecule leaves the surface it will, on the average, have a zero velocity component in the direction of flow. After undergoing one or several gas phase collisions, it will soon acquire a drift velocity equal to the linear flow velocity. These collisions and corresponding momentum exchanges will occur within one, or at most a few, mean free path lengths from the surface with the net effect of decreasing the linear velocity of those molecules flowing near the surface. As the area to volume ratio of the channels in the powder bed increases, the viscous drag effect will also increase, and the rate of flow, for a given pressure gradient, will decrease. It is, therefore, clear that the surface area of the particles which constitute the channel walls is related to the gas flow rate and pressure gradient. A general filtration spectrum is shown in Table 4.1.

4.2 SIMPLE LINE FLOW DIAGRAM OF THE APPARATUS

Since the two important parameters that have to be measured in permeametry are flow rate and pressure drop, two types of electronic measuring devices, operating under constant gas flow rate conditions, have been developed. Fig. 4.1 shows the general layout of the assembly, whilst Fig. 4.2 shows diagrammatically the simple line flow.

To convert the non-electrical physical parameters of pressure drop across the packed bed into electrical signals, two sensitive multitron absolute pressure transducers (obtained from Appleby and Ireland) were used to provide a rapid response

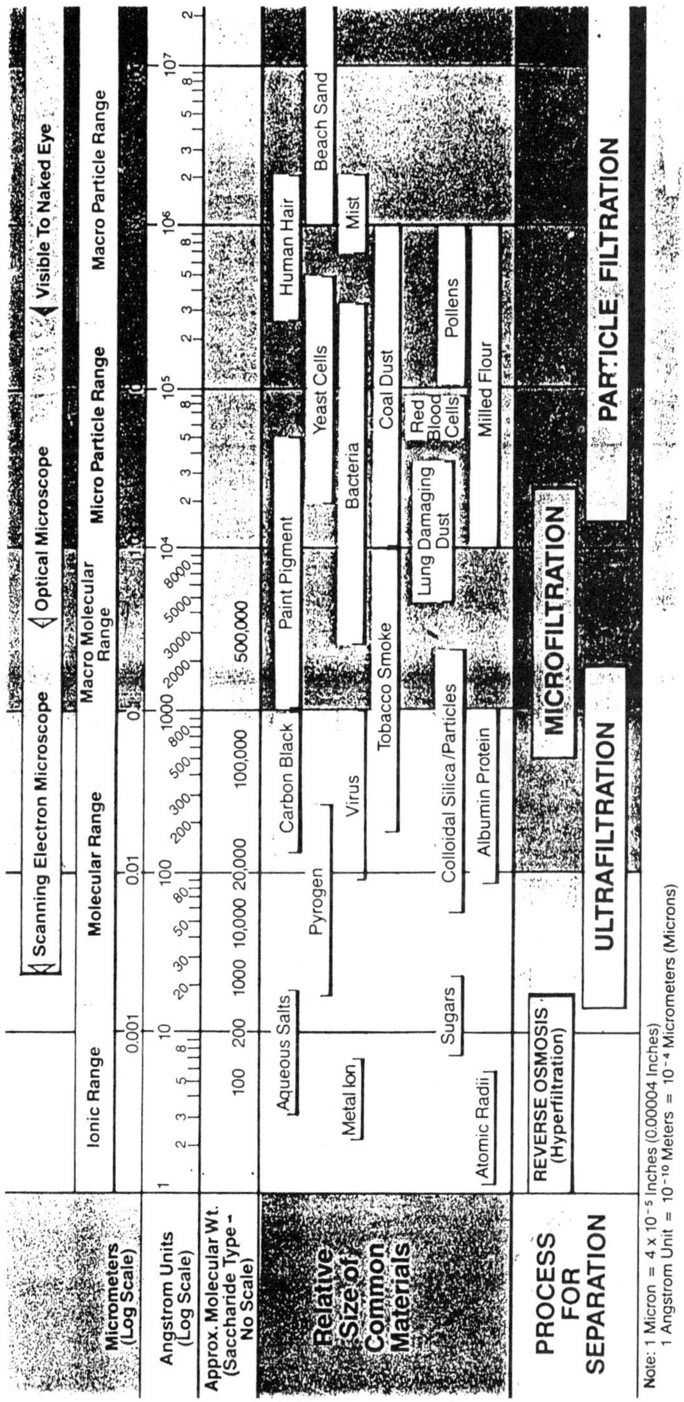

Table 4.1 – The filtration spectrum

Fig. 4.1 — General layout of the electronic permeameter.

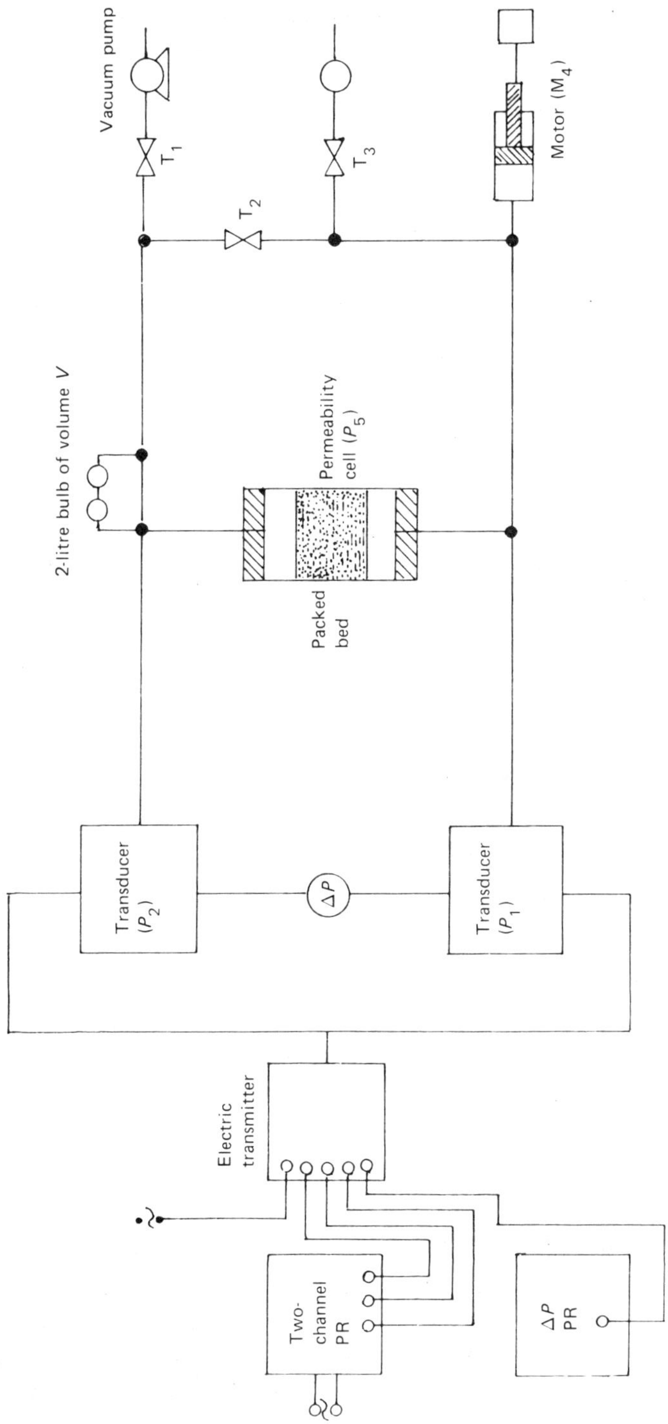

Fig. 4.2 — Simple linear flow diagram of the experimental assembly.

system. The transducer, based on the linear variable differential transformer (LVDT) principle, consist of a primary and secondary winding. Displacement of a movable magnetic core from its central zero position produces an output (at carrier frequency) which is proportional in magnitude to the core displacement. General components and mechanisms of LVDT pressure transducers are diagrammatically shown in Figs 4.3, 4.4, and 4.5.

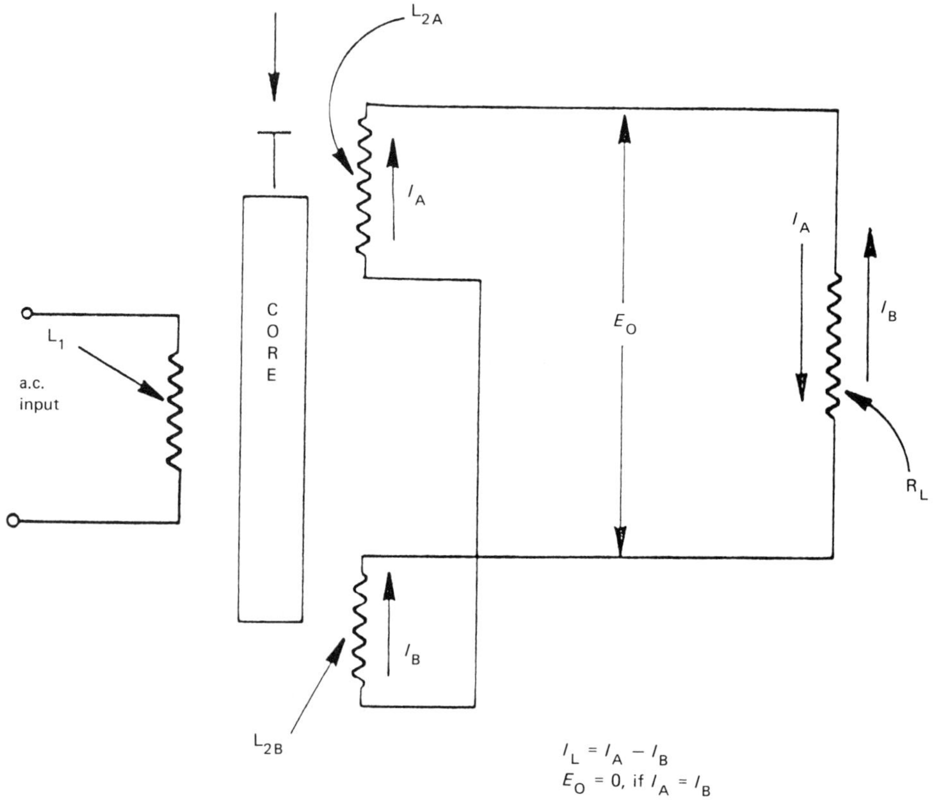

Fig. 4.3 — Circuit of the LVDT.

4.3 THE PRESSURE RANGE OF THE TRANSDUCERS

The two sensitive low voltage (1 volt) absolute multitron electronic pressure transducers used in monitoring the inlet (P_1) and outlet (P_2) pressures and pressure drops across the packed bed are shown in Fig. 4.6. The range of pressure measurements for the two transducers is 0–1 atm (0–760 mm Hg). Figs 4.7 and 4.8 show the

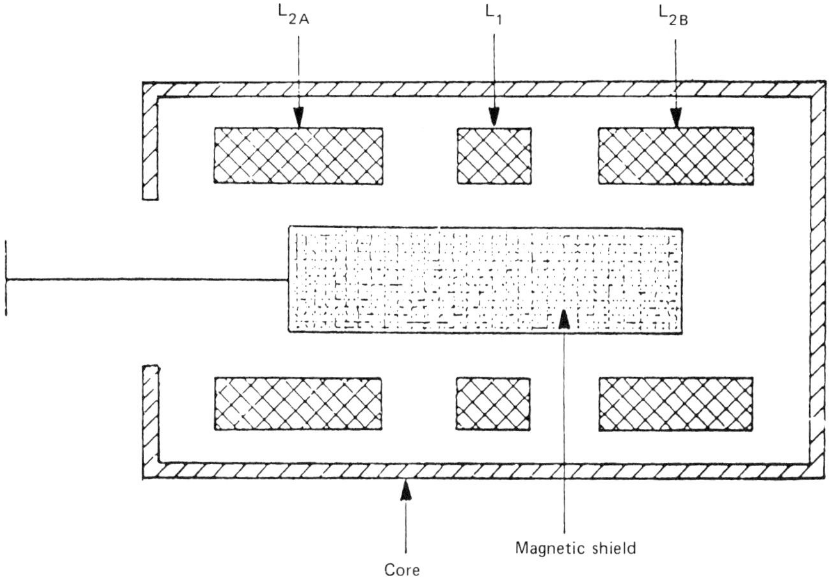

Fig. 4.4 — Construction of the LVDT and essential components.

working calibration charts for the pressure and pressure drop instruments as appearing on the two pen-recorders, at a constant air flow rate, $dV/dt = 4.167 \times 10^{-8}$ m³/sec.

4.4 THE ELECTRONIC PRESSURE SENSOR

The electronic units are built on a printed circuit board, PCB, (Figs 4.9, 4.10, and 4.11). Another section of the card is an amplifier that provides a current output signal for external use and isolates the transducer circuits from the external load.

The prime sensing source of the pressure transducer is a precision capsular element made by welding together two annular corrugated metal nesting diaphragms formed into shells of opposite curvature and fastened together with a hermetically sealed joint around their peripheries. The capsule is located in the housing (Fig. 4.12) by a fitting on one diaphragm that permits pressure to be connected to the inside of the capsule. A fitting on the other diaphragm accepts the stem of the LVDT ferrite core which is located in a non-magnetic tube on which the LVDT coil is mounted.

Differential pressure is detected by connecting the higher pressure to the housing chamber, i.e. outside the capsule, and the lower pressure inside the capsule. Evacuation of the capsule creates an absolute pressure aneroid capsule for barometric or scientific studies.

Any pressure changes in the packed bed will cause the pressure-sensitive elements to deflect. This deflection causes a movement of the ferromagnetic core and provides an a.c. signal proportional to the displacement of the sensing element.

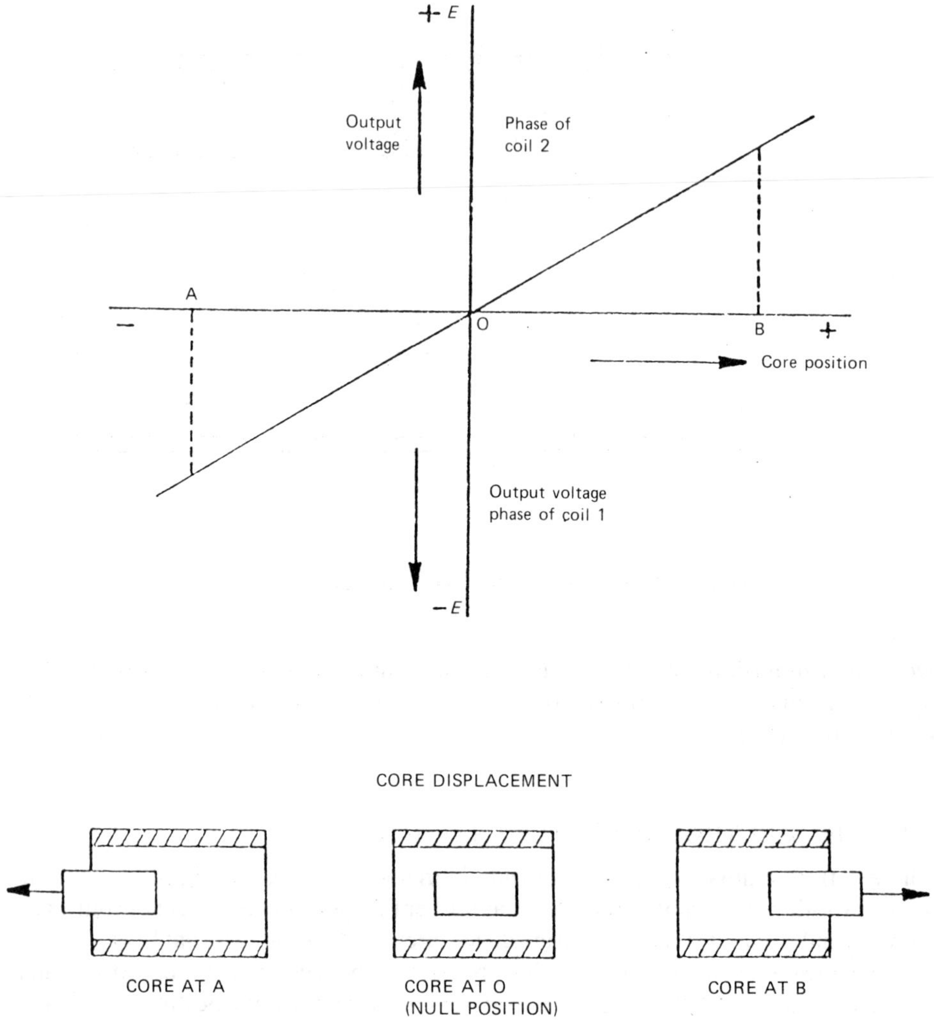

Fig. 4.5 — Linear characteristics of LVDT core movements and relative positions.

The signal is then transmitted to an electronic transmitter (Fig. 4.13) which provides the excitation, demodulation, amplification, and conversion to a voltage based output two-channel pen recorder, PR. A typical graphical output from the two pen recorders used is shown in Fig. 4.14. It can be seen from the graphical output in Fig. 4.14 of the two-channel (P_1 and P_2) pen recorders that

$P_1 = 0.131\ \text{V} = 99.56\ \text{mm Hg}$
$P_2 = 0.103\ \text{V} = 78.28\ \text{mm Hg}$
$\Delta P = 21.28\ \text{mm Hg}$

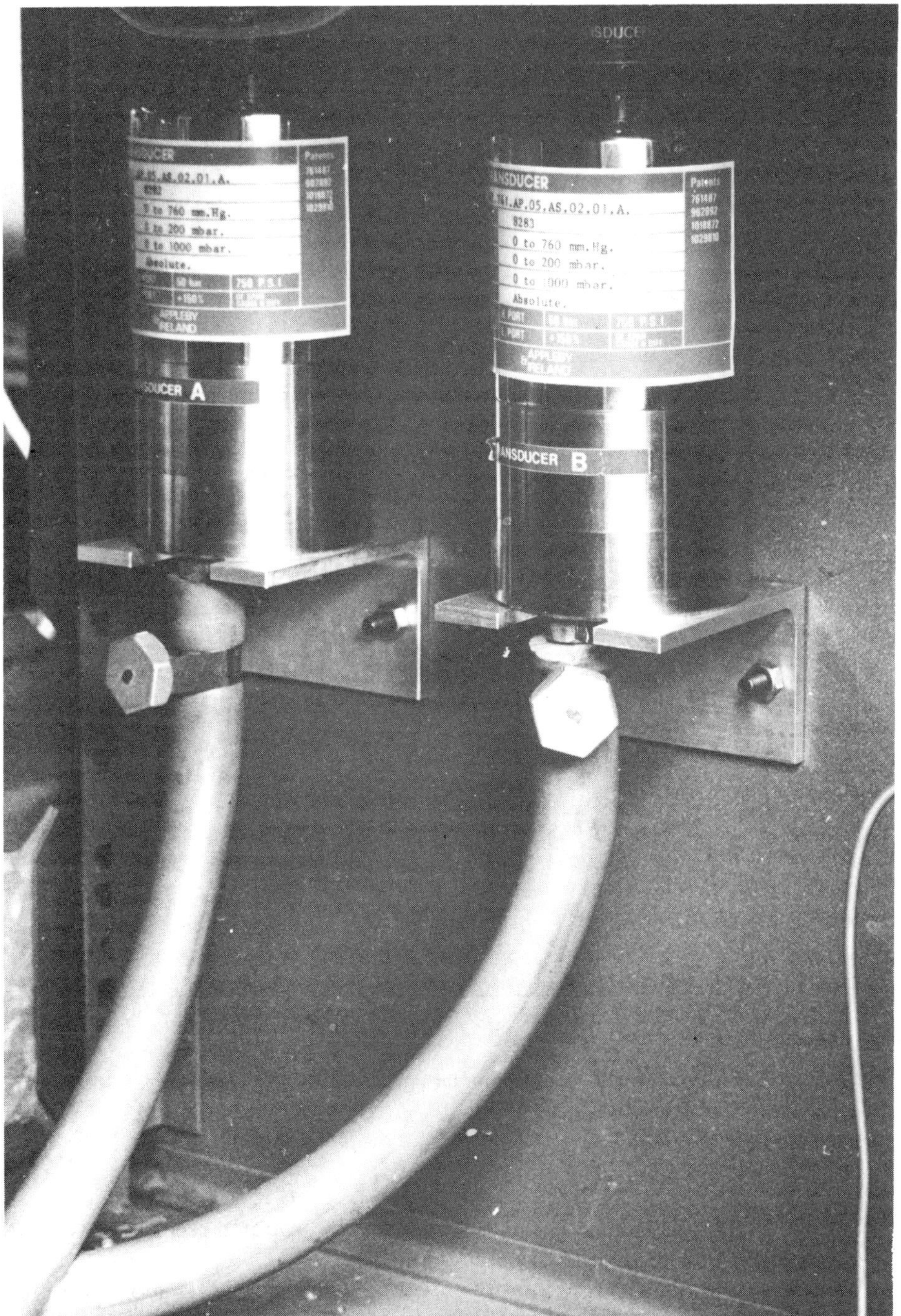

Fig. 4.6 — The two Multitron electronic pressure transducers used in monitoring the input and output pressures and the pressure drop.

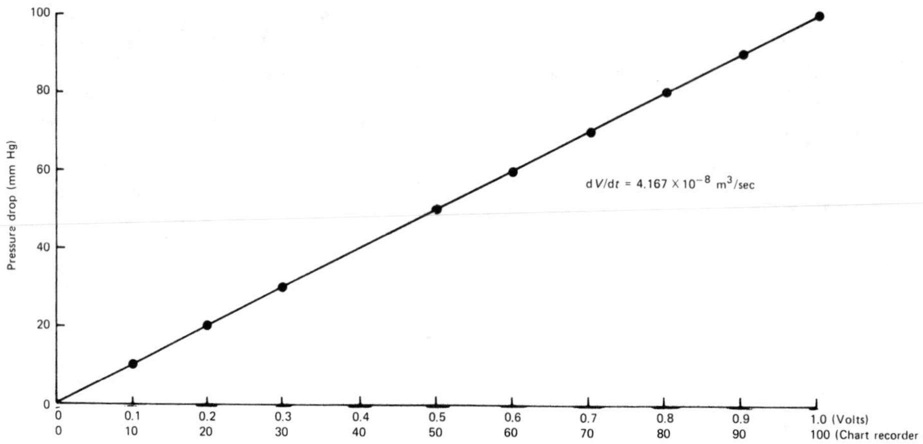

Fig. 4.7 — Calibration chart for the two-channel pen recorder.

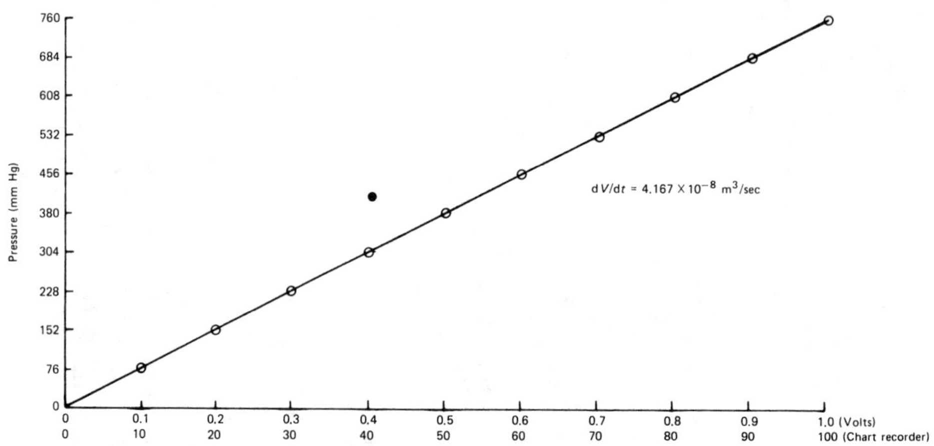

Fig. 4.8 — Calibration chart for the pressure drop pen recorder.

Now, from the same Fig. 4.14, the amplified pressure drop pen recorder gives the true value of $\Delta P = 21$ mm Hg.

It can, therefore, be seen that this pressure drop device provides a more accurate account than a manometric fluid in terms of accuracy and reproducibility of the pressure drop across the packed bed of powder.

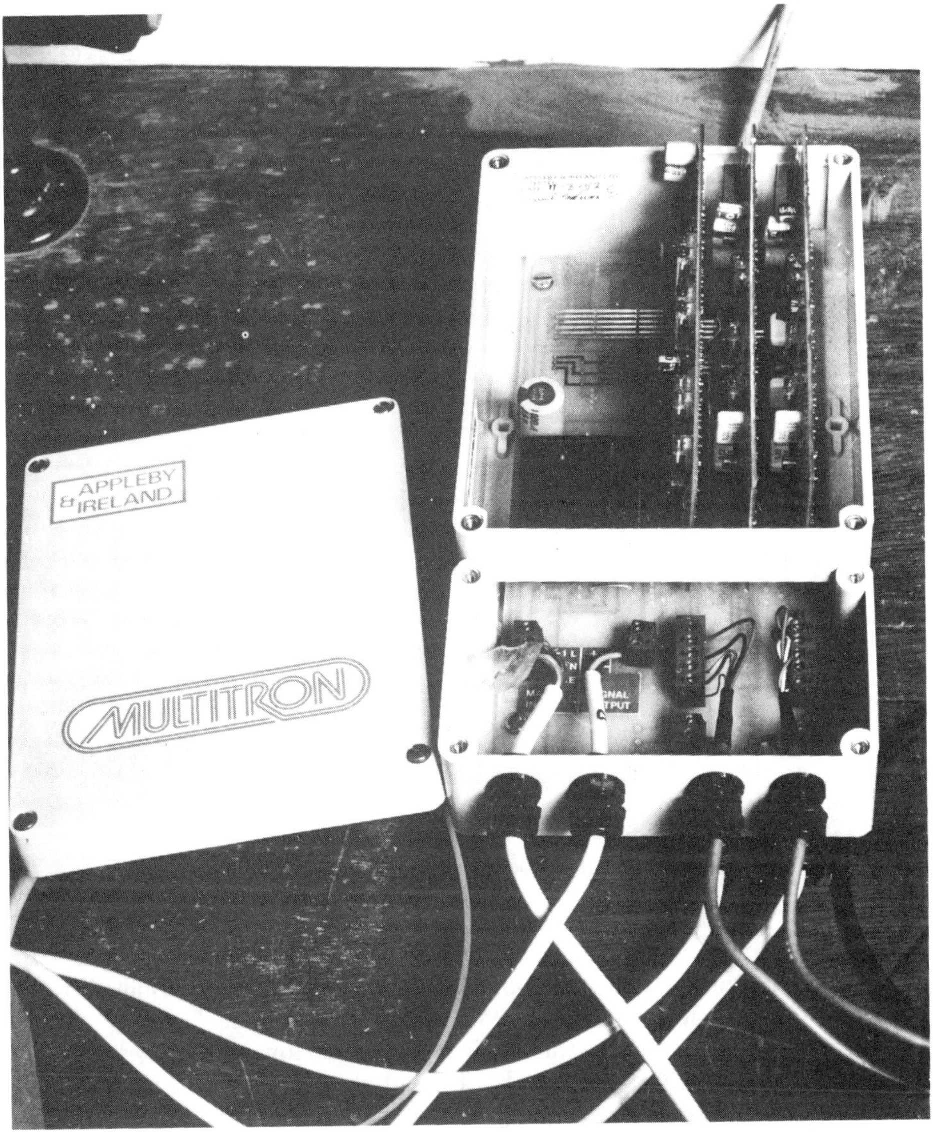

Fig. 4.9 — Printed circuit board (PCB) and transmitter.

4.5 GAS USED

Air was used as the fluid transport medium and was dried by passing it through a column of silica gels before its entry into the system.

4.6 POWDERS USED

Four different powders were used for the experimental analysis. Dicalcium phosphate dihydrate [$Ca(H_2PO_4)_2$] is a white crystalline water insoluble powder with a

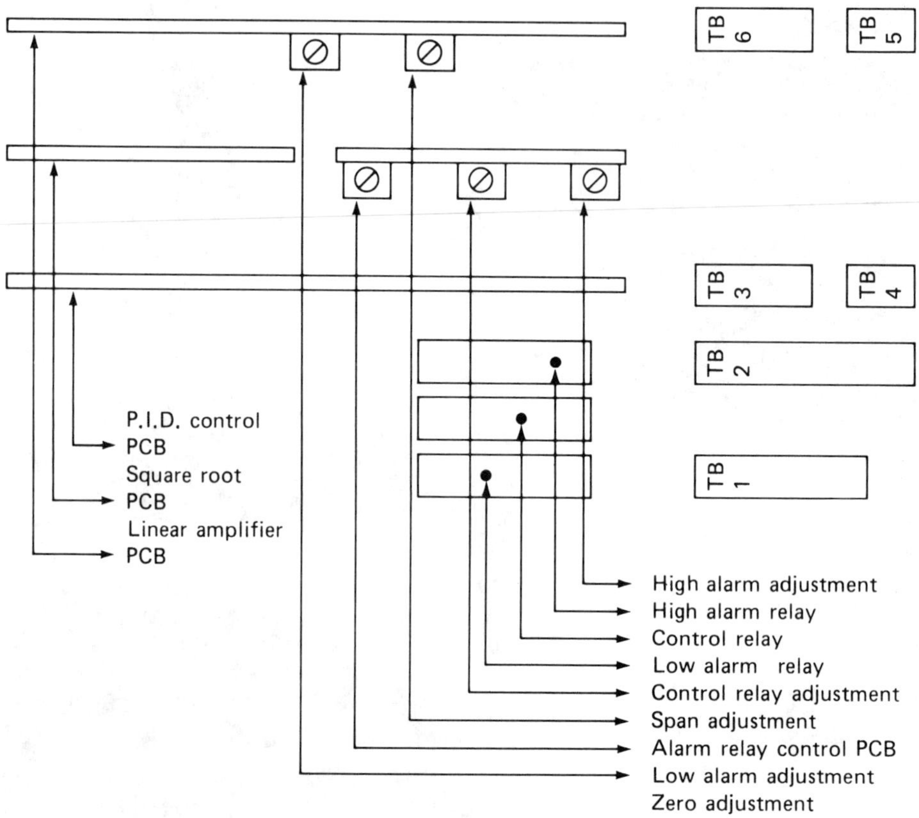

Fig. 4.10 — PCB and adjustment potentiometer layout.

mean particle size as determined by photosedimentation of 18 μm, supplied by Albright and Wilson Limited, London. The particle density as determined by air pycnometer (Beckman air pycnometer, model-930) is 2420 kg m^{-3}.

The spherical bronze powder was obtained as a sub-38 μm powder (by screening) from G.K.N. Bound Brook Limited, Lichfield, England. It was then further classified into seven size ranges by using the Bahco centrifuge. The mean sizes were determined by microscopy, and the results are given in Table 4.2. 625 particles were measured in each size range to give a standard deviation of the mean of 2%.

Cyclone tin powder (<25 μm) was obtained from Ronald Britton Limited, Rochdale, England, and copper powder (irregular <38 μm) was obtained from McKechnie Metal Powders Limited, Widnes, England.

4.7 PACKING OF THE CELL

All the powders, before being packed into the cell, were first dried at 100°C in an oven for about 24 hours and cooled at room temperature in a desiccator. The

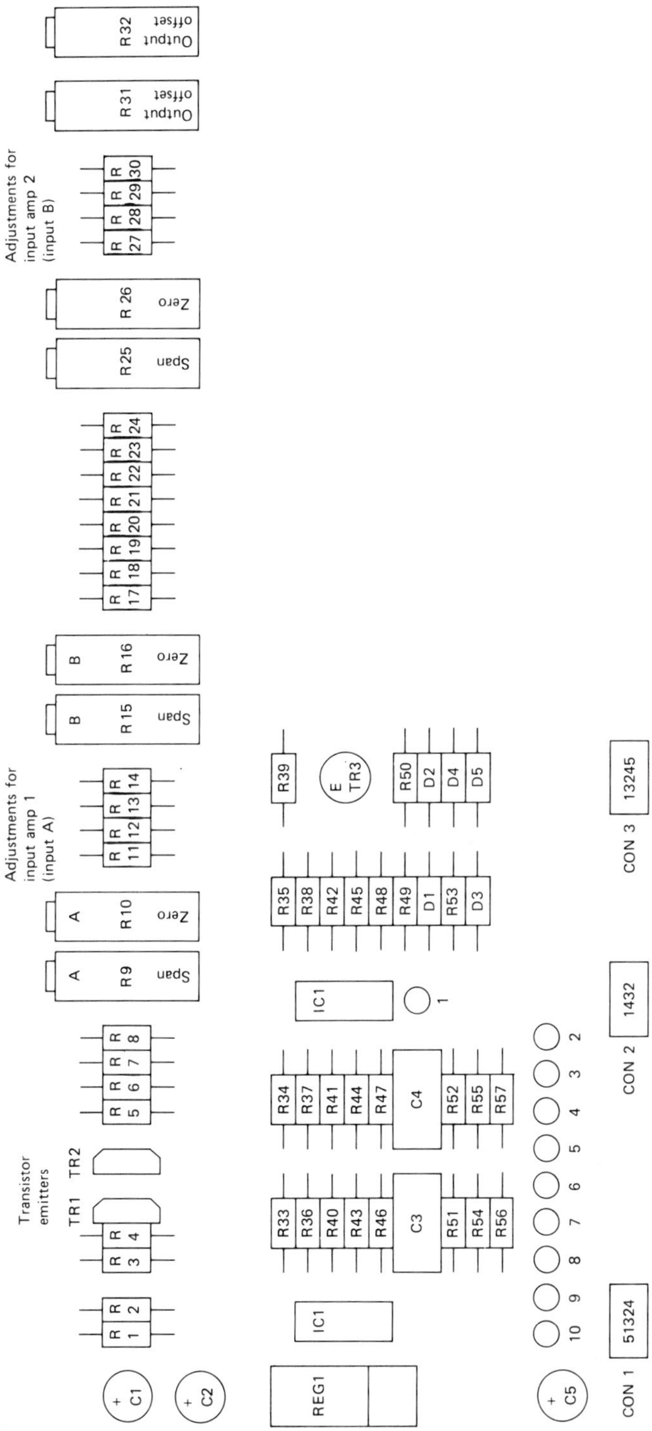

Fig. 4.11 — PID control module and PCB assembly.

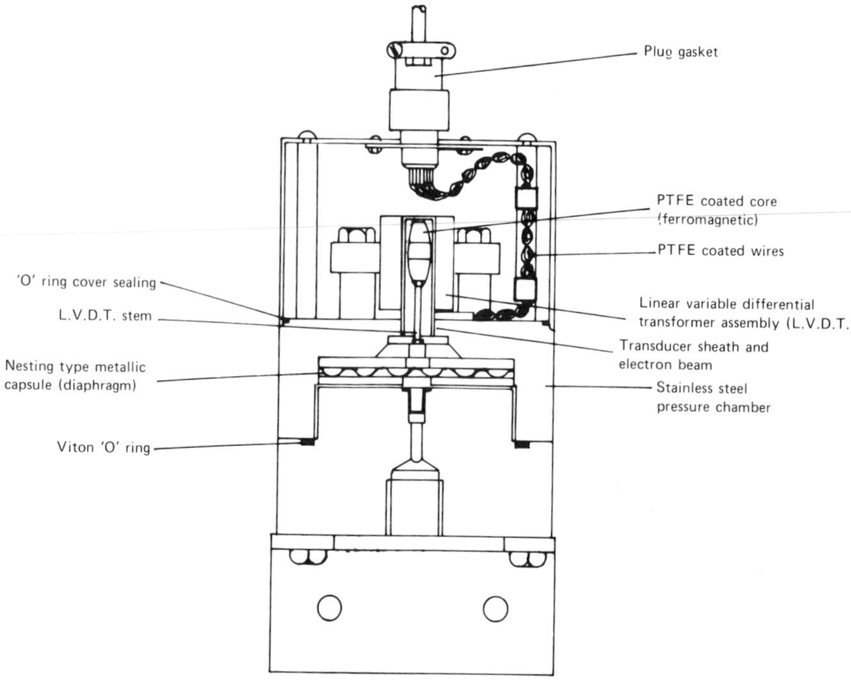

Fig. 4.12 — Multitron pressure transducer.

plungers or porous support were placed in the cylindrical bed and then covered with a Whatman No. 1 filter paper. Small amounts of powder were then carefully placed on top of the cylindrical filter disc in increments, and each increment was loosely packed by gentle vibration of the container. From the length and volume of the plug, measured with micrometric calipers, the porosity of the powder could be calculated. The porosity, ε, is given as;

$$\varepsilon = 1 - \frac{w}{\rho_s A L}$$

where w=mass of powder bed
ρ_s=powder density
A=cross-sectional area of the bed
L=length of the bed.

4.8 OPERATING PROCEDURE FOR MEASURING PERMEAMETRY THROUGH THE PACKED BED

The operating procedure will be described in more detail. All the stopcocks and flow tubes were constructed of Pyrex glass, and before starting the experiment, T_1 was closed and T_2 and T_3 opened to the atmosphere (see Fig. 4.2). This state of the

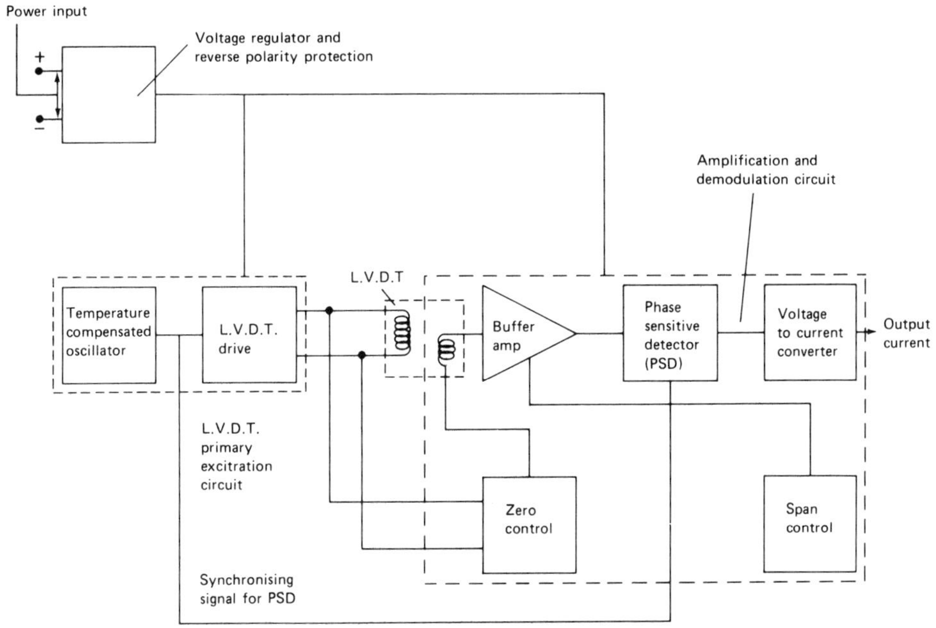

Fig. 4.13 — Transducer, demodulation, and signal output PCB.

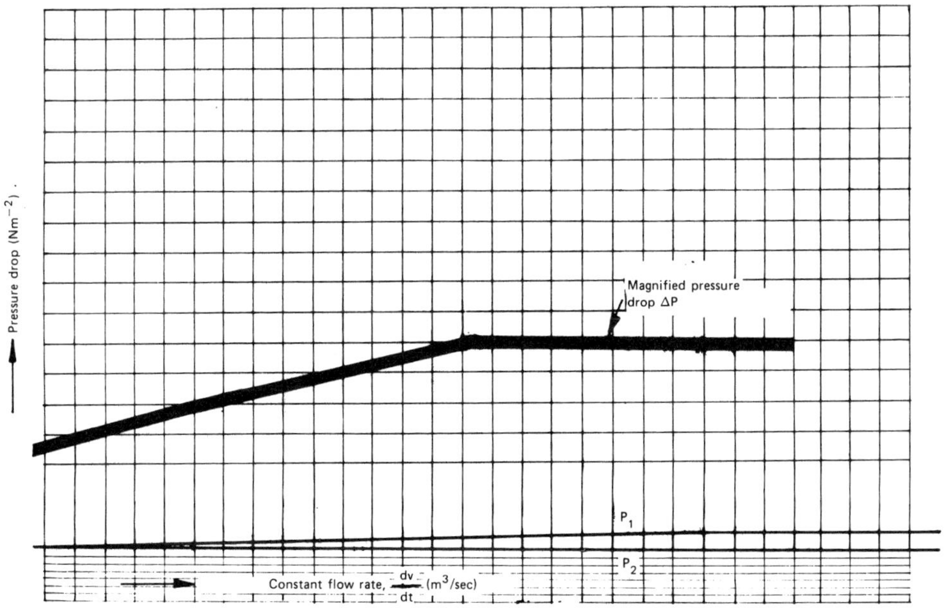

Fig. 4.14 — Typical output from the two pen recorders.

Table 4.2 — Mean sizes of spherical bronze powder

Size range (μm)	Mean size (microscopy) (μm)
8.1–15	13.375
15.0–18.5	14.663
18.5–25	15.240
8.1–38	17.153
15.0–38	19.576
18.5–38	20.841
21.0–38	21.708

apparatus enables the graduated syringe pipette tube on the electronic flow meter M_4, which is motor driven, to draw air and to be set at the starting point of 20 millilitres. The motor speed and choice of driving cogs were selected to give an optimum flow rate of 4.167×10^{-8} m³/sec.

Fig. 4.15 shows a close-up of Fig. 4.1 and the electronic flow meter. The dimensions and design characteristics of this flow meter (Microscal) is diagrammatically shown in Fig. 4.16, in which the main point of emphasis is the bevel gear.

The horizontally mounted electronic motor drives the bevel gear, giving a vertical drive to a small spur gear which in turn meshes with an idler spur gear of larger diameter. The two intermeshed spur gears of different diameters, running between the two joined parallel racks, produce a mesh-rack and pinion action, the gears rotating on fixed axis (one driven, one idling), thus imparting a linear motion to the rack assembly.

In the procedure, we next close T_3, and the vacuum pump is switched on. T_1 and T_2 are then gradually and carefully opened to minimize the risk of disturbing the powder bed during removal of desorbed atmospheric vapours and gases. This is an important preliminary to the accurate measurement of the pressure drop across packed beds. In their usual state, all solid surfaces are covered with absorbed film which must be removed (degassed) before any quantitative measurements are made.

When the required pressure is reached the system is then isolated from the vacuum pumping system by closing T_1, and the system is allowed to stabilize.

Bringing the two pens to a steady base line and setting the pressure drop PR to the zero base line is carried out by using the fine and coarse regulators. After this process, T_2 is then closed.

The apparatus is now ready for permeability measurements. The motor, M_4, is now switched on to admit the adsorbate, giving a constant flow rate through the packed bed, P_5. A red light on one side of the electronic flowmeter indicates that the adsorbate is flowing steadily. The time required for the adsorbate to flow across the adsorbent at a constant rate and pressure is 480 seconds.

Fig. 4.15 — Electronic flow meter Microscal (Fig. 4.1) at higher magnification.

ITEM No.	DESCRIPTION	No. OFF	MAT'	REMARKS
1	Vertical shaft	1	MS	1/4″ diameter × 2″
2	Bush bracket	1	BRM	1 1/2″ diameter × 1/2″
3	Straight mitre bevels	2	Drill	B24–24 davel
4	Horizontal shaft	1	MS	1/4″ diameter × 1″
5	Coupling	1	MS	1/2″ diameter × 3/4″
6	Motor bracket	1	MS	3″ × 3″ × 1/2″

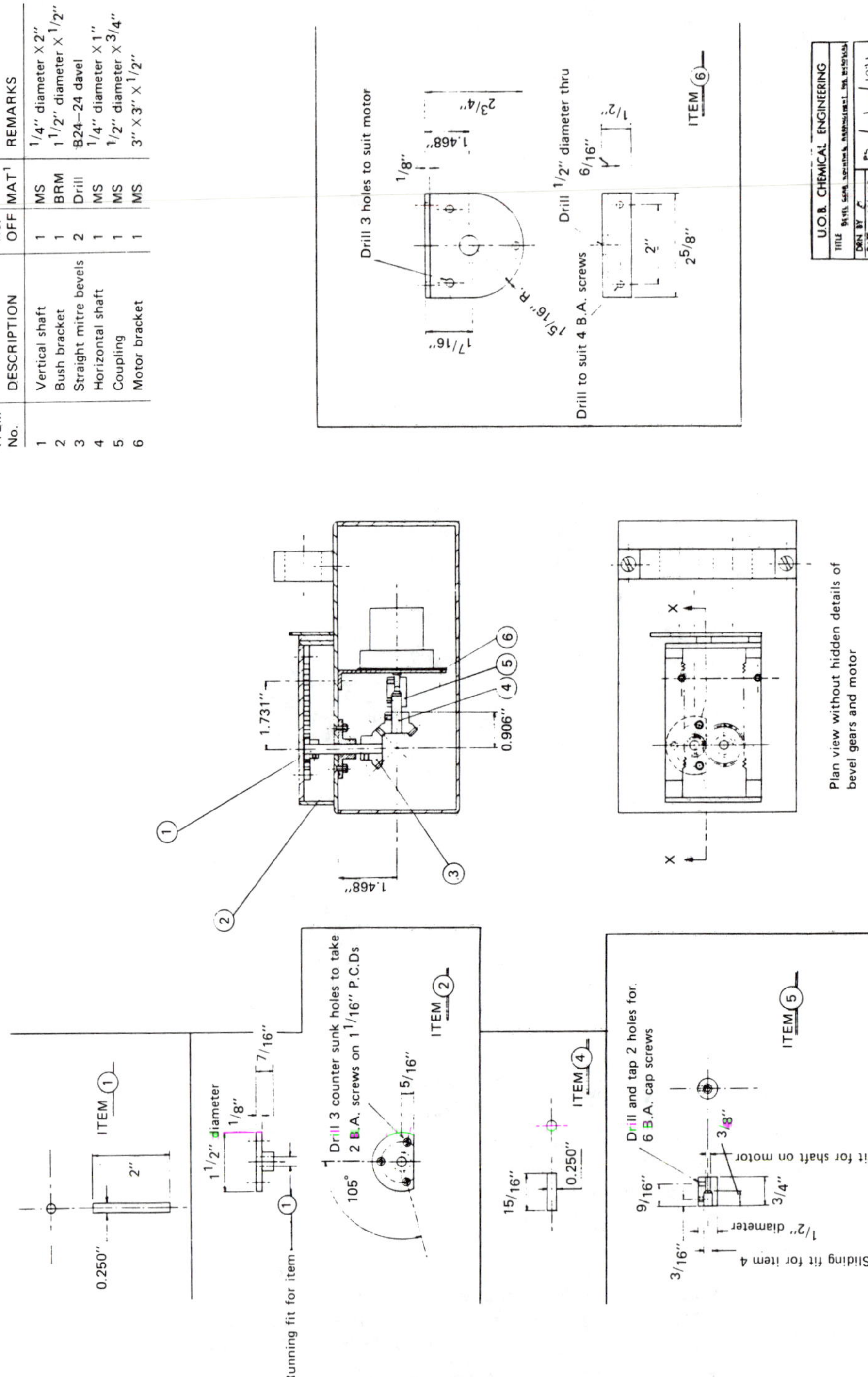

Fig. 4.16 — Dimensions and design characteristics of flow meter Microscal.

At the end of the 480 seconds, the flow meter is switched off from the power source to avoid breakage of the syringe. The pressure recorder P_1 firstly describes a steady increase in pressure on contact with the adsorbent and then deflect towards the original base line of P_2 when the flow is complete. The readings of P_1 and P_2 are then noted and compared with the amplified values of ΔP from the second PR.

After the experiment at one inlet pressure, the sample is again degassed in an identical manner as stated above and the process repeated at a different inlet pressure.

REFERENCES

[1] Poiseuille, J. M. L., Recherches experimentales sur le mouvement des liquides dans les tubes de tres petit diametre. *Mem. Savants Etrange* (*Paris*), **9**, 433–545 (1846); Engl. transl. *Rheol. Mem.*, **1**, (Jan. 1940).

[2] Lowell, S., *Introduction to powder surface area*. John Wiley & Sons, New York, Chichester (1979).

5

Experimental results: surface areas and coefficients of slip

5.1 INTRODUCTION

The outlet pressures and the pressure drops across packed beds of powder were determined as a constant and known amount of gas passed through the bed. The experimental procedures were described in Chapter 4.

5.2 METHODS OF CALCULATION

The data were determined by using the Carman–Arnell equation in the form given below. A full description of the terms used is given under *Steady state permeametry* in Appendix A1.1).

$$\frac{P_1}{\Delta P} \times \frac{dV}{dt} = \frac{A}{k\mu L} \frac{\varepsilon^3}{(1-\varepsilon)^2} \frac{1}{S_V^2} \overline{P} + \frac{8}{3}\left(\frac{\delta k_0}{k}\right) \frac{A}{LS_V}\left[\frac{2RT}{\pi M}\right]^{\frac{1}{2}} \frac{\varepsilon^2}{(1-\varepsilon)} \tag{5.1}$$

That is,

$$y = \frac{K_1}{S_V^2}\overline{P} + \frac{K_2}{S_V}\left(\frac{\delta k_0}{k}\right). \tag{5.2}$$

From the slope of the graph of the left-hand function against $\overline{P}$, a value of the specific surface area, S_V, can be found, and the intercept yields a value of $(\delta k_0/k)$, where

$$S_V^2 = \frac{K_1}{\text{Slope}} \tag{5.3}$$

and

$$\left(\frac{\delta k_0}{k}\right) = \frac{S_V \times \text{Intercept}}{K_2} \qquad (5.4)$$

These values were obtained by using regressional analysis.

If the contribution due to the second (slip) term is assumed negligible, the equation degenerates to the Carman–Kozeny equation, and the surface area given by this viscous flow term is:

$$S_K^2 = \frac{A}{k\mu L} \frac{\varepsilon^2}{(1-\varepsilon)^2} \frac{\bar{P}\Delta P}{P_1(\mathrm{d}V/\mathrm{d}t)} . \qquad (5.5)$$

If the contribution due to the viscous flow term is assumed negligible, the equation degenerates to the slip flow equation which is a form of Knudsen diffusion equation:

$$S_M = \frac{8}{3}\left(\frac{\delta k_0}{k}\right)\frac{A}{L}\left[\frac{2RT}{\pi M}\right]^{\frac{1}{2}} \frac{\varepsilon^2}{(1-\varepsilon)} \frac{\Delta P}{P_1(\mathrm{d}V/\mathrm{d}t)} . \qquad (5.6)$$

where R = universal gas constant (8.314 kJ/kmol K)
T = temperature in Kelvin (298 K)
M = molecular weight of the gas (29 kg/mol) for air.

Inserting equations (5.6) in the composite equation gives a quadratic in S_V:

$$S_V^2 - S_M S_V - S_K^2 = 0 , \qquad (5.7)$$

which has the solution:

$$S_V = S_M \pm \sqrt{\frac{S_M^2 + S_K^2}{2}} \qquad (5.8)$$

where S_V is the volume specific surface,
S_K is the volume specific surface using the viscous flow term,
S_M is the volume specific surface using the molecular flow term.

The Klinkenberg factor, B, is obtained from the relationship

$$B = \frac{8}{3}\delta k_0 \left[\frac{2RT}{\pi M}\right]^{\frac{1}{2}} \left(\frac{1-\varepsilon}{\varepsilon}\right) \mu S_V . \qquad (5.9)$$

The experimental and calculated data for each table are determined for a constant gas flow rate.

Values of the bed constants and other variables which may affect the results obtained are listed at the head of each experimental sheet (see below).

Columns one, two, and three contain the chart recorder readings for the inlet pressure (P_1 in volts), and the conversions into metre mercury and Newton per square metre. Columns four, five, and six give the final pressures, P_2. The pressure drop across the packed bed is given in column seven. Multiplying the constant gas flow rate by the ratio $P_1/\Delta P$, gives the values contained in column eight. The mean pressure, $\overline{P}$, is shown in column nine. Values of the calculated S_K, S_M, and S_V using equations (5.5), (5.6), and (5.8) are given in columns ten, eleven, and twelve.

5.3 RESULTS AND DISCUSSION

5.3.1 Preliminary tests

5.3.1.1 Dicalcium phosphate dihydrate powder (not pre-dried)

These preliminary tests were carried out in order to find out the effect of not drying the powder first in an oven before carrying out an experiment. For this test, the packed bed was compacted to a porosity of $\varepsilon = 0.6585$. The results obtained are shown in Table 5.1 and graphically in Fig. 5.1. Fig. 5.1 shows generally that at sub-atmospheric pressures the gas flow rate is a linear function of the mean pressure.

A regression analysis on the 10 experimental points from Fig. 5.1 at all pressures, gives a slope of 11.5665×10^{-12}, with a standard error of the slope of 0.0925×10^{-12}. The line intercepts the y-axis at a value of 27.2738×10^{-12}, with a correlation coefficient, $r = 0.9999$. The high correlation coefficient indicates good experimental technique and instrument design.

From the regression analysis, $S_V = 1.9771 \times 10^6\,\text{m}^2/\text{m}^3$, and $(\delta k_0/k) = 0.4002$.

The tabulated data for the same 10 values gives $S_V = 1.9757 \times 10^6\,\text{m}^2/\text{m}^3$, with a standard deviation, $s = 0.0151 \times 10^6\,\text{m}^2/\text{m}^3$.

The value of the Klinkenberg factor calculated from equation (5.9) is $B = 2.3578$ $(\text{Nm}^{-2} \times 10^4)$.

To obtain another value for the variables of interest for this preliminary test, the above powder bed was further compacted from a porosity of $\varepsilon = 0.6585$ to $\varepsilon = 0.6292$. The results, as before, are shown in Table 5.2, and graphically in Fig. 5.2.

Simple comparison of the results obtained at these two levels of compaction are shown and explained by Figs 5.3a,b,c, from where it can be seen that all the variables of interest in this study increased with decreasing porosity. It is thought that these graphs require no further comments at this stage.

5.3.1.2 Preliminary test with DCP, using modified Griffin permeameter

This test, using the modified Griffin permeameter, is an attempt to compare the S_V values previously obtained, with the apparatus designed.

In the equipment described in this book, a constant pressure drop is maintained across the bed. The relevant equations are much simpler than those required for equipment using a variable pressure drop. Comparisons between these two apparatuses were carried out to test the accuracy of the new permeameter. Experiments were first carried out at decreasing porosities in order to select an optimum porosity.

The results of the first group of experiments carried out using only the viscous flow equation are shown in Table 5.3.

The variation of the measured surface area, S_K, with porosity is plotted in Fig. 5.4, from which it can be seen that with decreasing porosity, S_K increases and tends to a constant value. The low values at high porosities are probably due to non-uniform packing of the bed resulting in channelling, i.e. preferential flow through the wider voids.

The working equations used for the Griffin permeameter are as follows:

Table 5.1 — Determination of specific surface area and Carman-Arnell constant. Data: Dicalcium phosphate dihydrate $[Ca(H_2PO_4)_2]$ (not dried)

$L = 7.256 \times 10^{-2}$ m
$A = 1.246 \times 10^{-4}$ m^2

$dV/dt = 4.167 \times 10^{-8}$ m^3/sec.
$V = 20 \times 10^{-6}$ m^3

$w = 7.4716 \times 10^{-3}$ kg.
$\rho_s = 2420$ kg m^{-3}

$\varepsilon = 0.6585$

$t = 480$ sec.

$\left[\frac{2RT}{\pi M}\right]^{\frac{1}{2}} = 231.74$ m/s

$\mu = 186 \times 10^{-7}$ Nsm^{-2}

$g = 9.81$ ms^{-2}

P_1			P_2			ΔP	$\frac{P_1}{\Delta P} \times \frac{dV}{dt}$	$\bar{P}$	(m^2/m^3 × 10^6)		
Volts	m Hg	(Nm^{-2} × 10^{-4})	Volts	m Hg	(Nm^{-2} × 10^{-4})	(Nm^{-2} × 10^{-4})	(m^3/ sec.) × 10^{-8}	(Nm^{-2} × 10^{-4})	S_K	S_M	S_V
0.921	0.6999	9.3378	0.8924	0.6782	9.0483	0.2895	134.4063	9.1930	1.7592	0.4012	1.9712
0.827	0.6285	8.3852	0.7990	0.6072	8.1010	0.2842	122.9456	8.2431	1.7417	0.4386	1.9748
0.727	0.5525	7.3712	0.6992	0.5314	7.0897	0.2815	109.1147	7.2304	1.7315	0.4942	1.9962
0.623	0.4735	6.3172	0.5970	0.4537	6.0531	0.2641	99.6735	6.1852	1.6756	0.5410	1.9678
0.522	0.3967	5.2926	0.4970	0.3777	5.0391	0.2535	86.9991	5.1658	1.6391	0.6200	1.9781
0.422	0.3210	4.2826	0.3990	0.3032	4.0452	0.2374	75.1710	4.1639	1.5831	0.7174	1.9819
0.323	0.2450	3.2687	0.3020	0.2290	3.0552	0.2135	63.7971	3.1619	1.4975	0.8454	1.9787
0.217	0.1650	2.2014	0.1994	0.1515	2.0213	0.1801	50.9341	2.1114	1.3695	1.0589	1.9977
0.117	0.0889	1.1861	0.1050	0.0798	1.0646	0.1215	40.6788	1.1254	1.1188	1.3592	1.9635
0.020	0.0152	0.2028	0.0170	0.0129	0.1746	0.0282	29.9669	0.1887	0.5338	1.8000	1.9465

See section 5.4.1.

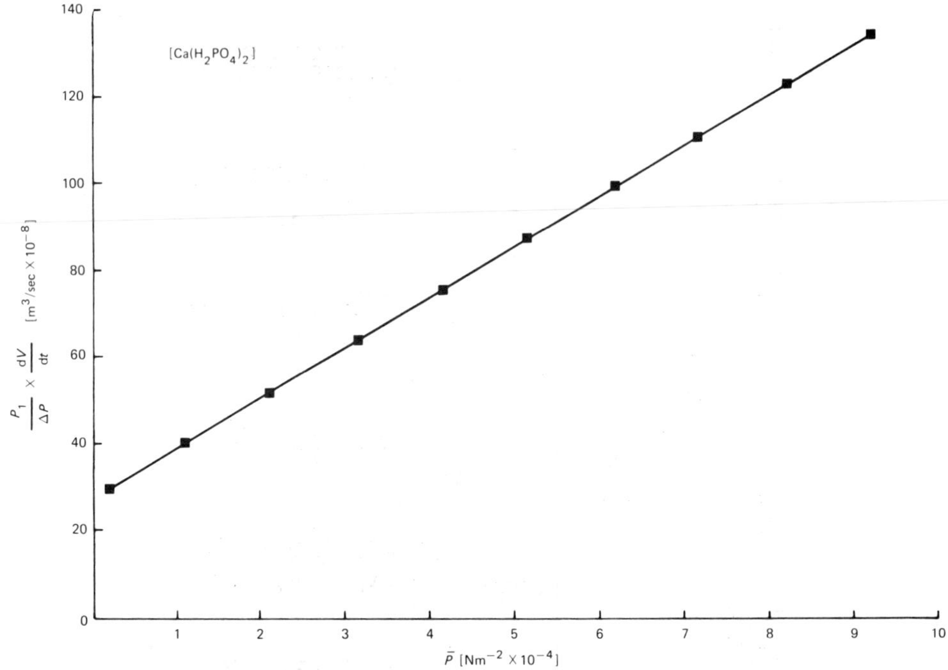

Fig. 5.1 — Evaluation of surface area and the Carman–Arnell constant.

$$S_K^2 = \frac{2A\rho fgt}{k\mu LaI_1} \frac{\varepsilon^3}{(1-\varepsilon)^2} \tag{5.10}$$

where

$$I_1 = \left(\frac{2V_0}{a\phi} + 1\right) \ln\left(\frac{h_1(\phi - h_2)}{h_2(\phi - h_1)} + 4\ln\left(\frac{\phi - h_1}{\phi - h_2}\right)\right. \tag{5.11}$$

$$S_M = \frac{8}{3}\left(\frac{\delta k_0}{k}\right)\frac{2At}{La\phi I_2}\left[\frac{2RT}{\pi M}\right]^{\frac{1}{2}} \frac{\varepsilon^2}{(1-\varepsilon)} \tag{5.12}$$

where

$$I_2 = \left(\frac{2V_0}{a\phi} + 1\right) \ln\frac{h_1}{h_2} - \frac{4}{\phi}(h_1 - h_2) \tag{5.13}$$

Equation (5.7), with values of S_K and S_M corrected for gas expansion using I_1 and I_2, may be written:

$$\frac{La(1-\varepsilon)I_2P_1}{2A\rho fg\varepsilon^2 t} = \frac{\varepsilon I_2 P_1}{k(1-\varepsilon)\mu I_1 S_V^2} + \frac{8}{3}\frac{\delta k_0}{t}\left[\frac{2RT}{\pi M}\right]^{\frac{1}{2}} \frac{1}{S_V} \tag{5.14}$$

Table 5.2 — Determination of specific surface area and Carman-Arnell constant. Data: Dicalcium phosphate dihydrate [$Ca(H_2PO_4)_2$] (not dried)

$L = 6.682 \times 10^{-2}$ m $dV/dt = 4.167 \times 10^{-8}$ m³/sec. $w = 7.4716 \times 10^{-3}$ kg.
$A = 1.246 \times 10^{-4}$ m² $V = 20 \times 10^{-6}$ m³ $\rho_s = 2420$ kg m^{-3}

$\varepsilon = 0.6292$ $t = 480$ sec. $\left[\frac{2RT}{\pi M}\right]^{\frac{1}{2}} = 231.74$ m/s

$\mu = 186 \times 10^{-7}$ Nsm^{-2} $g = 9.81$ ms^{-2}

	P_1			P_2		$\frac{P_1}{\Delta P} \times \frac{dV}{dt}$	ΔP	$\bar{P}$		(m²/m³ × 10⁶)	
Volts	m Hg	(Nm^{-2} × 10^{-4})	Volts	m Hg	(Nm^{-2} × 10^{-4})	(m³/ sec.) × 10^{-8}	(Nm^{-2} × 10^{-4})	(Nm^{-2} × 10^{-4})	S_K	S_M	S_V
0.958	0.7281	9.7140	0.907	0.6893	9.1963	78.1135	0.5177	9.4551	2.0970	0.9817	2.6445
0.851	0.6468	8.6293	0.803	0.6103	8.1424	73.8515	0.4869	8.3858	2.0321	1.0394	2.6172
0.750	0.5700	7.6047	0.703	0.5343	7.1284	66.5312	0.4763	7.3665	2.0066	1.1537	2.6647
0.649	0.4932	6.5801	0.603	0.4598	6.1345	61.5334	0.4456	6.3573	1.9383	1.2474	2.6599
0.545	0.4142	5.5261	0.505	0.3838	5.1205	56.7733	0.4056	5.3233	1.8465	1.3520	2.6424
0.438	0.3329	4.4414	0.402	0.3055	4.0758	50.6217	0.3656	4.2586	1.7491	1.5163	2.6645
0.330	0.2508	3.3461	0.300	0.2280	3.0419	45.8356	0.3042	3.1940	1.5920	1.6747	2.6361
0.222	0.1687	2.2507	0.199	0.1512	2.0172	40.1656	0.2335	2.1339	1.3899	1.9111	2.6422
0.114	0.0866	1.1554	0.100	0.0760	1.0139	34.0251	0.1415	1.0846	1.0767	2.2560	2.6874
0.050	0.0380	0.5069	0.0435	0.0331	0.4416	32.3469	0.0653	0.4742	0.7301	2.3730	2.6000

See section 5.4.2.

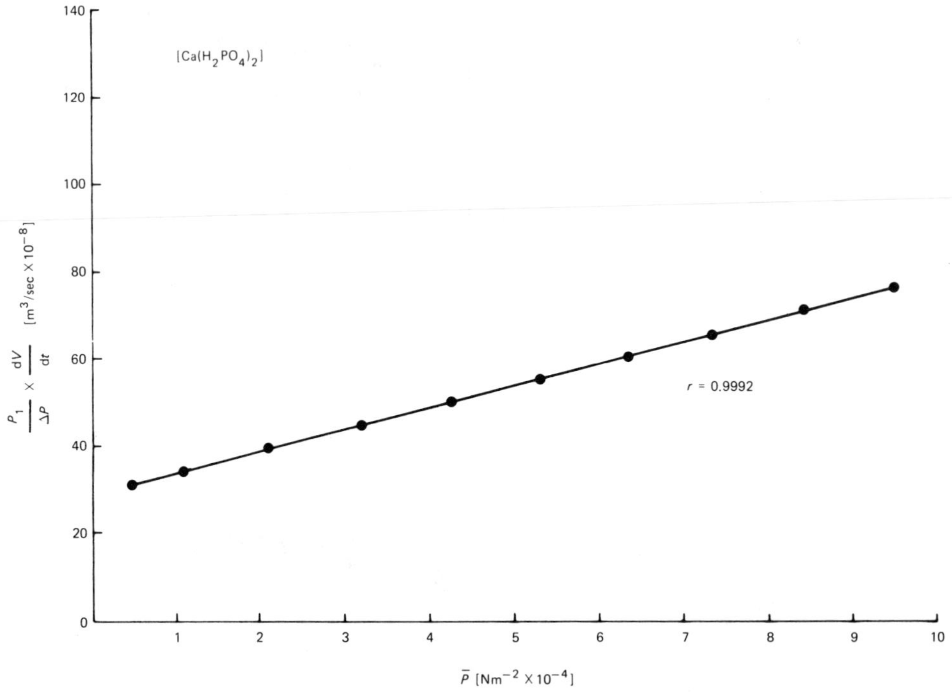

Fig. 5.2 — Evaluation of surface area and the Carman–Arnell constant.

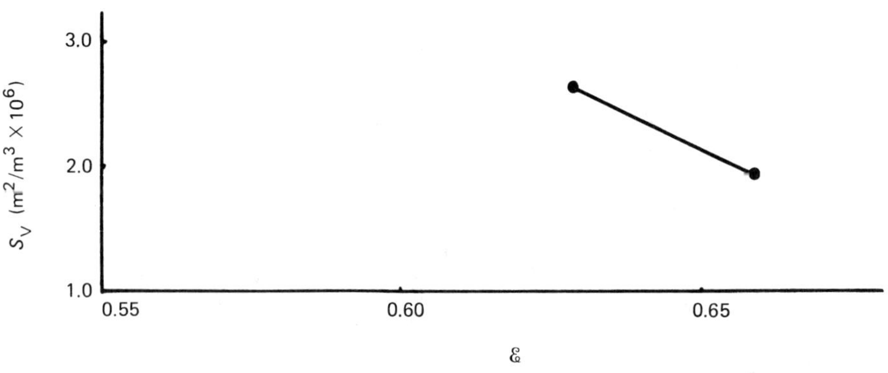

Fig. 5.3a — Variation of measured surface area with porosity.

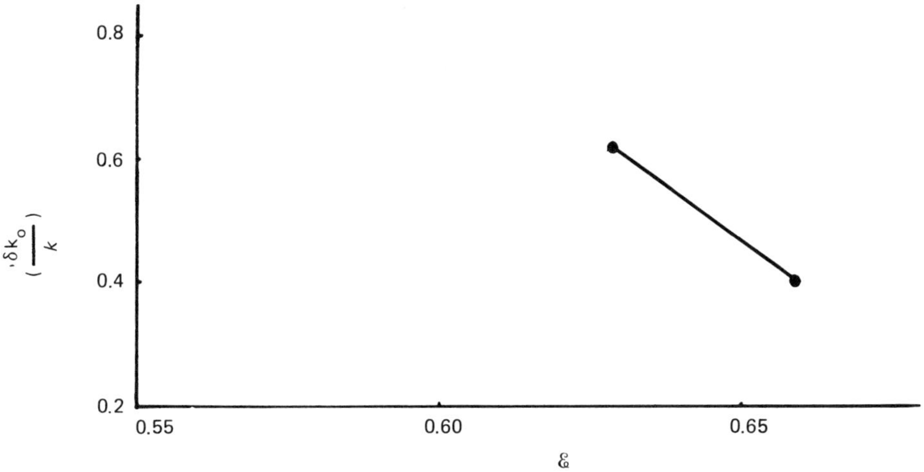

Fig. 5.3b — Variation of $(\delta k_0/k)$ with porosity.

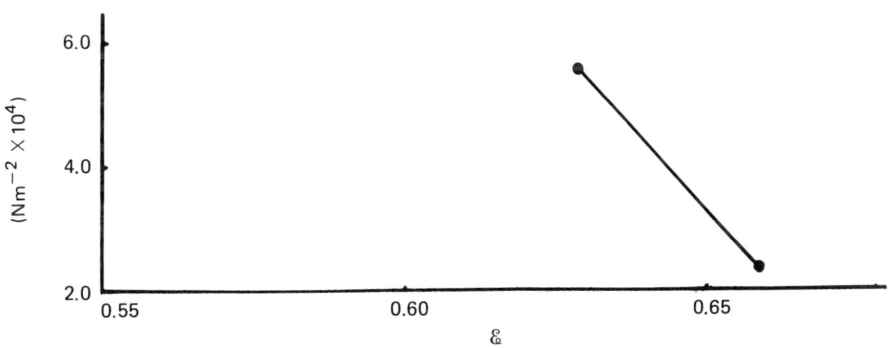

Fig. 5.3c — Variation of the Klinkenberg factor *B* with porosity.

From the slope of the graph of the left-hand function against P_1 a value of S_V can be found, and the intercept yields a value for $(\delta k_0/k)$.

The preliminary tests indicated that the surface area was independent of porosity for porosities smaller than 0.61.

To carry out the main comparative test, a new packed bed was prepared at a porosity within the required range (i.e. 0.609). The results obtained are shown in Table 5.4 and graphically in Fig. 5.5.

From the summary of results (Table 5.5) there appears to be a fair agreement

Table 5.3 — Surface areas at decreasing porosity using the Griffin permeameter

Weight of powder (g)	Length of bed (cm)	Time for fixed† volume of air to pass through the bed (s)	Bed porosity (ε)	S_K‡ ($m^2/m^3 \times 10^6$)
11.987	2.826	180	0.6540	1.7738
	2.715	238	0.6399	1.9352
	2.601	301	0.6242	2.0526
	2.456	385	0.6020	2.1366
	2.382	430	0.5896	2.1551
	2.301	485	0.5752	2.1678

† Corrected for time of passage through bed support (5 sec).
‡ Uncorrected value using Carman–Kozeny equation.

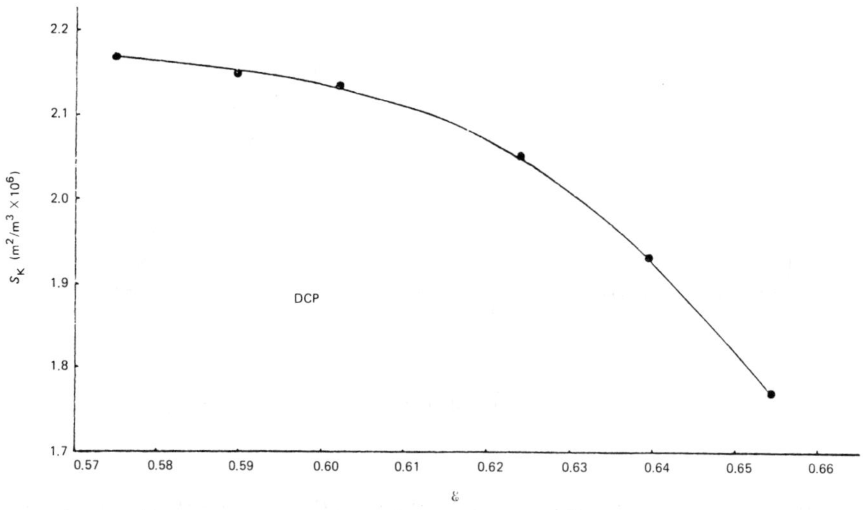

Fig. 5.4 — Variation of measured surface area, using the viscous flow equation (Griffin permeameter), with porosity.

between the surfaces $S_V = 2.6 \times 10^6\,m^2/m^3$ obtained from the two equipments, but for some unknown reasons (owing to the powder being undried) there is no consistency in values of $(\delta k_0/k)$ and B. It was not considered necessary to investigate

Table 5.4 — Determination of specific surface area and Carman-Arnell constant using modified Griffin permeameter. Data: Dicalcium phosphate dihydrate [$Ca(H_2PO_4)_2$]

$L = 4.186 \times 10^{-2}$ m	$\rho_s = 2420$ kg m^{-3}	$h_1 = 11.1$ cm	$g = 9.81$ ms^{-2}
$w = 20.0602$ g	$\rho_f = 860$ kg m^3	$h_2 = 7.2$ cm	$\mu = 186 \times 10^{-7}$ Nsm^{-2}
$A = 5.067 \times 10^{-4}$ m^2	$\varepsilon = 0.6092$	$h_3 = 4.5$ cm	$V = 1 \times 10^{-4}$ m^3
			$a = 4.75 \times 10^{-4}$ m^2

Equation (4.14) becomes:

$$2.4489 \times 10^{-6} y = \frac{16762}{S_V^2} \times + \frac{618}{S_V}\left(\frac{\delta k_0}{k}\right)$$

$$y = \frac{6.8446 \times 10^9}{S_V^2} \times + \frac{2.5235 \times 10^8}{S_V}\left(\frac{\delta k_0}{k}\right)$$

P_1	P_1	ϕ	I_2	I_1	t	y	x	($m^2/m^3 \times 10^6$)		
m Hg	(Nm$^{-2}\times 10^{-4}$)	m oil	m	m	sec.	(Nm$^{-2}\times$ s^{-1})	(Nm$^{-2}\times 10^{-4}$)	S_K	S_M	S_V
0.658	8.7788	10.4056	0.9287	0.9209	744	109.5816	8.8532	2.3587	0.4605	2.6001
0.605	8.0717	9.5674	0.9309	0.9220	733	102.5095	8.1496	2.3412	0.4923	2.6002
0.512	6.8309	8.0967	0.9358	0.9255	718	89.0300	6.9070	2.3171	0.5668	2.6177
0.429	5.7236	6.7842	0.9421	0.9300	680	79.2971	5.7981	2.2550	0.6364	2.5955
0.301	4.0158	4.7600	0.9580	0.9417	625	61.5542	4.0853	2.1619	0.8198	2.6103
0.196	2.6150	3.0995	0.9856	0.9630	555	46.4386	2.6764	2.0372	1.0867	2.6517
0.093	1.2408	1.4707	1.0642	1.0327	389	33.9450	1.2787	1.7056	1.4867	2.6039
0.032	0.4270	0.5061	1.3191	1.2188	240	21.6845	0.3945	1.3397	2.3273	2.9381

See section 5.4.3.

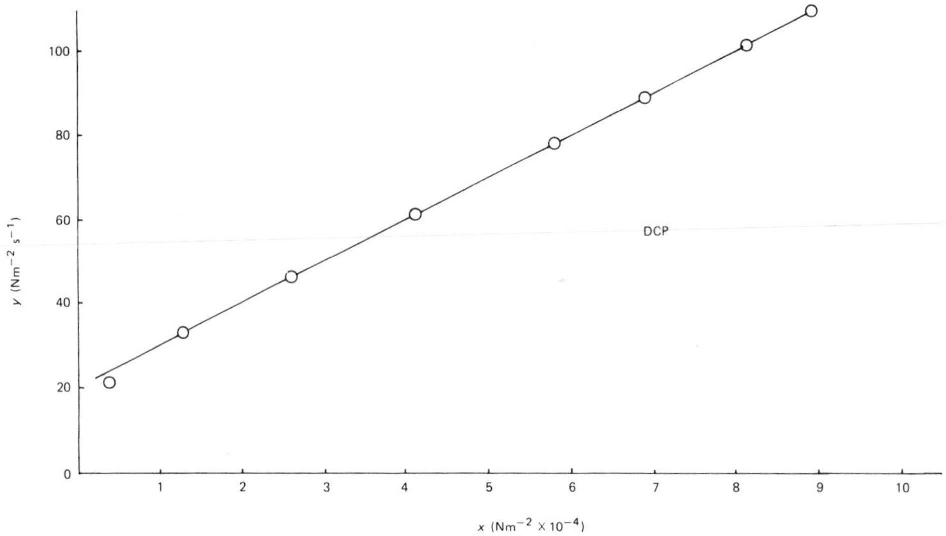

Fig. 5.5 — Evaluation of surface area and the Carman–Arnell constant.

Table 5.5 — Summary of results

	ε	S_V ($m^2/m^3 \times 10^6$)	Standard deviation s ($m^2/m^3 \times 10^6$)	$\left(\frac{\delta k_o}{k}\right)$	B ($Nm^{-2} \times 10^4$)
New equipment	0.6585	1.9760	0.015	0.400	2.358
	0.6292	2.6455	0.025	0.623	5.5886
Griffin	0.6092	2.6522	0.117	0.200	1.9556

this inconsistency further, since the anomaly disappears when the powder is pre-dried.

All subsequent powders discussed in this book were, however, dried in the oven as described in section 4.7. These preliminary tests are for comparative purposes only, and should be seen as such.

5.3.1.3 Preliminary test with spherical bronze ($<38\,\mu$m)

The last comparative experiment was carried out with spherical bronze powders $<38\,\mu$m. The results obtained are shown in Table 5.6, and graphically in Fig. 5.6. Again with this powder, there is fair agreement of $S_V = 0.21 \times 10^6\,m^2/m^3$ obtained with the new equipment designed, compared with $S_V = 0.24 \times 10^6\,m^2/m^3$ when using

Table 5.6 — Determination of specific surface area and Carman-Arnell constant. Data: Spherical bronze powders ($<38\,\mu m$)

$L = 7.310 \times 10^{-2}\,m$ $\quad$ $dV/dt = 4.167 \times 10^{-8}\,m^3/sec.$ $\quad$ $w = 0.0536\,kg$
$A = 1.246 \times 10^{-4}\,m^2$ $\quad$ $V = 20 \times 10^{-6}\,m^3$ $\quad$ $\rho_s = 8800\,kg\,m^{-3}$

$\varepsilon = 0.3317$ $\quad$ $t = 480\,sec.$ $\quad$ $\left[\frac{2RT}{\pi M}\right]^{\frac{1}{2}} = 231.74\,m/s$

$\mu = 186 \times 10^{-7}\,Nsm^{-2}$ $\quad$ $g = 9.81\,ms^{-2}$

P_1			P_2			ΔP	$\frac{P_1}{\Delta P} \times \frac{dV}{dt}$	$\bar{P}$	($m^2/m^3 \times 10^6$)		
Volts	m Hg	($Nm^{-2} \times 10^{-4}$)	Volts	m Hg	($Nm^{-2} \times 10^{-4}$)	($Nm^{-2} \times 10^{-4}$)	(m^3/ sec.) $\times 10^{-8}$	($Nm^{-2} \times 10^{-4}$)	S_K	S_M	S_V
0.9118	0.6930	9.2457	0.9000	0.6840	9.1256	0.1201	320.7896	9.1857	0.2072	0.0012	0.2078
0.8118	0.6170	8.2318	0.8000	0.6030	8.1117	0.1201	285.6113	8.1718	0.2070	0.0013	0.2077
0.7120	0.5411	7.2191	0.7000	0.5320	7.0977	0.1214	247.7923	7.1584	0.2081	0.0015	0.2088
0.6119	0.4650	6.2038	0.6000	0.4560	6.0833	0.1200	215.4270	6.1438	0.2067	0.0017	0.2076
0.5120	0.3891	5.1912	0.5000	0.3800	5.0698	0.1214	178.1856	5.1305	0.2077	0.0021	0.2087
0.4119	0.3130	4.1760	0.4000	0.3040	4.0553	0.1202	144.7703	4.1159	0.2064	0.0026	0.2077
0.3120	0.2370	3.1620	0.3000	0.2280	3.0419	0.1201	109.7090	3.1020	0.2059	0.0034	0.2076
0.2113	0.1610	2.1480	0.2000	0.1520	2.0279	0.1201	74.5272	2.0880	0.2049	0.0051	0.2075
0.1120	0.0851	1.1354	0.1000	0.0760	1.0140	0.1214	38.9721	1.0747	0.2033	0.0097	0.2082
0.0225	0.0171	0.2281	0.0100	0.0076	0.1014	0.1267	7.5019	0.1650	0.1816	0.0504	0.2085

See section 5.4.4.

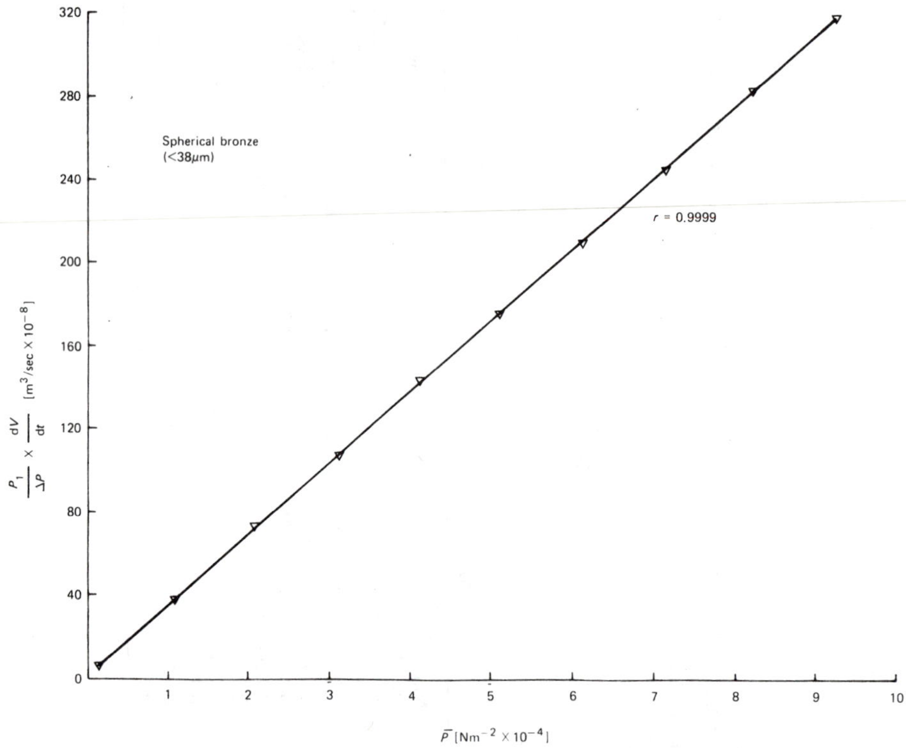

Fig. 5.6 — Evaluation of surface area and the Carman–Arnell constant.

the modified Griffin apparatus by permeametry, and $S_V = 0.2394 \times 10^6\,m^2/m^3$ by microscopy, obtained by Maghafarti [1].

It may, therefore, be concluded from these series of preliminary tests that the new instrument performed satisfactorily.

5.4 CALCULATIONS

5.4.1 Table 5.1

From equation (5.1)

$$\frac{P_1}{\Delta P} \times \frac{dV}{dt} = \frac{45.2111}{S_V^2}\bar{P} + \frac{1.3474}{S_V}\left(\frac{\delta k_0}{k}\right)$$

A regression analysis on the 10 experimental points gives:

Slope	$= 11.5665 \times 10^{-12}$
SE of slope	$= 0.0925 \times 10^{-12}$
Intercept	$= 27.2738 \times 10^{-8}$
Correlation coefficient, r	$= 0.9999$

Therefore,

$$S_V^2 = \frac{45.2111}{11.5665} \times 10^{-12}$$

$$S_V = 1.9771 \times 10^6 \, m^2/m^3$$

From the table, $S_V = 1.9757 \times 10^6 \, m^2/m^3$, with a standard deviation, $s = 0.0151 \times 10^6 \, m^2/m^3$.

$$\left(\frac{\delta k_0}{k}\right) = \frac{1.9771 \times 10^6 \times 17.2738 \times 10^{-8}}{1.3474}$$

$$= 0.4002$$

From equation (5.9), the Klinkenberg factor, $B = 2.3578 \, (Nm^{-2} \times 10^4)$.

5.4.1.1 *Worked example*

$$B = \frac{8}{3}[5 \times 0.4002] \times 231.74 \times \frac{0.3415}{0.6585} \times 186 \times 10^{-7} \times 1.9771 \times 10^6$$

$$= \underline{2.358 \, (Nm^{-2} \times 10^4)}$$

5.4.2 Table 5.2

From equation (5.1)

$$\frac{P_1}{\Delta P} \times \frac{dV}{dt} = \frac{36.3655}{S_V^2} \bar{P} + \frac{1.2310}{S_V} \left(\frac{\delta k_0}{k}\right)$$

A regression analysis on the 10 experimental points gives:

Slope	$= 5.1962 \times 10^{-12}$
SE of slope	$= 0.0753 \times 10^{-12}$
Intercept	$= 29.0209 \times 10^{-8}$
Correlation coefficient, r	$= 0.9992$

Therefore,

$$S_V^2 = \frac{36.3655}{5.1962} \times 10^{-12}$$

$$S_V = 2.6455 \times 10^6 \, m^2/m^3$$

From the table, $S_V = 2.6459 \times 10^6 \, m^2/m^3$, with a standard deviation, $s = 0.0251 \times 10^6 \, m^2/m^3$.

$$\left(\frac{\delta k_0}{k}\right)=\frac{2.6455\times 10^6\times 29.0209\times 10^{-8}}{1.2310}$$

$$=0.6237$$

From equation (5.9), the Klinkenberg factor, $B=5.5886\ (\text{Nm}^{-2}\times 10^4)$.

5.4.3 Table 5.4

A regression analysis on the 8 experimental points gives:

$$\text{Slope} = 10.2151\times 10^{-4}$$
$$\text{Intercept} = 19.3012$$

This gives:

$$S_V=2.600\times 10^6\,\text{m}^2/\text{m}^3$$

and

$$\left(\frac{\delta k_0}{k}\right)=0.200$$

The tabulated data for the same 8 values gives:

$$S_V=2.6522\times 10^6\,\text{m}^2/\text{m}^3$$

with

$$s=0.1169\times 10^6\,\text{m}^2/\text{m}^3$$

If the final value is considered an outsider:

$$S_V=2.6113\times 10^6\,\text{m}^2/\text{m}^3$$

$$s=0.0193\times 10^6\,\text{m}^2/\text{m}^3.$$

5.4.4 Table 5.6

From equation (5.1)

$$\frac{P_1}{\Delta P}\times\frac{\text{d}V}{\text{d}t}=\frac{1.4978}{S_V^2}\overline{P}+\frac{0.1730}{S_V}\left(\frac{\delta k_0}{k}\right)$$

A regression analysis on the 10 experimental points gives:

Slope	$= 34.6418 \times 10^{-12}$
SE of slope	$= 0.1094 \times 10^{-12}$
Intercept	$= 1.8128 \times 10^{-8}$
Correlation coefficient, r	$= 0.9999$

Therefore,

$$S_V^2 = \frac{1.4978}{34.6418} \times 10^{-12}$$

$$S_V = 0.2080 \times 10^6\,\mathrm{m^2/m^3}$$

From the table, $S_V = 0.2080 \times 10^6\,\mathrm{m^2/m^3}$, with a standard deviation, $s = 0.0005 \times 10^6\,\mathrm{m^2/m^3}$.

$$\left(\frac{\delta k_0}{k}\right) = \frac{0.208 \times 10^6 \times 1.8128 \times 10^{-8}}{0.173}$$

$$= 0.0218$$

From equation (5.9), the Klinkenberg factor $B = 0.0525\ (\mathrm{Nm^{-2}} \times 10^4)$.

REFERENCE

[1] Pourghaffar-Maghafarti, R., MPhil thesis, University of Bradford (1982).

6

Determination of surface area and Carman–Arnell constant, using dicalcium phosphate dihydrate powder (dry)

6.1 INTRODUCTION

Further tests with dicalcium phosphate dihydrate powder were carried out at four different porosities by gently compacting the powder beds. A known powder weight was hand-compacted stepwise until a constant bed volume was obtained. The results so obtained at various porosities are shown in Tables 6.1–6.4 and as graphs in Figs 6.1 and 6.2. Values of all the regression lines, i.e. the slopes and intercepts, standard errors of the slopes, the correlation coefficients, the calculated specific surfaces, S_V, the calculated coefficients of slip, i.e. Carman and Arnell $\left(\frac{\delta k_0}{k}\right)$, and the Klinkenberg factors B, are shown below each experimental table.

6.2 RESULTS AND DISCUSSION

It is generally observed from all the graphs evaluating the surface areas of the powders that the flow rate of gas through the packed beds is a linear function of the mean pressure.

A statistical analysis of the entire results obtained at different porosities shows that the correlation coefficients and the parameters for the other regression lines are statistically highly significant at the significance levels tested. It should further be noted that the tables of the distribution of Student's t function with 8 degrees of freedom ($v = n - 2$) gives at 5% (5% is the complement of 95% level of confidence) significance level, $t = 2.3060$, 1% significance level, $t = 3.3554$, and 0.1% significance level, $t = 5.0413$. The high correlations obtained from these calculations are probably due to the sensitivity of the instrument used.

Table 6.1 — Determination of specific surface area and Carman–Arnell constant. Data: Dicalcium phosphate dihydrate $[Ca(H_2PO_4)_2$

$L = 7.1986 \times 10^{-2}$ m
$A = 1.246 \times 10^{-4}$ m^2
$\varepsilon = 0.6559$

$dV/dt = 4.167 \times 10^{-8}$ m^3/sec.
$V = 20 \times 10^{-6}$ m^3
$t = 480$ sec.

$w = 7.4701 \times 10^{-3}$ kg
$\rho_s = 2420$ kg m^{-3}

$\mu = 186 \times 10^{-7}$ Nsm^{-2} $\quad g = 9.81$ ms^{-2} $\quad \left[\frac{2RT}{\pi M}\right]^{1/2} = 231.74$ m/s

P_1			P_2			ΔP	$\frac{P_1}{\Delta P} \times \frac{dV}{dt}$	$\bar{P}$	(m^2/m^3 × 10^6)		
Volts	m Hg	(Nm^{-2} × 10^{-4})	Volts	m Hg	(Nm^{-2} × 10^{-4})	(Nm^{-2} × 10^{-4})	(m^3/sec.) × 10^{-8}	(Nm^{-2} × 10^{-4})	S_K	S_M	S_V
0.932	0.7083	9.4498	0.902	0.6855	9.1457	0.3041	129.4880	9.2977	1.7847	0.1766	1.8752
0.838	0.6369	8.4973	0.808	0.6141	8.1972	0.3001	117.9882	8.3472	1.7715	0.1938	1.8710
0.738	0.5609	7.4833	0.708	0.5381	7.1863	0.2970	104.9929	7.3348	1.7603	0.2178	1.8726
0.632	0.4803	6.4079	0.603	0.4583	6.1144	0.2935	90.9769	6.2612	1.7472	0.2514	1.8774
0.530	0.4028	5.3740	0.502	0.3815	5.0898	0.2842	78.7947	5.2319	1.7162	0.2902	1.8674
0.430	0.3268	4.3600	0.403	0.3063	4.0865	0.2735	66.4282	4.2233	1.6793	0.3443	1.8603
0.330	0.2503	3.3461	0.304	0.2310	3.0819	0.2642	52.7753	3.2140	1.6436	0.4333	1.8745
0.225	0.1710	2.2814	0.203	0.1543	2.0376	0.2438	38.9934	2.1595	1.5673	0.5865	1.8878
0.120	0.0912	1.2167	0.102	0.0775	1.0340	0.1827	27.7503	1.1253	1.3412	0.8241	1.8151
0.0231	0.0176	0.2348	0.016	0.0122	0.1628	0.0720	13.5891	0.1988	0.8056	1.6829	2.0064

See section 6.3.1.

Table 6.2 — Determination of specific surface area and Carman–Arnell constant. Data: Dicalcium phosphate dihydrate $[Ca(H_2PO_4)_2$

$L = 6.5940$ $\quad$ $dV/dt = 4.167 \times 10^{-8}$ m³/sec. $\quad$ $w = 7.4701 \times 10^{-3}$ kg

$A = 1.246 \times 10^{-4}$ m² $\quad$ $V = 20 \times 10^{-6}$ m³ $\quad$ $\rho_s = 2420$ kg m^{-3}

$\varepsilon = 0.6243$ $\quad$ $t = 480$ sec.

$\mu = 186 \times 10^{-7}$ Nsm^{-2} $\quad$ $g = 9.81$ ms^{-2} $\quad$ $\left[\frac{2RT}{\pi M}\right]^{1/2} = 231.74$ m/s

P_1			P_2			$\frac{P_1}{\Delta P} \times \frac{dV}{dt}$	ΔP	$\bar{P}$	(m²/m³ × 10⁶)		
Volts	m Hg	(Nm^{-2} × 10^{-4})	Volts	m Hg	(Nm^{-2} × 10^{-4})	(m³/sec.) × 10^{-8}	(Nm^{-2} × 10^{-4})	(Nm^{-2} × 10^{-4})	S_K	S_M	S_V
0.960	0.7296	9.7340	0.907	0.6893	9.1964	75.4493	0.5376	9.4652	2.0957	0.3581	2.2824
0.858	0.6521	8.7000	0.805	0.6118	8.1624	67.4347	0.5376	8.4312	2.0922	0.4007	2.3021
0.755	0.5738	7.6554	0.703	0.5340	7.1244	60.0754	0.5310	7.3899	2.0752	0.4497	2.3122
0.651	0.4947	6.6001	0.600	0.4560	6.0837	53.2274	0.5167	6.3419	2.0424	0.5076	2.3119
0.550	0.4180	5.5768	0.501	0.3807	5.0791	46.6918	0.4977	5.3279	1.9987	0.5787	2.3088
0.440	0.3344	4.4614	0.394	0.2994	3.9945	39.8172	0.4669	4.2279	1.9281	0.6786	2.2970
0.340	0.2584	3.4475	0.297	0.2257	3.0112	32.9263	0.4363	3.2293	1.8530	0.8206	2.3082
0.232	0.1763	2.3521	0.195	0.1482	1.9772	26.1435	0.3749	2.1646	1.7026	1.0335	2.2960
0.130	0.0988	1.3182	0.104	0.0790	1.0540	20.7908	0.2642	1.1861	1.4133	1.2995	2.2053
0.040	0.0304	0.4056	0.028	0.0210	0.2802	13.4779	0.1254	0.3429	0.9438	2.0047	2.3791

See section 6.3.2.

Table 6.3 — Determination of specific surface area and Carman–Arnell constant. Data: Dicalcium phosphate dihydrate $[Ca(H_2PO_4)_2$

$L = 6.280 \times 10^{-2}$ m $dV/dt = 4.167 \times 10^{-8}$ m³/sec. $w = 7.4701 \times 10^{-3}$ kg
$A = 1.246 \times 10^{-4}$ m² $V = 20 \times 10^{-6}$ m³ $\rho_s = 2420$ kg m^{-3}
$\varepsilon = 0.6055$ $t = 480$ sec.

$\mu = 186 \times 10^{-7}$ Nsm^{-2} $g = 9.81$ ms^{-2} $\left[\frac{2RT}{\pi M}\right]^{1/2} = 231.74$ m/s

P_1			P_2			ΔP	$\frac{P_1}{\Delta P} \times \frac{dV}{dt}$	$\bar{P}$	(m²/m³ × 10⁶)		
Volts	m Hg	(Nm^{-2} × 10^{-4})	Volts	m Hg	(Nm^{-2} × 10^{-4})	(Nm^{-2} × 10^{-4})	(m³/sec.) × 10^{-8}	(Nm^{-2} × 10^{-4})	S_K	S_M	S_V
0.984	0.7478	9.9768	0.921	0.6999	9.3378	0.6390	65.0599	9.6573	2.1256	0.3704	2.3188
0.880	0.6688	8.9229	0.818	0.6217	8.2945	0.6284	59.1689	8.6087	2.1044	0.4073	2.3179
0.780	0.5928	7.9089	0.719	0.5464	7.2898	0.6191	53.2327	7.5993	2.0845	0.4527	2.3231
0.670	0.5092	6.7935	0.610	0.4640	6.1905	0.6030	46.9461	6.4920	2.0516	0.5133	2.3242
0.562	0.4271	5.6982	0.505	0.3838	5.1205	0.5777	41.1016	5.4093	2.0014	0.5863	2.3160
0.450	0.3420	4.5628	0.396	0.3009	4.0145	0.5483	34.6766	4.2886	1.9402	0.6950	2.3186
0.350	0.2660	3.5489	0.301	0.2288	3.0526	0.4963	29.7970	3.3007	1.8362	0.8088	2.2846
0.244	0.1854	2.4735	0.202	0.1535	2.0479	0.4256	24.2178	2.2607	1.6856	0.9951	2.2551
0.142	0.1079	1.4396	0.110	0.0836	1.1154	0.3242	18.5034	1.2775	1.4496	1.3024	2.2404
0.075	0.0570	0.7605	0.049	0.0372	0.4963	0.2642	11.9947	0.6284	1.2628	2.0091	2.6182

See section 6.3.3.

Table 6.4 — Determination of specific surface area and Carman–Arnell constant. Data: Dicalcium phosphate dihydrate $[Ca(H_2PO_4)_2$

$L = 5.634 \times 10^{-2}$ m
$A = 1.246 \times 10^{-4}$ m^2
$\varepsilon = 0.5603$

$dV/dt = 4.167 \times 10^{-8}$ m^3/sec.
$V = 20 \times 10^{-6}$ m^3
$t = 480$ sec.

$w = 7.4701 \times 10^{-3}$ kg
$\rho_s = 2420$ kg m^{-3}

$\mu = 186 \times 10^{-7}$ Nsm^{-2} $g = 9.81$ ms^{-2} $\left[\frac{2RT}{\pi M}\right]^{1/2} = 231.74$ m/s

P_1			P_2			ΔP	$\frac{P_1}{\Delta P} \times \frac{dV}{dt}$	$\bar{P}$	$(m^2/m^3 \times 10^6)$		
Volts	m Hg	$(Nm^{-2} \times 10^{-4})$	Volts	m Hg	$(Nm^{-2} \times 10^{-4})$	$(Nm^{-2} \times 10^{-4})$	$(m^3/sec.) \times 10^{-8}$	$(Nm^{-2} \times 10^{-4})$	S_K	S_M	S_V
0.991	0.7532	10.0489	0.901	0.6848	9.1363	0.9126	45.8340	9.5926	2.1271	0.4266	2.3511
0.885	0.6726	8.9736	0.797	0.6057	8.0810	0.8926	41.8922	8.5273	2.0989	0.4672	2.3455
0.785	0.5966	7.9596	0.701	0.5330	7.1111	0.8485	39.0897	7.5353	2.0426	0.5007	2.3082
0.680	0.5168	6.8949	0.599	0.4552	6.0731	0.8218	34.9611	6.4840	2.0035	0.5598	2.3029
0.571	0.4339	5.7889	0.492	0.3739	4.9884	0.8005	30.1341	5.3886	1.9673	0.6495	2.3187
0.451	0.3428	4.5735	0.379	0.2880	3.8424	0.7311	26.0672	4.2079	1.8691	0.7509	2.2819
0.351	0.2668	3.5595	0.281	0.2136	2.8498	0.7097	20.8996	3.2046	1.6587	0.9365	2.1918
0.250	0.1900	2.5349	0.192	0.1459	1.9465	0.5884	17.9519	2.2407	1.6436	1.0903	2.2768
0.159	0.1208	1.6117	0.110	0.0836	1.1154	0.4963	13.5320	1.3636	1.4768	1.4464	2.3676
0.080	0.0608	0.8112	0.047	0.0357	0.4763	0.3349	10.0934	0.6437	1.1748	1.9392	2.4929

See section 6.3.4.

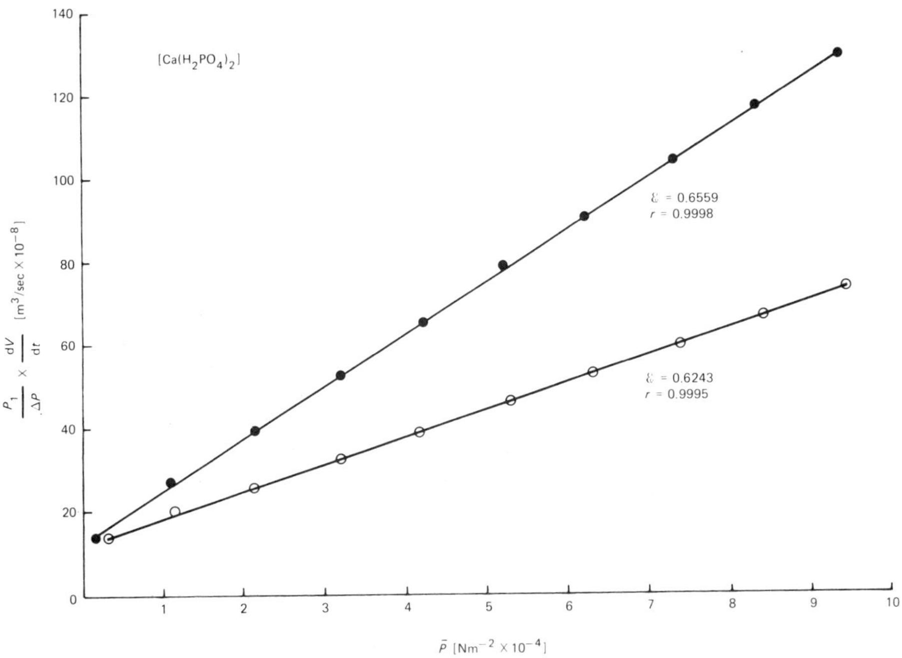

Fig. 6.1 — Evaluation of surface area and the Carman–Arnell constant.

Further statistical comparisons showed that with decreasing porosity, the values of the intercepts would tend to decrease.

Table 6.5 gives a comparative summary of the specific surfaces and the permeability coefficients at different porosities. Since dicalcium phosphate is a porous powder, from the table the surface areas generally increased with decreasing porosity (Fig. 6.3), rendering it difficult to assign an unequivocal value. However, the permeametry surface values indicate little significant changes after the initial compaction at a porosity of $\varepsilon = 0.6559$, even though it tended to increase with decreasing porosity, hence the interpretation must be handled with caution.

A similar comparison with the Carman–Arnell coefficient (Fig. 6.4) shows almost a general constancy of $\left(\dfrac{\delta k_0}{k}\right) = 0.2$ in value for this powder, except at a porosity of $\varepsilon = 0.6559$ where a slightly lower value of 0.1710 was obtained, probably owing to surface irregularities causing channelling in the non-homogeneous bed. The Carman–Arnell coefficient may, therefore, be taken as constant but considerably lower than the usually accepted value of 0.45 when considered specifically for this particular powder bed.

Fig. 6.5 indicates that with decreasing ε, the Klinkenberg factor B tends to increase significantly, which is in agreement with Klinkenberg's study. What Klinkenberg's work failed to point out (even though his work was not related to surface

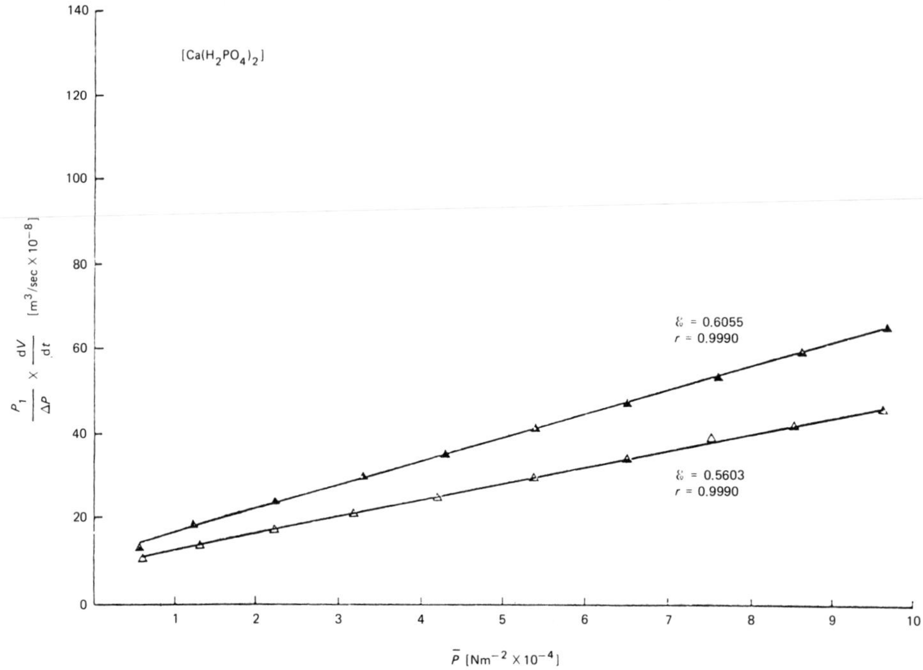

Fig. 6.2 — Evaluation of surface area and the Carman–Arnell constant.

Table 6.5 — Variation of the specific surface areas and the permeability coefficients of slip $[Ca(H_2PO_4)_2]$

ε	S_V ($m^2/m^3 \times 10^6$)	$\left(\frac{\delta k_0}{k}\right)$	Klinkenberg factor, B ($Nm^{-2} \times 10^4$)
0.6559	1.8724	0.1710	0.9693
0.6243	2.2995	0.2230	1.7736
0.6055	2.3172	0.2115	1.8352
0.5603	2.3279	0.2006	2.1062

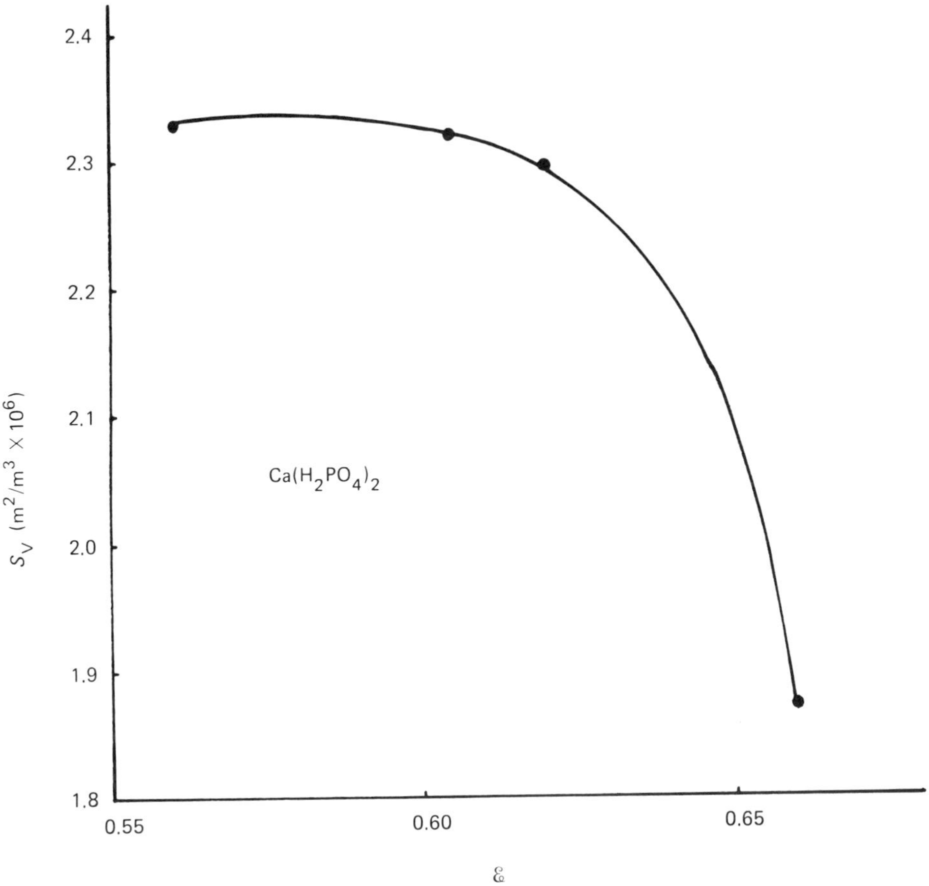

Fig. 6.3 — Variation for measured surface area with porosity.

areas) was whether the increase in B was linear, exponential, or otherwise. It can be shown from this study that

$$y = A_i - CX$$

or expressed numerically,

$$B = 8.4474 - 11.0815\ \varepsilon$$

where A_i=intercept
C=slope
X=a constant depending on the type of material.

The standard error of the slope is 3.6931, with r=0.9046. Even with the margin of error of the slope and the limited experimental data, it can be seen that B tends to increase with decreasing ε.

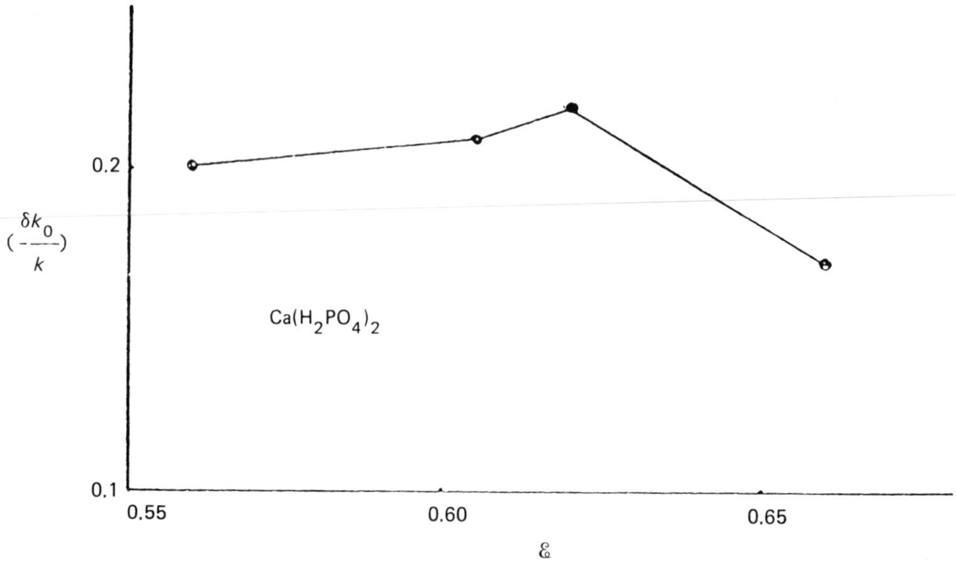

Fig. 6.4 — Variation of ($\delta k_0/k$) with porosity.

6.3.1 Calculations (Table 6.1)

From equation (5.1)

$$\frac{P_1}{\Delta P} \times \frac{dV}{dt} = \frac{44.3601}{S_V^2} \bar{P} + \frac{1.3373}{S_V} \left(\frac{\delta k_0}{k} \right)$$

A regression analysis on the 10 experimental points gives:

Slope $= 12.6524 \times 10^{-12}$
SE of slope $= 0.0801 \times 10^{-12}$
Intercept $= 12.2133 \times 10^{-8}$
Correlation coefficient, $r = 0.9998$

Therefore, $$S_V^2 = \frac{44.3601}{12.6524 \times 10^{-12}}$$

$$\doteq 1.8800 \times 10^6 \text{ m}^2/\text{m}^3$$

From the table, $S_v = 1.8808 \times 10^6$ m²/m³, with a standard deviation, $s = 0.0483 \times 10^6$ m²/m³.

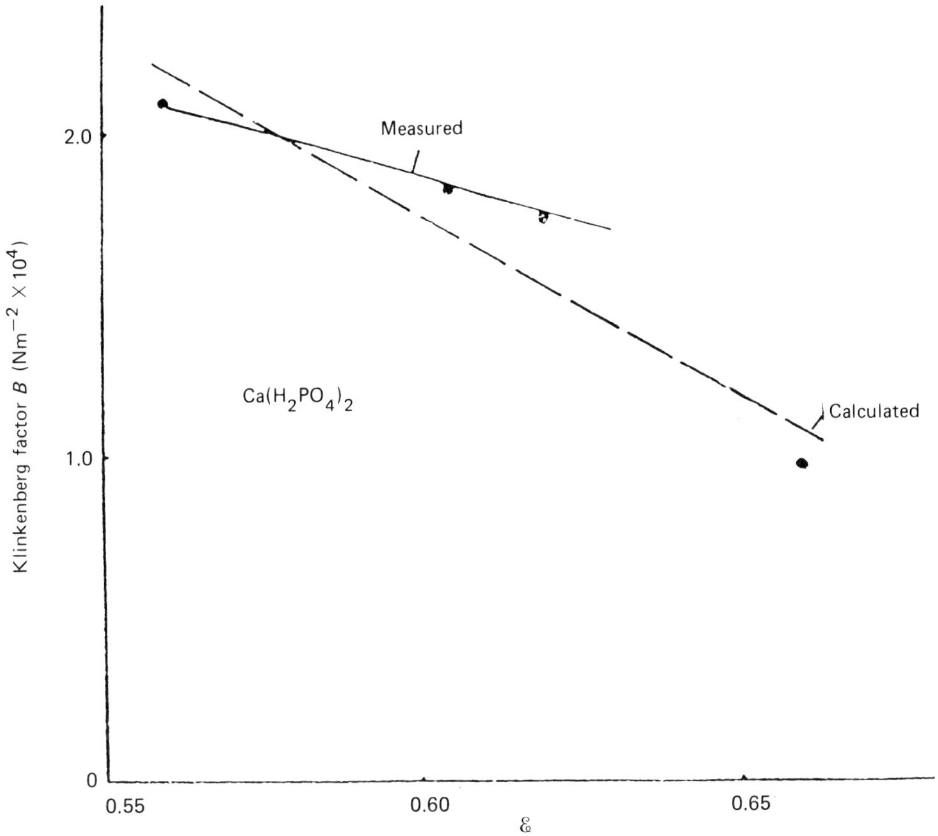

Fig. 6.5 — Variation of Klinkenberg factor B with porosity.

$$\left(\frac{\delta k_0}{k}\right) = \frac{1.8800 \times 10^6 \times 12.2133 \times 10^{-8}}{1.3373}$$

$$= 0.1710$$

From equation (5.9), the Klinkenberg factor $B = 0.9693$ ($Nm^{-2} \times 10^4$).

6.3.2 Calculations (Table 6.2)

From equation (5.1)

$$\frac{P_1}{\Delta P} \times \frac{dV}{dt} = \frac{35.0101}{S_V^2}\bar{P} + \frac{1.2116}{S_V}\left(\frac{\delta k_0}{k}\right)$$

A regression analysis on the 10 experimental points gives:

Slope $= 6.6212 \times 10^{-12}$
SE of slope $= 0.07167 \times 10^{-12}$
Intercept $= 11.7509 \times 10^{-8}$
Correlation coefficient, $r = 0.9995$

Therefore,

$$S_V^2 = \frac{35.0101}{6.6212 \times 10^{-12}}$$

$$S_V = 2.2995 \times 10^6 \text{ m}^2/\text{m}^3$$

From the table, $S_v = 2.3003 \times 10^6$ m²/m³, with a standard deviation, $s = 0.0422 \times 10^6$ m²/m³.

$$\left(\frac{\delta k_0}{k}\right) = \frac{2.2995 \times 10^6 \times 11.7509 \times 10^{-8}}{1.2116}$$

$$= 0.2230$$

From equation (5.9), the Klinkenberg factor $B = 1.7736$ ($\text{Nm}^{-2} \times 10^4$).

6.3.3 Calculations (Table 6.3)

From equation (5.1)

$$\frac{P_1}{\Delta P} \times \frac{dV}{dt} = \frac{30.4382}{S_V^2}\bar{P} + \frac{1.1394}{S_V}\left(\frac{\delta k_0}{k}\right)$$

A regression analysis on the 10 experimental points gives:

Slope $= 5.6686 \times 10^{-12}$
SE of slope $= 0.0942 \times 10^{-12}$
Intercept $= 10.3973 \times 10^{-8}$
Correlation coefficient, $r = 0.9990$

Therefore,

$$S_V^2 = \frac{30.4382}{5.6686 \times 10^{-12}}$$

$$S_V = 2.3172 \times 10^6 \text{ m}^2/\text{m}^3$$

From the table, $S_v = 2.3317 \times 10^6$ m²/m³, with a standard deviation, $s = 0.1051 \times 10^6$ m²/m³.

$$\left(\frac{\delta k_0}{k}\right) = \frac{2.3172 \times 10^6 \times 10.3973 \times 10^{-8}}{1.1394}$$

$$= 0.2115$$

From equation (5.9), the Klinkenberg factor $B = 1.8352$ ($\text{Nm}^{-2} \times 10^4$).

6.3.4 Calculations (Table 6.4)

From equation (5.1)

$$\frac{P_1}{\Delta P} \times \frac{\mathrm{d}V}{\mathrm{d}t} = \frac{21.6397}{S_\mathrm{V}^2}\bar{P} + \frac{0.9757}{S_\mathrm{V}}\left(\frac{\delta k_0}{k}\right)$$

A regression analysis on the 10 experimental points gives:

Slope $= 3.9932 \times 10^{-12}$
SE of slope $= 0.0754 \times 10^{-12}$
Intercept $= 8.4085 \times 10^{-8}$
Correlation coefficient, $r = 0.9990$

Therefore, $$S_\mathrm{V}^2 = \frac{21.6397}{3.9932 \times 10^{-12}}$$
$$S_\mathrm{V} = 2.3279 \times 10^6 \ \mathrm{m^2/m^3}$$

From the table, $S_\mathrm{v} = 2.3237 \times 10^6 \ \mathrm{m^2/m^3}$, with a standard deviation, $s = 0.0773 \times 10^6 \ \mathrm{m^2/m^3}$.

$$\left(\frac{\delta k_0}{k}\right) = \frac{2.3279 \times 10^6 \times 8.4085 \times 10^{-8}}{0.9757}$$
$$= 0.2006$$

From equation (5.9), the Klinkenberg factor $B = 2.1062 \ (\mathrm{Nm^{-2}} \times 10^4)$.

7

Determination of surface area and Carman–Arnell constant using spherical bronze powders

7.1 INTRODUCTION

The main experimental results for all the spherical bronze powders tested, showing the relationship between the gas flow rates and the mean pressures, are presented in Tables 7.1–7.7 and graphically in Figs 7.1–7.7. The results show generally that the experimental points lie accurately on a series of straight lines with no departure from linearity. A closer appraisal of the tables and graphs reveal that the S_V values for all the measured particle size ranges increases with increasing fineness of the powder sample.

7.2 RESULTS AND DISCUSSION

As explained in section 6.1, values of all the powder bed constants, the regression constants, the correlation coefficients, the calculated specific surface areas, and the coefficients of slip, are shown below each experimental table. The results show that all the correlation coefficients are statistically significant at the specified significance levels of 0.1%.

To illustrate the results on a comparative basis (see Table 7.8) the variables have been plotted against mean particle size in Figs 7.8–7.10. The results demonstrate quite clearly that S_V, $\left(\frac{\delta k_0}{k}\right)$, and the factor B, would decrease with increasing mean particle size.

To determine the effects of porosity on the variable parameters the results have further been plotted against ε, and these may be observed in Figs 7.11–7.13. The graphs highlight the previous trends found in this study, namely, that with increasing

Table 7.1 — Determination of specific surface area and Carman–Arnell constant. Data: Spherical bronze powders (13.375 μm)

$L = 7.8260 \times 10^{-2}$ m $dV/dt = 4.167 \times 10^{-8}$ m³/sec. $w = 0.05378$ kg
$A = 1.246 \times 10^{-4}$ m² $V = 20 \times 10^{-6}$ m³ $\rho_s = 8800$ kg m^{-3}
$\varepsilon = 0.3732$ $t = 480$ sec.

$\mu = 186 \times 10^{-7}$ Nsm^{-2} $g = 9.81$ ms^{-2} $\left[\frac{2RT}{\pi M}\right]^{1/2} = 231.74$ m/s

P_1			P_2			ΔP	$\frac{P_1}{\Delta P} \times \frac{dV}{dt}$	$\bar{P}$	(m²/m³ × 10⁶)		
Volts	m Hg	(Nm^{-2} × 10^{-4})	Volts	m Hg	(Nm^{-2} × 10^{-4})	(Nm^{-2} × 10^{-4})	(m³/sec.) × 10^{-8}	(Nm^{-2} × 10^{-4})	S_K	S_M	S_V
0.9675	0.7353	9.8100	0.9000	0.6840	9.1256	0.3762	108.6610	9.4678	0.4444	0.0089	0.4489
0.8370	0.6361	8.4866	0.8000	0.6080	8.1117	0.3749	94.3283	8.2992	0.4465	0.0103	0.4517
0.7368	0.5600	7.4713	0.7000	0.5320	7.0977	0.3736	88.3322	7.2845	0.4451	0.0116	0.4509
0.6366	0.4838	6.4547	0.6000	0.4560	6.0838	0.3709	72.5175	6.2693	0.4426	0.0134	0.4494
0.5357	0.4071	5.4314	0.5000	0.3800	5.0698	0.3616	62.5903	5.2506	0.4360	0.0155	0.4438
0.4355	0.3310	4.4160	0.4000	0.3040	4.0558	0.3602	51.0868	4.2359	0.4335	0.0190	0.4431
0.3353	0.2548	3.3994	0.3010	0.2280	3.0419	0.3575	39.6232	3.2207	0.4292	0.0244	0.4416
0.2337	0.1776	2.3695	0.2010	0.1520	2.0279	0.3416	28.9043	2.1987	0.4152	0.0335	0.4323
0.1326	0.1008	1.3448	0.1020	0.0760	1.0140	0.3308	16.9401	1.1794	0.3972	0.0572	0.4268
0.0508	0.0386	0.5150	0.0100	0.0076	0.1014	0.4136	5.1886	0.3082	0.3669	0.1867	0.4719

See section 7.3.1.

Table 7.2 — Determination of specific surface area and Carman–Arnell constant. Data: Spherical bronze powders (14.663 μm)

$L = 7.743 \times 10^{-2}$ m $dV/dt = 4.167 \times 10^{-8}$ m³/sec. $w = 0.05254$ kg
$A = 1.246 \times 10^{-4}$ m² $V = 20 \times 10^{-6}$ m³ $\rho_s = 8800$ kg m^{-3}
$\varepsilon = 0.3811$ $t = 480$ sec.

$\mu = 186 \times 10^{-7}$ Nsm^{-2} $g = 9.81$ ms^{-2} $\left[\frac{2RT}{\pi M}\right]^{1/2} = 231.74$ m/s

P_1			P_2			ΔP	$\frac{P_1}{\Delta P} \times \frac{dV}{dt}$	$\bar{P}$	(m²/m³ × 10⁶)		
Volts	m Hg	(Nm$^{-2} \times 10^{-4}$)	Volts	m Hg	(Nm$^{-2} \times 10^{-4}$)	(Nm$^{-2} \times 10^{-4}$)	(m³/sec.) × 10^{-8}	(Nm$^{-2} \times 10^{-4}$)	S_K	S_M	S_V
0.9281	0.7054	9.4111	0.9000	0.6840	9.1256	0.2855	137.3592	9.2684	0.4108	0.0058	0.4137
0.3282	0.6294	8.3972	0.8000	0.6080	8.1117	0.2855	122.5609	8.2545	0.4105	0.0065	0.4138
0.7281	0.5534	7.3832	0.7000	0.5320	7.0977	0.2855	107.7611	7.2405	0.4100	0.0074	0.4137
0.6282	0.4774	6.3693	0.6000	0.4560	6.0838	0.2855	92.9628	6.2266	0.4094	0.0086	0.4137
0.5282	0.4014	5.3553	0.5000	0.3800	5.0698	0.2855	78.1630	5.2126	0.4084	0.0102	0.4135
0.4281	0.3254	4.3413	0.4000	0.3040	4.0558	0.2855	63.3632	4.1986	0.4072	0.0126	0.4136
0.3232	0.2494	3.3274	0.3010	0.2280	3.0419	0.2855	48.5649	3.18479	0.4051	0.0165	0.4134
0.2280	0.1733	2.3121	0.2010	0.1520	2.0279	0.2842	33.9005	2.1700	0.4001	0.0236	0.4121
0.1280	0.0973	1.2982	0.1020	0.0760	1.0140	0.2842	19.0345	1.1561	0.3899	0.0421	0.4115
0.0432	0.0328	0.4376	0.0100	0.0076	0.1014	0.3362	5.4238	0.2695	0.3526	0.1476	0.4340

See section 7.3.2.

Table 7.3 — Determination of specific surface area and Carman–Arnell constant. Data: Spherical bronze powders (15,240 μm)

$L = 7.6840 \times 10^{-2}$ m $dV/dt = 4.167 \times 10^{-8}$ m³/sec. $w = 0.05163$ kg
$A = 1.246 \times 10^{-4}$ m² $V = 20 \times 10^{-6}$ m³ $\rho_s = 8800$ kg m^{-3}
$\varepsilon = 0.3872$ $t = 480$ sec.

$\mu = 186 \times 10^{-7}$ Nsm^{-2} $g = 9.81$ ms^{-2} $\left[\frac{2RT}{\pi M}\right]^{1/2} = 231.74$ m/s

P_1			P_2			ΔP	$\frac{P_1}{\Delta P} \times \frac{dV}{dt}$	$\overline{P}$	($m^2/m^3 \times 10^6$)		
Volts	m Hg	($Nm^{-2} \times 10^{-4}$)	Volts	m Hg	($Nm^{-2} \times 10^{-4}$)	($Nm^{-2} \times 10^{-4}$)	(m^3/sec.) $\times 10^{-8}$	($Nm^{-2} \times 10^{-4}$)	S_K	S_M	S_V
0.9250	0.7030	9.3791	0.9000	0.6840	9.1256	0.2535	154.1724	9.2524	0.4025	0.0051	0.4051
0.8250	0.6270	8.3652	0.8000	0.6080	8.1117	0.2535	137.5061	8.2385	0.4025	0.0057	0.4054
0.7250	0.5510	7.3512	0.7000	0.5320	7.0977	0.2535	120.8380	7.2245	0.4016	0.0065	0.4049
0.6250	0.4750	6.3373	0.6000	0.4560	6.0838	0.2535	104.1717	6.2106	0.4011	0.0076	0.4049
0.5250	0.3990	5.3233	0.5000	0.3800	5.0698	0.2535	87.5037	5.1966	0.4002	0.0090	0.4047
0.4250	0.3230	4.3093	0.4000	0.3040	4.0558	0.2535	70.8357	4.1826	0.3991	0.0111	0.4047
0.3250	0.2470	3.2954	0.3010	0.2280	3.0419	0.2535	54.1694	3.1687	0.3972	0.0146	0.4046
0.2250	0.1710	2.2814	0.2010	0.1520	2.0279	0.2535	37.5014	2.1547	0.3937	0.0211	0.4044
0.1250	0.0950	1.2675	0.1020	0.0760	1.0140	0.2535	20.8350	1.1408	0.3843	0.0379	0.4040
0.0376	0.0286	0.3816	0.0100	0.0076	0.1014	0.2802	5.6750	0.2415	0.3388	0.1391	0.4154

See section 7.3.3.

Table 7.4 — Determination of specific surface area and Carman–Arnell constant. Data: Spherical bronze powders (17.153 μm)

$L = 7.9800 \times 10^{-2}$ m $dV/dt = 4.167 \times 10^{-8}$ m³/sec. $w = 0.0544$ kg
$A = 1.246 \times 10^{-4}$ m² $V = 20 \times 10^{-6}$ m³ $\rho_s = 8800$ kg m^{-3}
$\varepsilon = 0.3783$ $t = 480$ sec.

$\mu = 186 \times 10^{-7}$ Nsm^{-2} $g = 9.81$ ms^{-2} $\left[\dfrac{2RT}{\pi M}\right]^{1/2} = 231.74$ m/s

P_1			P_2			ΔP	$\dfrac{P_1}{\Delta P} \times \dfrac{dV}{dt}$	$\overline{P}$	(m²/m³ × 10⁶)		
Volts	m Hg	(Nm^{-2} × 10^{-4})	Volts	m Hg	(Nm^{-2} × 10^{-4})	(Nm^{-2} × 10^{-4})	(m³/sec.)	(Nm^{-2} × 10^{-4})	S_K	S_M	S_V
0.9230	0.7015	9.3591	0.9000	0.6840	9.1256	0.2335	167.0209	9.2423	0.3606	0.0024	0.3618
0.8230	0.6255	8.3452	0.8000	0.6080	8.1117	0.2335	155.5904	8.2284	0.3603	0.0027	0.3616
0.7230	0.5495	7.3312	0.7000	0.5320	7.0977	0.2335	130.8313	7.2145	0.3599	0.0031	0.3615
0.6230	0.4735	6.3173	0.6000	0.4560	6.0838	0.2335	112.7374	6.2006	0.3595	0.0035	0.3613
0.5230	0.3975	5.3033	0.5000	0.3800	5.0698	0.2335	94.6418	5.1865	0.3588	0.0042	0.3609
0.4230	0.3215	4.2893	0.4000	0.3040	4.0558	0.2335	76.5461	4.1726	0.3579	0.0052	0.3605
0.3230	0.2455	3.2754	0.3010	0.2280	3.0419	0.2335	58.4522	3.1587	0.3563	0.0068	0.3597
0.2230	0.1695	2.2614	0.2010	0.1520	2.0279	0.2335	40.3566	2.1446	0.3534	0.0099	0.3584
0.1230	0.0935	1.2475	0.1020	0.0760	1.0140	0.2335	22.2627	1.1308	0.3455	0.0180	0.3546
0.0390	0.0296	0.3949	0.0100	0.0076	0.1014	0.2935	5.6066	0.2482	0.3225	0.0713	0.3601

See section 7.3.4.

Table 7.5 — Determination of specific surface area and Carman–Arnell constant. Data: Spherical bronze powders (19.576 μm)

$L = 7.9720 \times 10^{-2}$ m
$A = 1.246 \times 10^{-4}$ m^2
$\varepsilon = 0.3841$

$dV/dt = 4.167 \times 10^{-8}$ m^3/sec.
$V = 20 \times 10^{-6}$ m^3
$t = 480$ sec.

$w = 53.8412$ kg $\times 10^{-3}$ kg
$\rho_s = 8800$ kg m^{-3}

$\mu = 186 \times 10^{-7}$ Nsm^{-2}

$g = 9.81$ ms^{-2}

$\left[\frac{2RT}{\pi M}\right]^{1/2} = 231.74$ m/s

	P_1			P_2		ΔP	$\frac{P_1}{\Delta P} \times \frac{dV}{dt}$	$\bar{P}$	(m^2/m^3 $\times 10^6$)		
Volts	m Hg	(Nm$^{-2} \times 10^{-4}$)	Volts	m Hg	(Nm$^{-2} \times 10^{-4}$)	(Nm$^{-2} \times 10^{-4}$)	8^3/sec.) $\times 10^{-8}$	(Nm$^{-2} \times 10^{-4}$)	S_K	S_M	S_V
0.9158	0.6960	9.2857	0.9000	0.6840	9.1256	0.1601	241.6834	9.2057	0.3094	0.0019	0.3104
0.8158	0.6200	8.2718	0.8000	0.6080	8.1117	0.1601	215.2941	8.1918	0.3094	0.0022	0.3105
0.7158	0.5440	7.2578	0.7000	0.5320	7.0977	0.1601	188.9023	7.1777	0.3090	0.0025	0.3103
0.6158	0.4680	6.2439	0.6000	0.4560	6.0838	0.1601	162.5130	6.1639	0.3089	0.0028	0.3103
0.5158	0.3920	5.2299	0.5000	0.3800	5.0698	0.1601	136.1211	5.1498	0.3084	0.0034	0.3101
0.4158	0.3160	4.2159	0.4000	0.3040	4.0558	0.1601	109.7293	4.1359	0.3079	0.0042	0.3100
0.3158	0.2400	3.2020	0.3010	0.2280	3.0419	0.1601	83.3400	3.1219	0.3069	0.0056	0.3100
0.2158	0.1640	2.1880	0.2010	0.1520	2.0279	0.1601	56.9481	2.1080	0.3051	0.0081	0.3100
0.1158	0.0880	1.1741	0.1020	0.0760	1.0140	0.1601	30.5589	1.0941	0.3000	0.0151	0.3100
0.0337	0.0256	0.3416	0.0100	0.0076	0.1014	0.1601	5.9261	0.2215	0.2502	0.0520	0.3000

See section 7.3.5.

Table 7.6 — Determination of specific surface area and Carman–Arnell constant. Data: Spherical bronze powders (20.841 μm)

$L = 7.8120 \times 10^{-2}$ m $dV/dt = 4.167 \times 10^{-8}$ m³/sec. $w = 0.05263$ kg
$A = 1.246 \times 10^{-4}$ m² $V = 20 \times 10^{-6}$ m³ $\rho_s = 8800$ kg m^{-3}
$\varepsilon = 0.3855$ $t = 480$ sec.

$\mu = 186 \times 10^{-7}$ Nsm^{-2} $g = 9.81$ ms^{-2} $\left[\frac{2RT}{\pi M}\right]^{1/2} = 231.74$ m/s

P_1			P_2			ΔP	$\frac{P_1}{\Delta P} \times \frac{dV}{dt}$	$\bar{P}$	(m²/m³ × 10⁶)		
Volts	m Hg	(Nm^{-2} × 10^{-4})	Volts	m Hg	(Nm^{-2} × 10^{-4})	(Nm^{-2} × 10^{-4})	(m³/sec.) × 10^{-8}	(Nm^{-2} × 10^{-4})	S_K	S_M	S_V
0.9145	0.6950	9.2724	0.9000	0.6840	9.1256	0.1468	263.2023	9.1990	0.3018	0.0017	0.3027
0.8145	0.6190	8.2585	0.8000	0.6080	8.1117	0.1468	234.4221	8.1851	0.3016	0.0019	0.3026
0.7145	0.5430	7.2445	0.7000	0.5320	7.0977	0.1468	205.6392	7.1711	0.3015	0.0021	0.3026
0.6145	0.4670	6.2306	0.6000	0.4560	6.0838	0.1468	176.8590	6.1572	0.3013	0.0025	0.3026
0.5145	0.3910	5.2166	0.5000	0.3800	5.0698	0.1468	148.0761	5.1432	0.3010	0.0029	0.3025
0.4145	0.3150	4.2026	0.4000	0.3040	4.0558	0.1468	119.2932	4.1292	0.3005	0.0037	0.3024
0.3145	0.2390	3.1887	0.3010	0.2280	3.0419	0.1468	90.5130	3.11153	0.2995	0.0048	0.3020
0.2145	0.1630	2.1747	0.2010	0.1520	2.0279	0.1468	61.7301	2.1013	0.2980	0.0071	0.3020
0.1145	0.0370	1.1608	0.1020	0.0760	1.0140	0.1468	32.9499	1.0874	0.2933	0.0132	0.3000
0.0311	0.0236	0.3149	0.0100	0.0076	0.1014	0.2135	6.1461	0.2082	0.2970	0.0710	0.3346

See section 7.3.6.

Table 7.7 — Determination of specific surface area and Carman–Arnell constant. Data: Spherical bronze powders (20.841 μm)

$L = 7.8680 \times 10^{-2}$ m
$A = 1.246 \times 10^{-4}$ m^2
$\varepsilon = 0.3990$

$dV/dt = 4.167 \times 10^{-8}$ m^3/sec.
$V = 20 \times 10^{-6}$ m^3
$t = 480$ sec.

$w = 0.05185$ kg
$\rho_s = 8800$ kg m^{-3}

$\mu = 186 \times 10^{-7}$ Nsm^{-2}

$g = 9.81$ ms^{-2}

$\left[\frac{2RT}{\pi M}\right]^{1/2} = 231.74$ m/s

P_1			P_2			ΔP	$\frac{P_1}{\Delta P} \times \frac{dV}{dt}$	$\overline{P}$	(m^2/m$^3 \times 10^6$)		
Volts	m Hg	(Nm$^{-2} \times 10^{-4}$)	Volts	m Hg	(Nm$^{-2} \times 10^{-4}$)	(Nm$^{-2} \times 10^{-4}$)	(m^3/sec.) $\times 10^{-8}$	(Nm$^{-2} \times 10^{-4}$)	S_K	S_M	S_V
0.9125	0.6935	9.2523	0.9000	0.6840	9.1256	0.1267	304.2962	9.1889	0.3007	0.0009	0.3012
0.8125	0.6175	8.2384	0.8000	0.6080	8.1117	0.1267	270.9504	8.1751	0.3005	0.0010	0.3010
0.7125	0.5415	7.2244	0.7000	0.5320	7.0977	0.1267	237.6012	7.1610	0.3003	0.0012	0.3009
0.6125	0.4655	6.2105	0.6000	0.4560	6.0838	0.1267	204.2554	6.1472	0.3002	0.0013	0.3009
0.5125	0.3895	5.1965	0.5000	0.3800	5.0698	0.1267	170.9062	5.1332	0.2998	0.0016	0.3006
0.4125	0.3135	4.1825	0.4000	0.3040	4.0558	0.1267	137.5570	4.1191	0.2994	0.0020	0.3000
0.3125	0.2375	3.1686	0.3010	0.2280	3.0419	0.1267	104.2112	3.1053	0.2987	0.0026	0.3000
0.2125	0.1615	2.1546	0.2010	0.1520	2.0279	0.1267	70.8620	2.0912	0.2973	0.0038	0.3000
0.1125	0.0855	1.1407	0.1020	0.0760	1.0140	0.1267	37.5162	1.0774	0.2932	0.0073	0.3000
0.0336	0.0255	0.3402	0.0100	0.0076	0.1014	0.2008	5.9346	0.2208	0.3059	0.0386	0.3258

See section 7.3.7.

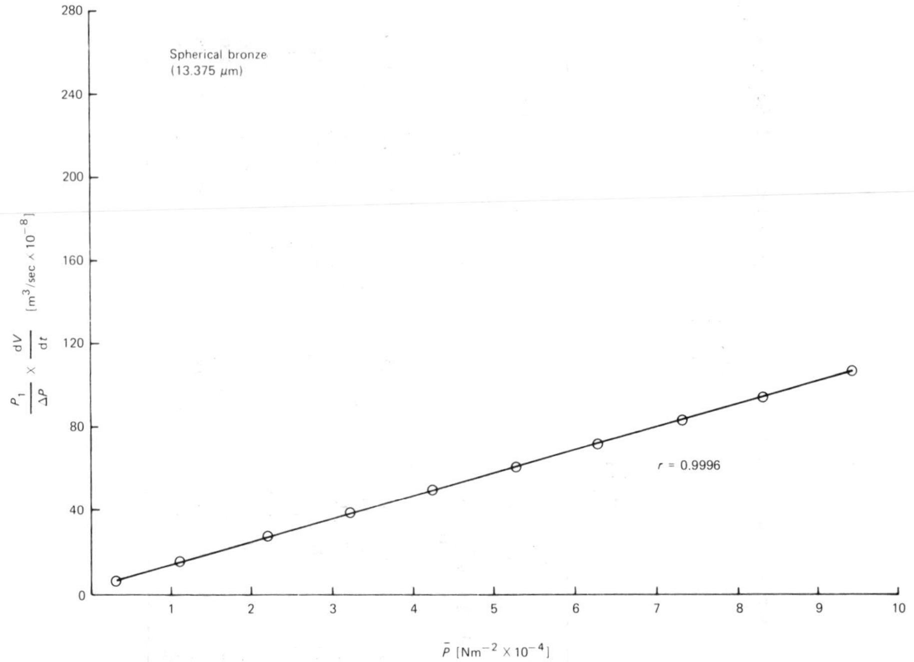

Fig. 7.1 — Evaluation of surface area and the Carman–Arnell constant.

mean particle size and/or increasing ε, the specific surfaces, $\left(\frac{\delta k_0}{k}\right)$, and the factor B, would collectively decrease. The experimental scatter found is evidently related to the limited scope and range of the particle size used.

Two micrographs of the spherical bronze powders used for this study are presented in Figs A1 and A2 (in Appendix A) at (×40 magnifications). The micrographs show differences typical of the powders, especially the large compared with the smaller particles. Consequently these differences would cause a change and variation in the values of any measured surface areas of the the powder.

When Fig. A1 is closely observed, it reveals possible dense structures of interlocked particles in which large and small particles can be seen. It is, therefore, reasonable to suggest that the clusters and thick powder particle concentrations visible in Fig. A1 when compared with the isolated patches of powder particles in Fig. A2 form an important distinctive feature in explaining the differences in values of surface areas obtained.

7.3.1 Calculations (Table 7.1)

From equation (5.1)

$$\frac{P_1}{\Delta P} \times \frac{dV}{dt} = \frac{2.2658}{S_V^2}\,\bar{P} + \frac{0.2187}{S_V}\left(\frac{\delta k_0}{k}\right)$$

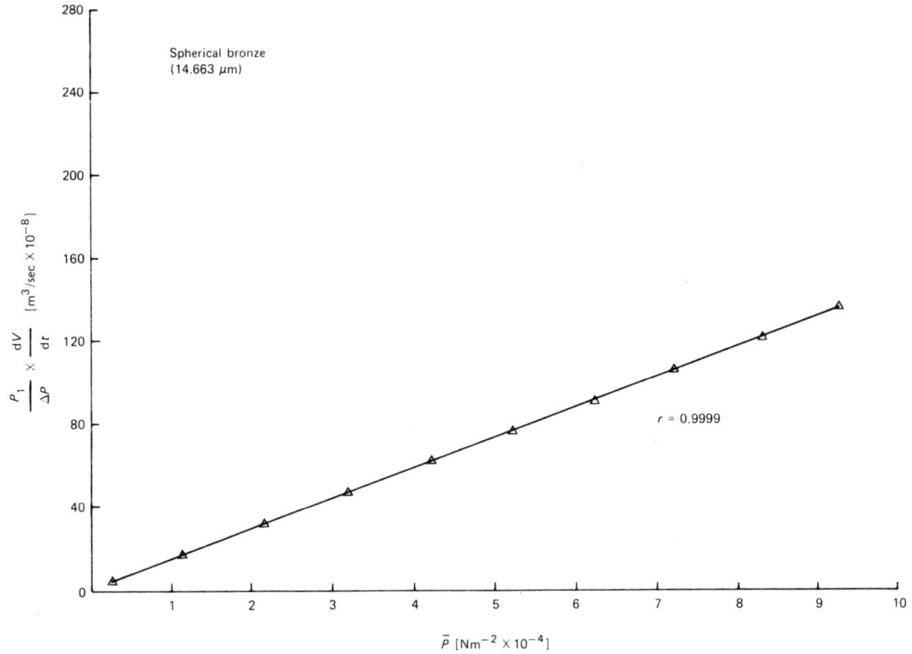

Fig. 7.2 — Evaluation of surface area and the Carman–Arnell constant.

A regression analysis on the 10 experimental points gives:

Slope	$= 11.0501 \times 10^{-12}$
SE of slope	$= 0.1073 \times 10^{-12}$
Intercept	$= 3.5923 \times 10^{-8}$
Correlation coefficient, r	$= 0.9996$

Therefore,

$$S_V^2 = \frac{2.2658}{11.0501 \times 10^{-12}}$$

$$S_V = 0.4528 \times 10^6 \text{ m}^2/\text{m}^3$$

From the table, $S_V = 0.4500 \times 10^6 \text{ m}^2/\text{m}^3$, with a standard deviation, $s = 0.0122 \times 10^6 \text{ m}^2/\text{m}^3$.

$$\left(\frac{\delta k_0}{k}\right) = \frac{0.4528 \times 10^6 \times 3.5923 \times 10^{-8}}{0.2187} = 0.0744$$

From equation (5.9), the Klinkenberg factor $B = 0.3252$ ($\text{Nm}^{-2} \times 10^4$).

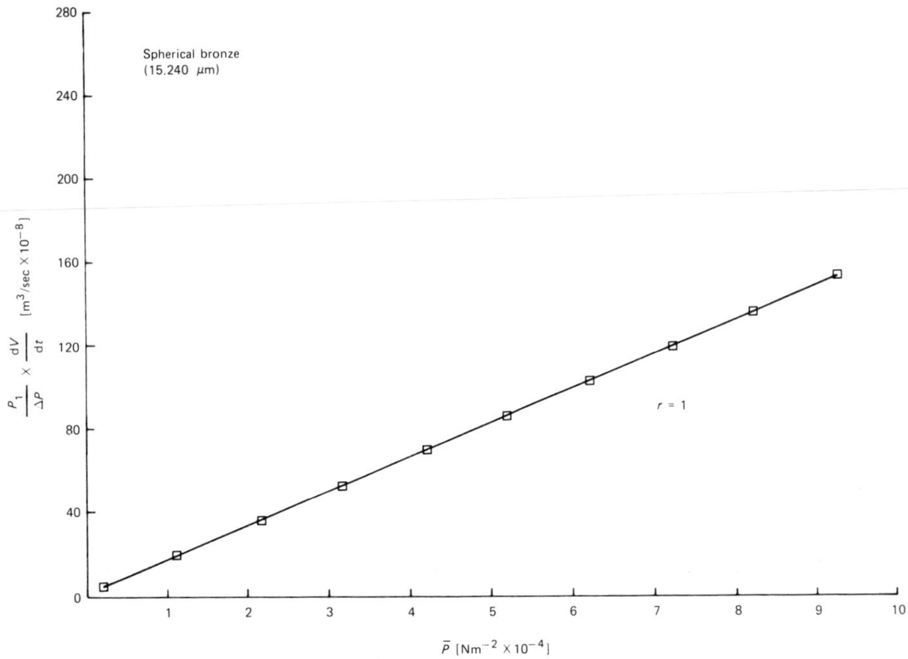

Fig. 7.3 — Evaluation of surface area and the Carman–Arnell constant.

7.3.2 Calculations (Table 7.2)

From equation (5.1)

$$\frac{P_1}{\Delta P} \times \frac{dV}{dt} = \frac{2.5029}{S_V^2} \bar{P} + \frac{0.2334}{S_V} \left(\frac{\delta k_0}{k} \right)$$

A regression analysis on the 10 experimental points gives:

Slope	$= 14.6192 \times 10^{-12}$
SE of slope	$= 0.0217 \times 10^{-12}$
Intercept	$= 1.9938 \times 10^{-8}$
Correlation coefficient, r	$= 0.9999$

Therefore,

$$S_V^2 = \frac{2.5029}{14.6192 \times 10^{-12}}$$

$$S_V = 0.4138 \times 10^6 \ \mathrm{m^2/m^3}$$

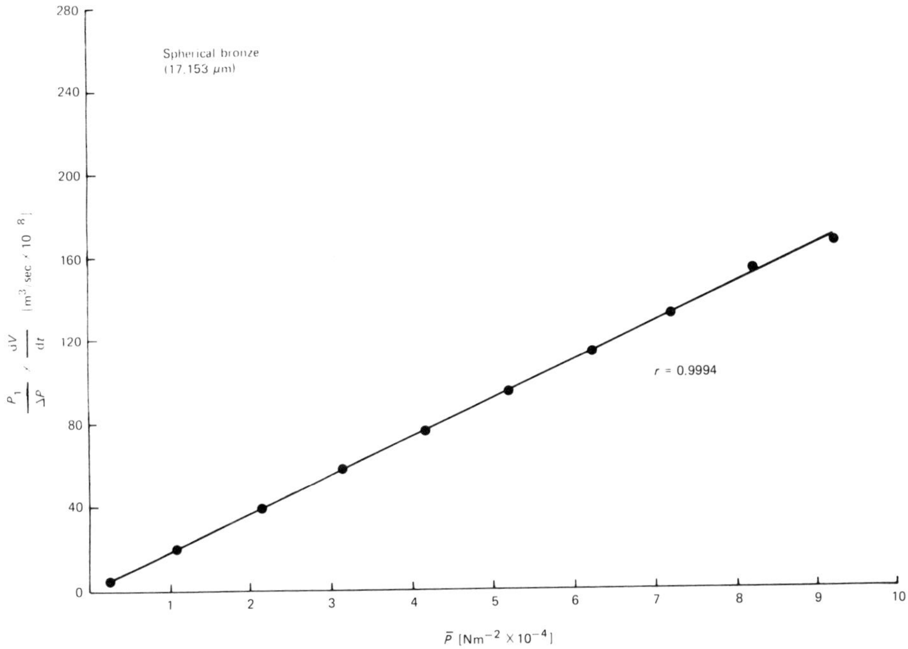

Fig. 7.4 — Evaluation of surface area and the Carman–Arnell constant.

From the table, $S_V = 0.4153 \times 10^6\ \mathrm{m^2/m^3}$, with a standard deviation, $s = 0.0066 \times 10^6\ \mathrm{m^2/m^3}$.

$$\left(\frac{\delta k_0}{k}\right) = \frac{0.4138 \times 10^6 \times 1.9338 \times 10^{-8}}{0.2334} = 0.0343$$

From equation (5.9), the Klinkenberg factor $B = 1.1325\ (\mathrm{Nm}^{-2} \times 10^4)$.

7.3.3 Calculations (Table 7.3)

From equation (5.1)

$$\frac{P_1}{\Delta P} \times \frac{\mathrm{d}V}{\mathrm{d}t} = \frac{2.6978}{S_V^2}\,\bar{P} + \frac{0.2452}{S_V}\left(\frac{\delta k_0}{k}\right)$$

A regression analysis on the 10 experimental points gives:

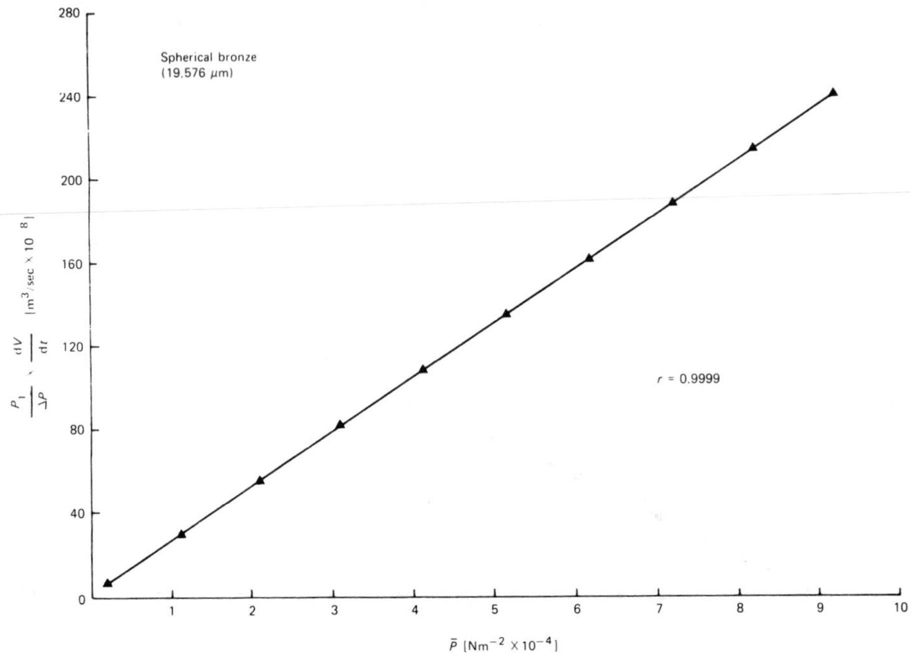

Fig. 7.5 — Evaluation of surface area and the Carman–Arnell constant.

Slope	$= 16.4579 \times 10^{-12}$
SE of slope	$= 0.0118 \times 10^{-12}$
Intercept	$= 1.9505 \times 10^{-8}$
Correlation coefficient, r	$= 1$

Therefore,

$$S_V^2 = \frac{2.6978}{16.4579 \times 10^{-12}}$$

$$S_V = 0.4050 \times 10^6 \text{ m}^2/\text{m}^3$$

From the table, $S_V = 0.4058 \times 10^6 \text{ m}^2/\text{m}^3$, with a standard deviation, $s = 0.0034 \times 10^6 \text{ m}^2/\text{m}^3$.

$$\left(\frac{\delta k_0}{k}\right) = \frac{0.4050 \times 10^6 \times 1.9505 \times 10^{-8}}{0.2452} = 0.0322$$

From equation (5.9), the Klinkenberg factor $B = 0.1186$ (Nm$^{-2} \times 10^4$).

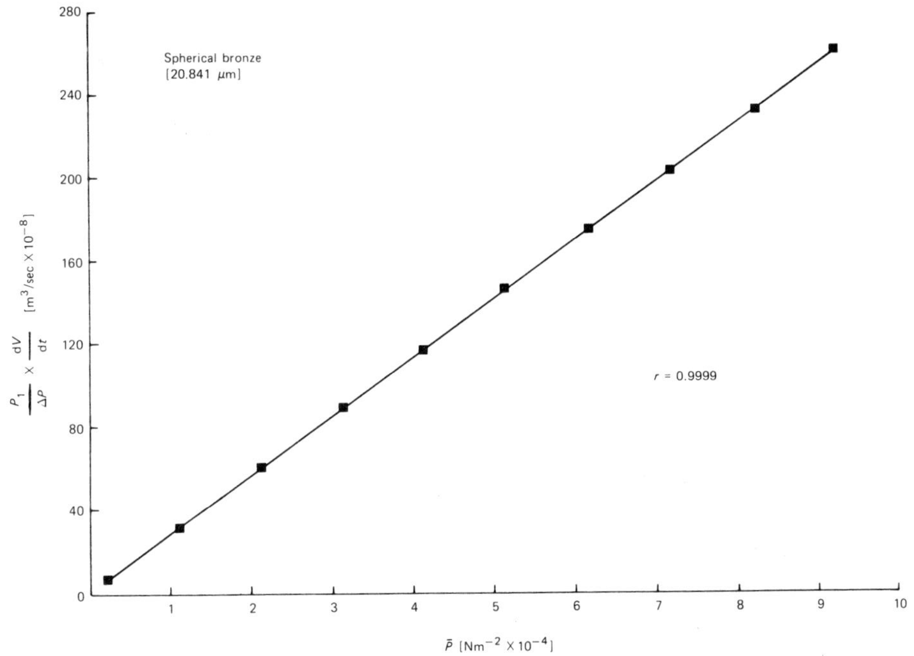

Fig. 7.6 — Evaluation of surface area and the Carman–Arnell constant.

7.3.4 Calculations (Table 7.4)

From equation (5.1)

$$\frac{P_1}{\Delta P} \times \frac{dV}{dt} = \frac{2.3501}{S_V^2}\,\bar{P} + \frac{0.2221}{S_V}\left(\frac{\delta k_0}{k}\right)$$

A regression analysis on the 10 experimental points gives:

Slope	$= 18.1757 \times 10^{-12}$
SE of slope	$= 0.2215 \times 10^{-12}$
Intercept	$= 1.1108 \times 10^{-8}$
Correlation coefficient, r	$= 0.9994$

Therefore,

$$S_V^2 = \frac{2.3501}{18.1757 \times 10^{-12}}$$

$$S_V = 0.3596 \times 10^6\ m^2/m^3$$

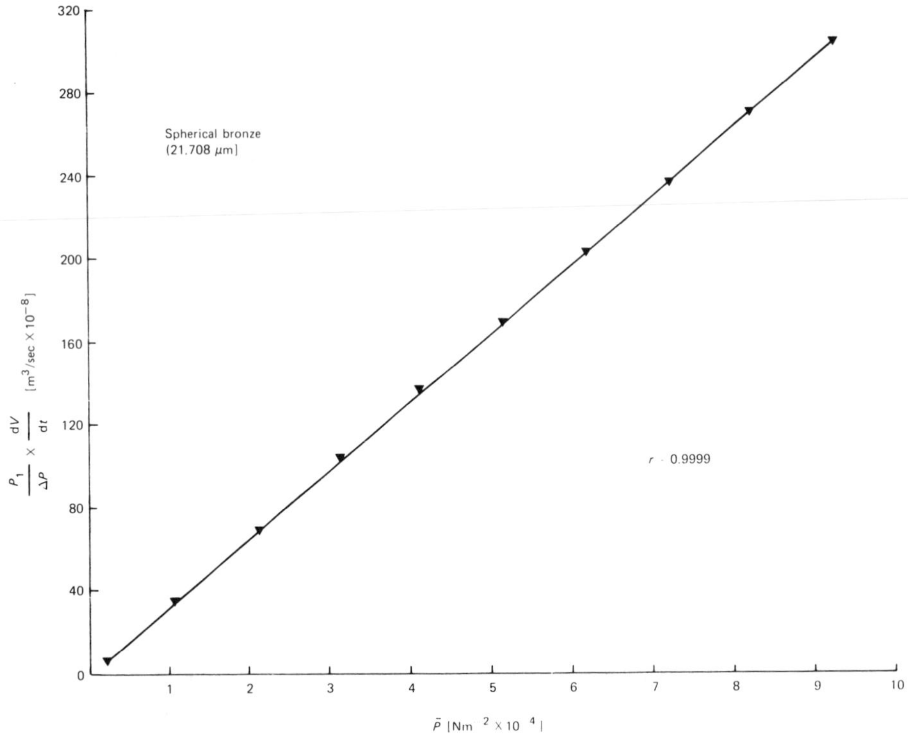

Fig. 7.7 — Evaluation of surface area and the Carman–Arnell constant.

Table 7.8 — Variation of the permeability coefficients of slip with mean particle size (Spherical bronze)

Mean particle size range (μm)	S_V ($m^2/m^3 \times 10^6$)	$\left(\frac{\delta k_0}{k}\right)$ ($\times 10^{-1}$)	Klinkenberg factor, B ($Nm^{-2} \times 10^4$)
13.375	0.4528	0.744	0.3252
14.663	0.4138	0.343	0.1325
15.240	0.4050	0.322	0.1186
17.153	0.3596	0.180	0.0612
19.576	0.3101	0.200	0.0571
20.841	0.3023	0.183	0.0507
21.708	0.3010	0.105	0.0274

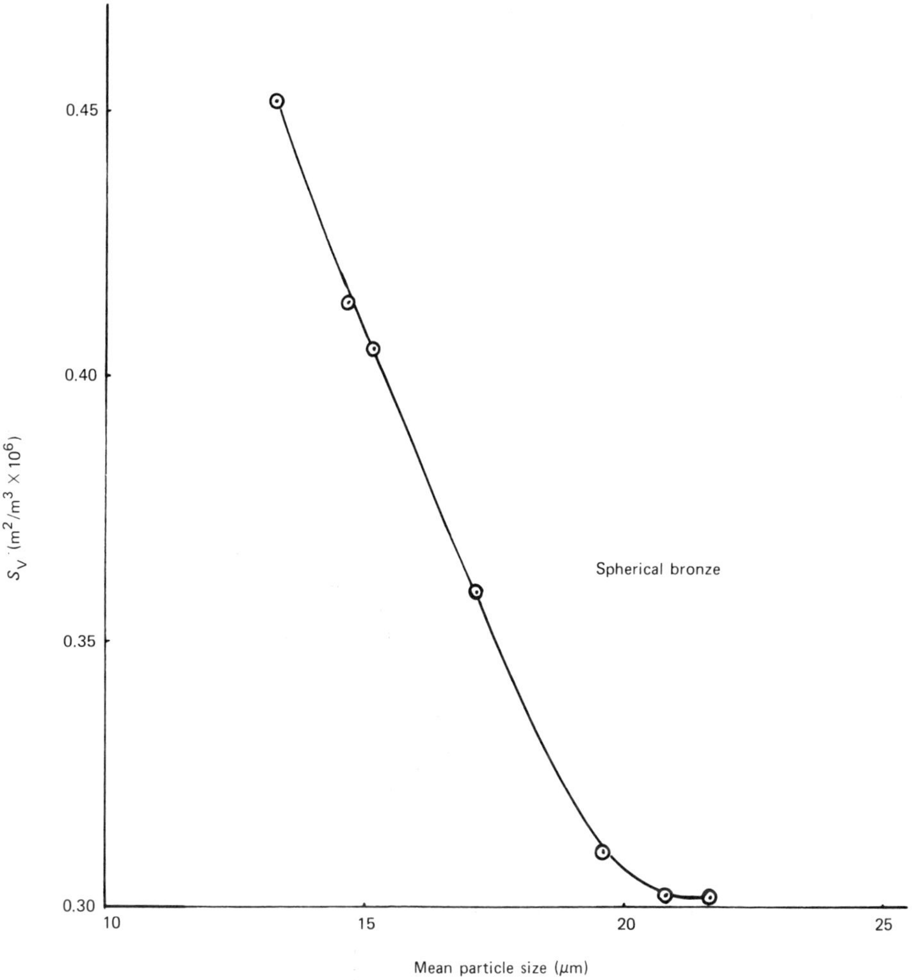

Fig. 7.8 — Variation of measured surface area with mean particle size.

From the table, $S_V = 0.3600 \times 10^6\ m^2/m^3$, with a standard deviation, $s = 0.0021 \times 10^6\ m^2/m^3$.

$$\left(\frac{\delta k_0}{k}\right) = \frac{0.36 \times 10^6 \times 1.1108 \times 10^{-8}}{0.2221} = 0.0180$$

From equation (5.9), the Klinkenberg factor $B = 0.0612\ (Nm^{-2} \times 10^4)$.

7.3.5 Calculations (Table 7.5)

From equation (5.1)

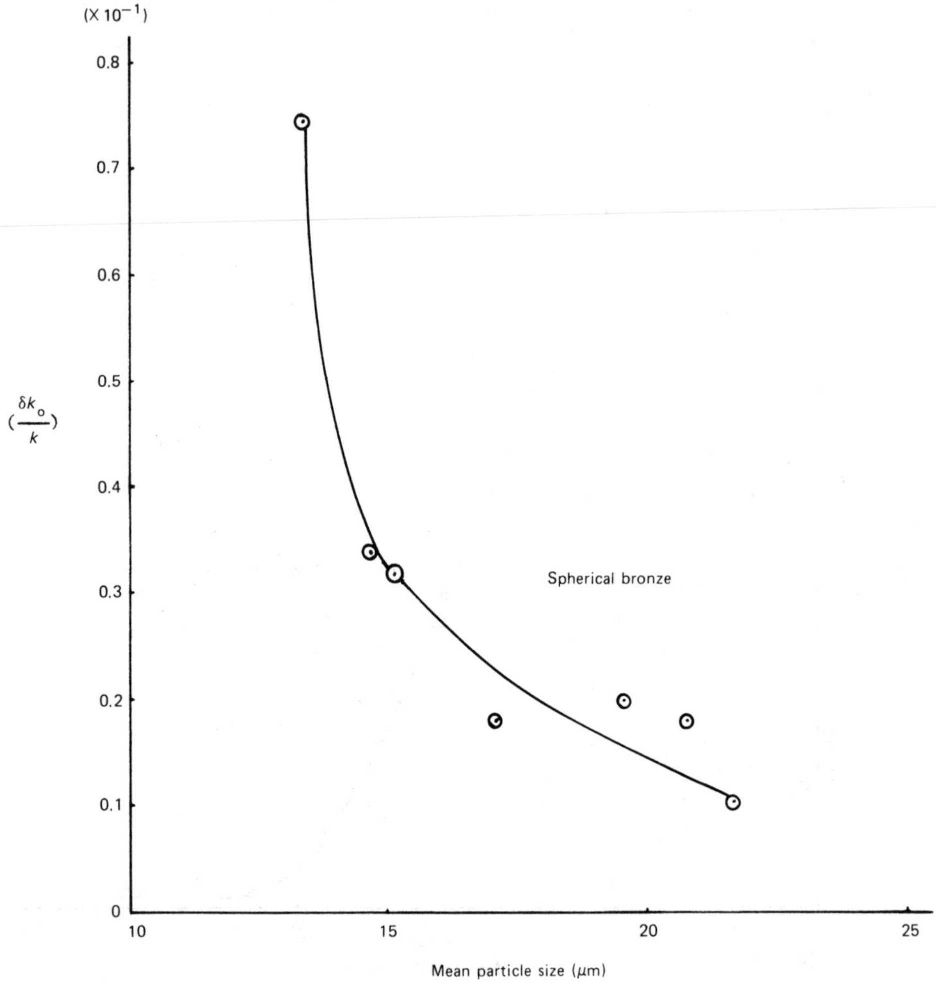

Fig. 7.9 — Variation of ($\delta k_0/k$) with mean particle size.

$$\frac{P_1}{\Delta P} \times \frac{dV}{dt} = \frac{2.5123}{S_V^2}\bar{P} + \frac{0.2313}{S_V}\left(\frac{\delta k_0}{k}\right)$$

A regression analysis on the 10 experimental points gives:

Slope	$= 26.1295 \times 10^{-12}$
SE of slope	$= 0.0606 \times 10^{-12}$
Intercept	$= 1.4156 \times 10^{-8}$
Correlation coefficient, r	$= 0.9999$

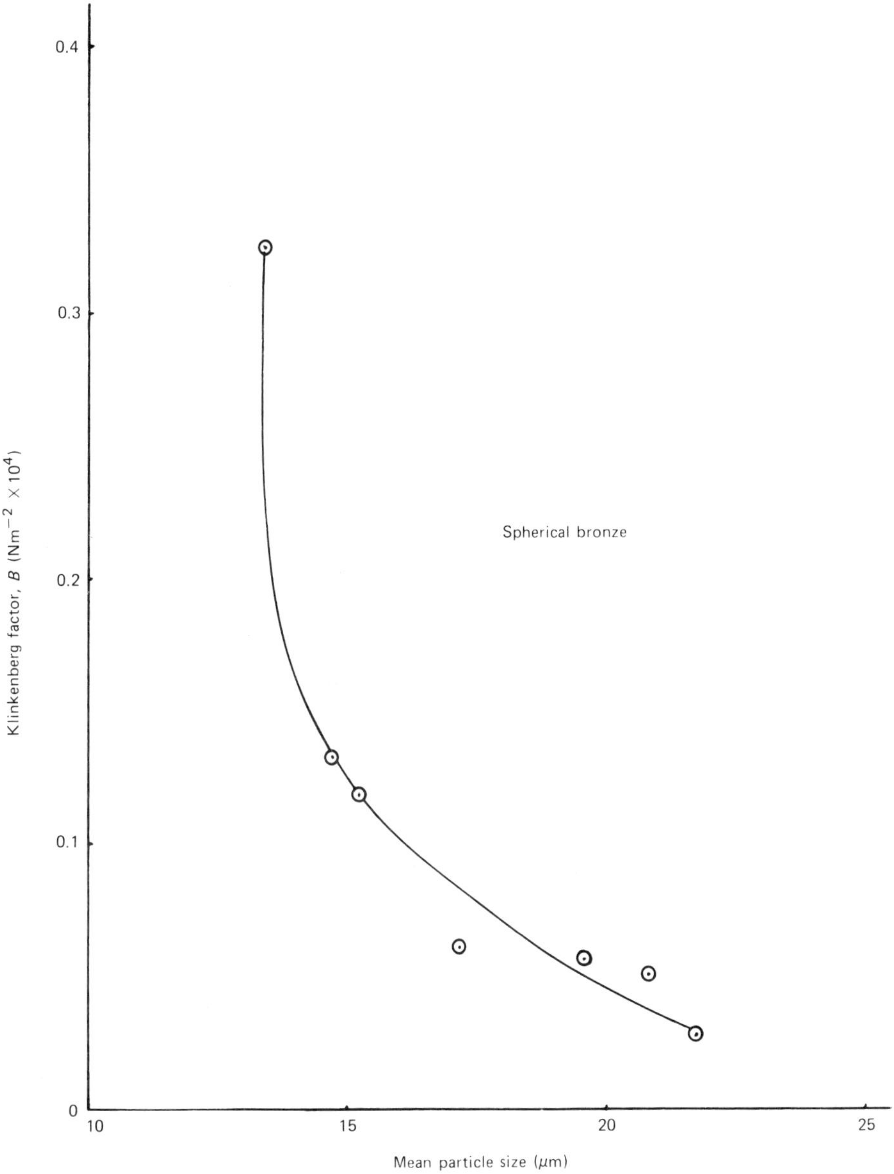

Fig. 7.10 — Relationship between Klinkenberg factor B with mean particle size.

Therefore,

$$S_V^2 = \frac{2.5123}{26.1295 \times 10^{-12}}$$

$$S_V = 0.3101 \times 10^6 \text{ m}^2/\text{m}^3$$

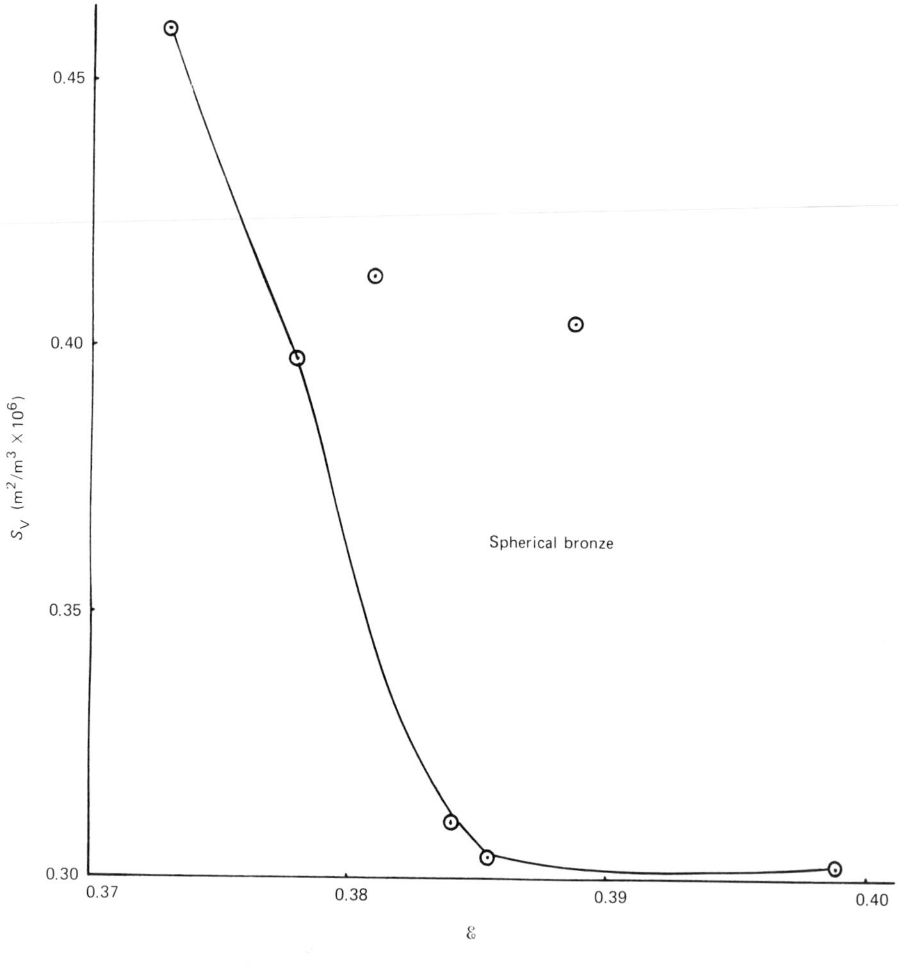

Fig. 7.11 — Variation of measured surface area with porosity.

From the table, $S_V = 0.3100 \times 10^6$ m^2/m^3, with a standard deviation, $s = 0.0032 \times 10^6$ m^2/m^3.

$$\left(\frac{\delta k_0}{k}\right) = \frac{0.3101 \times 10^6 \times 1.4156 \times 10^{-8}}{0.2313} = 0.0190$$

From equation (5.9), the Klinkenberg factor $B = 0.0571$ (Nm$^{-2} \times 10^4$).

7.3.6 Calculations (Table 7.6)

From equation (5.1)

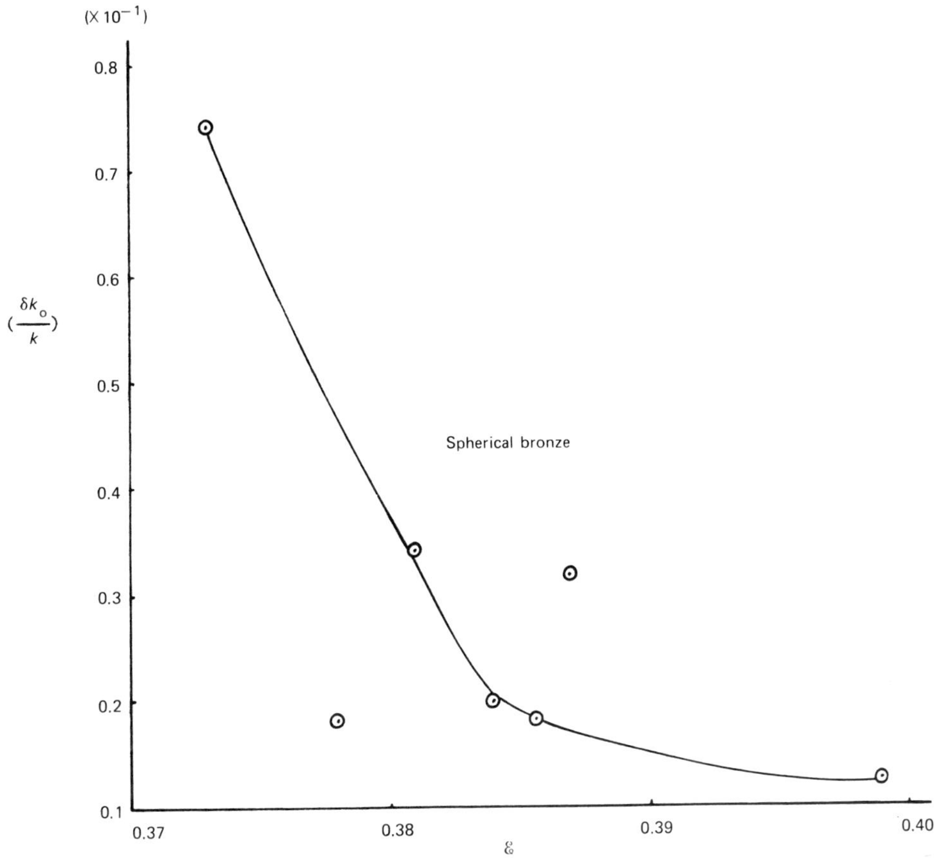

Fig. 7.12 — Variation of $(\delta k_0/k)$ with porosity.

$$\frac{P_1}{\Delta P} \times \frac{dV}{dt} = \frac{2.6025}{S_V^2}\,\overline{P} + \frac{0.2384}{S_V}\left(\frac{\delta k_0}{k}\right)$$

A regression analysis on the 10 experimental points gives:

Slope	$= 28.4837 \times 10^{-12}$
SE of slope	$= 0.0582 \times 10^{-12}$
Intercept	$= 1.4424 \times 10^{-8}$
Correlation coefficient, r	$= 0.9999$

Therefore,

$$S_V^2 = \frac{2.6025}{28.4837 \times 10^{-12}}$$

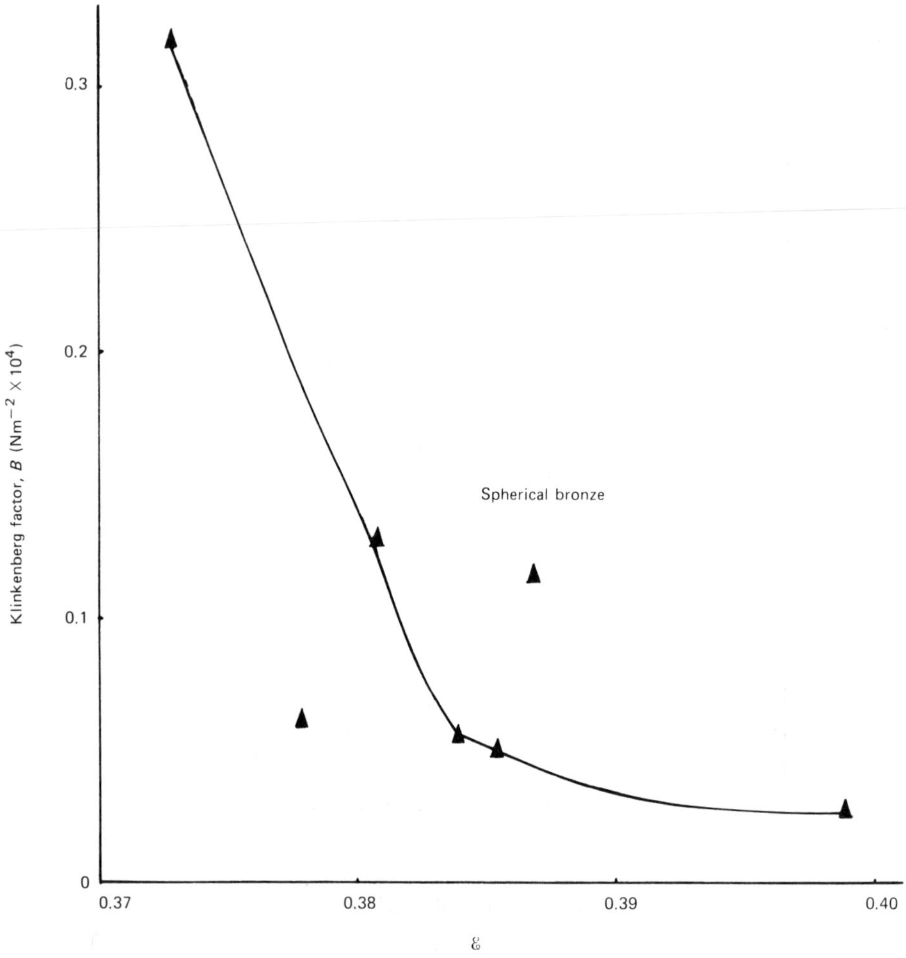

Fig. 7.13 — Variation of Klinkenberg factor B with porosity.

$$S_V = 0.3023 \times 10^6 \text{ m}^2/\text{m}^3$$

From the table, $S_V = 0.3054 \times 10^6 \text{ m}^2/\text{m}^3$, with a standard deviation, $s = 0.0103 \times 10^6 \text{ m}^2/\text{m}^3$.

$$\left(\frac{\delta k_0}{k}\right) = \frac{0.3023 \times 10^6 \times 1.4424 \times 10^{-8}}{0.2384} = 0.0183$$

From equation (5.9), the Klinkenberg factor $B = 0.0507$ ($\text{Nm}^{-2} \times 10^4$).

7.3.7 Calculations (Table 7.7)

From equation (5.1)

$$\frac{P_1}{\Delta P} \times \frac{\mathrm{d}V}{\mathrm{d}t} = \frac{2.9936}{S_V^2}\,\bar{P} + \frac{0.2592}{S_V}\left(\frac{\delta k_0}{k}\right)$$

A regression analysis on the 10 experimental points gives:

Slope	$= 33.0695 \times 10^{-12}$
SE of slope	$= 0.1076 \times 10^{-12}$
Intercept	$= 0.9033 \times 10^{-8}$
Correlation coefficient, r	$= 0.9999$

Therefore,

$$S_V^2 = \frac{2.9936}{33.0695 \times 10^{-12}}$$

$$S_V = 0.3010 \times 10^6 \ \mathrm{m^2/m^3}$$

From the table, $S_V = 0.303 \times 10^6$ m^2/m^3, with a standard deviation, $s = 0.0080 \times 10^6$ m^2/m^3.

$$\left(\frac{\delta k_0}{k}\right) = \frac{0.301 \times 10^6 \times 0.9033 \times 10^{-8}}{0.2592} = 0.0105$$

From equation (5.9), the Klinkenberg factor $B = 0.0274$ (Nm$^{-2} \times 10^4$).

8

Determination of surface area and Carman–Arnell constant using cyclone tin and copper powders

8.1 INTRODUCTION

The results of the effect of the mean pressures, $\overline{P}$, on the flow rate of two different powders, are presented firstly for cyclone tin powder in Table 8.1, and graphically in Fig. 8.1. This is followed by the results obtained for copper powders in Table 8.2 and in Fig. 8.2, graphically. Values of the change in measured surface with increasing mean pressure in the bed, S_K/S_V, for the two powders are shown in Table 8.3 and as graphs in Fig. 8.3. The graphs of S_K/S_V against $\overline{P}$ show a distinct difference between the two powder grades at very low pressures; this is probably due to changes in flow from mainly streamline to mainly diffusion.

8.2 RESULTS AND DISCUSSION

A general analysis of the results for the two powder types from Table 8.1 and 8.2, Figs 8.1 and 8.2 suggests that the powders responded in different ways to the values of the coefficients of slip. For cyclone tin which has a particle size of $<25\ \mu m$, $\left(\frac{\delta k_0}{k}\right)$ was 0.1604, but for copper with particle size $<38\ \mu m$ $\left(\frac{\delta k_0}{k}\right)$ equals 0.0103, a decrease from the value of tin of 93.6%, which explains the fact that $\left(\frac{\delta k_0}{k}\right)$ depends to a greater degree on the surface properties, i.e. particle size of the powder, rather than on any other conceivable variable factor.

Further comparison with the Klinkenberg factor, B, showed a rapid decrease

Table 8.1 — Determination of specific surface area and Carman-Arnell constant. Data: Cyclone tin powder ($<25\ \mu m$)

$L = 7.2250 \times 10^{-2}$ m
$A = 1.246 \times 10^{-4}$ m^2
$\varepsilon = 0.3796$
$\mu = 186 \times 10^{-7}$ Nsm^{-2}

$dV/dt = 4.167 \times 10^{-8}$ m^3/sec.
$V = 20 \times 10^{-6}$ m^3
$t = 480$ sec.
$g = 9.81$ ms^{-2}

$w = 0.04077$ kg.
$\rho_s = 7300$ kg m^{-3}
$\left[\frac{2\ RT}{\pi\ M}\right]^{\frac{1}{2}} = 231.74$ m/s

P_1			P_2			ΔP	$\frac{P_1}{\Delta P} \times \frac{dV}{dt}$	$\bar{P}$	(m^2/m^3 × 10^6)		
Volts	m Hg	(Nm^{-2} × 10^{-4})	Volts	m Hg	(Nm^{-2} × 10^{-4})	(Nm^{-2} × 10^{-4})	(m^3/sec.) × 10^{-8}	(Nm^{-2} × 10^{-4})	S_K	S_M	S_V
0.9692	0.7366	9.8274	0.9000	0.6840	9.1256	0.7018	58.3511	9.4765	0.6544	0.0680	0.6893
0.8686	0.6601	8.8068	0.8000	0.6080	8.1117	0.6951	52.7952	8.4593	0.6500	0.0752	0.6887
0.7674	0.5832	7.7808	0.7000	0.5320	7.0977	0.6831	47.4639	7.4393	0.6429	0.0837	0.6861
0.6645	0.5050	6.7375	0.6000	0.4560	6.0838	0.6537	42.9481	6.4107	0.6274	0.0924	0.6753
0.5620	0.4271	5.6982	0.5000	0.3800	5.0698	0.6284	37.7855	5.3840	0.6130	0.1051	0.6678
0.4592	0.3490	4.6562	0.4000	0.3040	4.0558	0.6004	32.3158	4.3560	0.5962	0.1229	0.6608
0.3553	0.2700	3.6023	0.3010	0.2280	3.0419	0.5604	26.7858	3.3221	0.5719	0.1482	0.6508
0.2511	0.1908	2.5456	0.2010	0.1520	2.0279	0.5177	20.4897	2.2868	0.5425	0.1938	0.6480
0.1513	0.1150	1.5343	0.1020	0.0760	1.0140	0.5203	12.2880	1.2742	0.5229	0.3231	0.7088
0.0771	0.0586	0.7818	0.0100	0.0076	0.1014	0.6804	4.7880	0.4416	0.4932	0.8293	1.0589

See section 8.4.1.

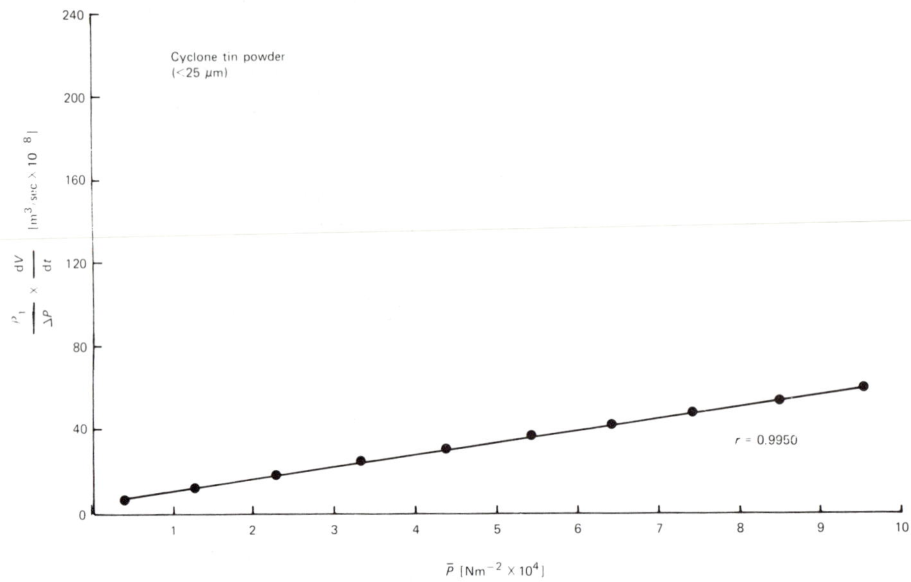

Fig. 8.1 — Evaluation of surface area and the Carman–Arnell constant.

from $B = 1.0245$ ($Nm^{-2} \times 10^4$) for tin, to $B = 0.0296$ for copper, a decrease from the value of tin of 97.11%. The reason advanced for this response is again that of the finer particle size of tin compared with copper. Thus, this observation is in agreement with the results of the study carried out in section 7.1 with the smooth spherical bronze powders where, with decreasing particle size, the coefficient B increased.

8.3 CONCLUSION

It may be concluded from this study, therefore, that the permeability coefficients of slip may be explained by the overriding effects of the nature of the solid surface and the size of the inter-particle voids, i.e. particle size of the powder used. With decreasing particle size the slip coefficients, $\left(\frac{\delta k_0}{k}\right)$, and the Klinkenberg factor, B, would increase.

8.4.1 Calculations (Table 8.1)

From equation (5.1)

$$\frac{P_1}{\Delta P} \times \frac{dV}{dt} = \frac{2.6353}{S_V^2} \bar{P} + \frac{0.2475}{S_V} \left(\frac{\delta k_0}{k}\right)$$

A regression analysis on the 10 experimental points gives:

Table 8.2 — Determination of specific surface area and Carman-Arnell constant.
Data: Copper powder ($< 38\ \mu m$)

$L = 6.97 \times 10^{-2}$ m
$A = 1.246 \times 10^{-4}\ m^2$
$\varepsilon = 0.5310$
$\mu = 186 \times 10^{-7}\ Nsm^{-2}$

$dV/dt = 4.167 \times 10^{-8}\ m^3/sec.$
$V = 20 \times 10^{-6}\ m^3$
$t = 480$ sec.
$g = 9.81\ ms^{-2}$

$w = 36.3716 \times 10^{-3}$ kg.
$\rho_s = 8.930\ kg\ m^{-3}$
$\left[\frac{2\ RT}{\pi\ M}\right]^{\frac{1}{2}} = 231.74$ m/s.

	P_1			P_2		ΔP	$\frac{P_1}{\Delta P} \times \frac{dV}{dt}$	$\bar{P}$		$(m^2/m^3 \times 10^6)$	
Volts	m Hg	$(Nm^{-2} \times 10^{-4})$	Volts	m Hg	$(Nm^{-2} \times 10^{-4})$	$(Nm^{-2} \times 10^{-4})$	$(m^3/sec.) \times 10^{-8}$	$(Nm^{-2} \times 10^{-4})$	S_K	S_M	S_V
0.9099	0.6915	9.2257	0.9000	0.6840	9.1256	0.1001	384.0509	9.1757	0.5596	0.0018	0.5605
0.8099	0.6155	8.2117	0.8000	0.6080	8.1117	0.1000	342.1815	8.1617	0.5591	0.0020	0.5601
0.7099	0.5395	7.1977	0.7000	0.5320	7.0977	0.1000	299.9282	7.1477	0.5589	0.0023	0.5600
0.6099	0.4635	6.1838	0.6000	0.4560	6.0838	0.1000	257.6789	6.1338	0.5585	0.0027	0.5598
0.5099	0.3875	5.1698	0.5000	0.3800	5.0698	0.1000	215.4256	5.1198	0.5581	0.0032	0.5597
0.4099	0.3115	4.1559	0.4000	0.3040	4.0558	0.1001	173.0034	4.1059	0.5577	0.0040	0.5597
0.3099	0.2355	3.1419	0.3010	0.2280	3.0419	0.1000	130.9229	3.0919	0.5563	0.0052	0.5589
0.2099	0.1595	2.1279	0.2010	0.1520	2.0279	0.1000	88.6696	2.0779	0.5542	0.0077	0.5581
0.1099	0.0835	1.1140	0.1020	0.0760	1.0140	0.1000	46.4204	1.0640	0.5481	0.0147	0.5555
0.0251	0.0191	0.2548	0.0100	0.0076	0.1014	0.1534	6.9215	0.1781	0.5807	0.0988	1.6322

See section 8.4.2.

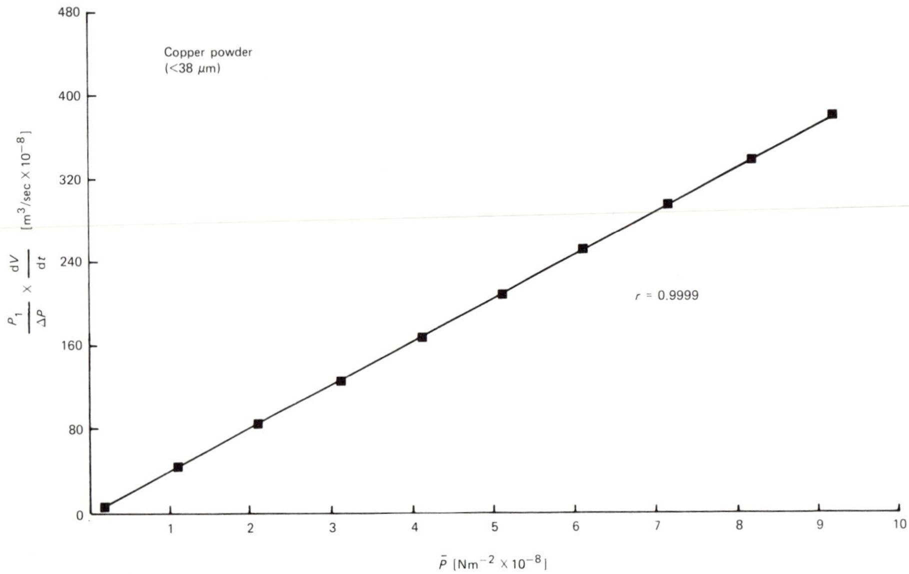

Fig. 8.2 — Evaluation of surface area and the Carman–Arnell constant.

Table 8.3 — Values of S_K/S_V at various mean pressures, $\bar{P}$, and porosities (cyclone tin and copper powders)

Cyclone tin powder (< 20 μm) ε = 0.3796	Copper powder (< 38 μm) ε = 0.5310
0.9494	0.9984
0.9438	0.9982
0.9370	0.9980
0.9291	0.9977
0.9179	0.9971
0.9022	0.9964
0.8788	0.9953
0.8372	0.9930
0.7377	0.9867
0.4658	0.9185

Slope $= 5.6844 \times 10^{-12}$
SE of slope $= 0.2064 \times 10^{-12}$
Intercept $= 5.8326 \times 10^{-8}$
Correlation coefficient, $r = 0.9950$

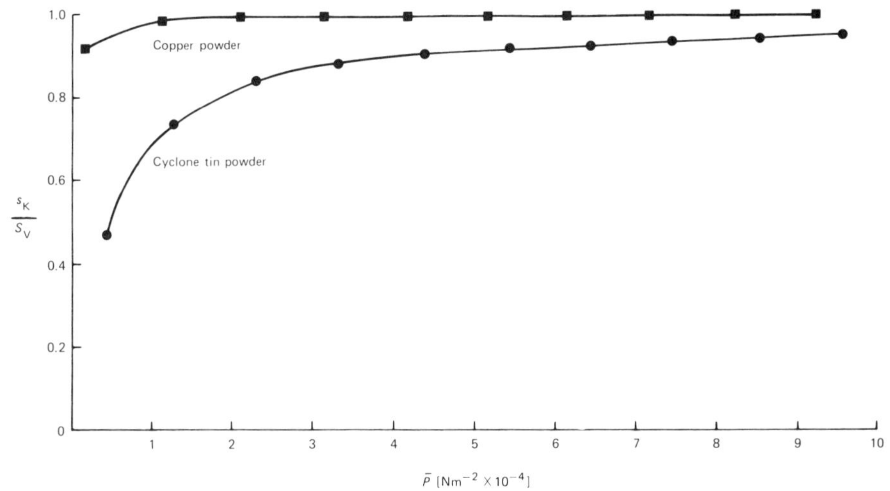

Fig. 8.3 — Change in measured surface with increasing mean pressure in the bed.

Therefore,

$$S_V^2 = \frac{2.6353}{5.6844 \times 10^{-12}}$$

$$S_V = 0.6809 \times 10^6 \ m^2/m^3$$

From the table, $S_V = 0.6800 \times 10^6 \ m^2/m^3$, with a standard deviation, $s = 0.0189 \times 10^6 \ m^2/m^3$ on the 9 acceptable points. The increase in the last point may be due to more surface area being available for flow.

$$\left(\frac{\delta k_0}{k}\right) = \frac{0.6809 \times 10^6 \times 5.8326 \times 10^{-8}}{0.2475} = 0.1604$$

From equation (5.9), the Klinkenberg factor $B = 1.0245$ (Nm$^{-2} \times 10^4$).

8.4.2 Calculations (Table 8.2)

From equation (5.1)

$$\frac{P_1}{\Delta P} \times \frac{dV}{dt} = \frac{13.106}{S_V^2} \qquad \bar{P} + \frac{0.6643}{S_V}\left(\frac{\delta k_0}{k}\right)$$

A regression analysis on the 10 experimental points gives:

Slope $= 41.7875 \times 10^{-12}$
SE of slope $= 0.0839 \times 10^{-12}$
Intercept $= 1.2258 \times 10^{-8}$
Correlation coefficient, $r = 0.9999$
Therefore,

$$S_V^2 = \frac{13.106}{41.7875 \times 10^{-12}}$$

$$S_V = 0.5600 \times 10^6 \text{ m}^2/\text{m}^3$$

From the table, $S_V = 0.5664 \times 10^6 \text{ m}^2/\text{m}^3$, with a standard deviation, $s = 0.0219 \times 10^6 \text{ m}^2/\text{m}^3$.

$$\left(\frac{\delta k_0}{k}\right) = \frac{0.56 \times 10^6 \times 1.2258 \times 10^{-8}}{0.6643} = 0.0103$$

From equation (5.9), the Klinkenberg factor $B = 0.0296$ ($\text{Nm}^{-2} \times 10^4$).

9

General discussion of results, and conclusions

9.1 GENERAL DISCUSSION

Specific surface is not a unique property of a powder but increases with increasing resolution of the measuring technique.

In permeametry an envelope surface is deduced from measurements of pressure drops across a bed of powder and of flow rates through the bed. The assumptions made in deriving specific surface from these two variables are so sweeping that the derived surface is highly suspect. Nevertheless, this technique gives a useful measure of powder fineness.

The Carman–Kozeny equation applies in the viscous flow region, but as the powder becomes finer this equation breaks down. The controlling factor is the ratio of hydraulic diameter and mean free path of the gas molecules ($^{r}H/\lambda$), and as this approaches unity, 'slip flow' occurs. This effect increases with:

(a) increasing fineness, i.e. ^{r}H reduced
(b) decreasing porosity, i.e. ^{r}H reduced (see Appendix A2.1).
(c) decreasing pressure, i.e. λ increased.

To account for the enhanced flow that arises because of this breakdown a second term is introduced into the equation, and this term includes an unknown 'constant' $\left(\frac{\delta k_0}{k}\right)$, the coefficient of slip. It was considered likely that this so-called constant was a function of ($^{r}H/\lambda$), and an attempt has been made to investigate this relationship. A related parameter, the Klinkenberg factor, was also evaluated.

To carry out the investigation an electronic permeameter, operating under constant gas flow rate conditions, was designed. Inlet and outlet pressures were measured, using pressure transducers, and a particularly useful feature was the incorporation of an electronic gauge to amplify the difference between these two pressures, i.e. the pressure drop.

The first group of experiments was carried out to find out how compacting a powder to different porosities in a cylindrical tube affected both the measured specific surface areas and the coefficients of slip.

The powder used for this purpose was dicalcium phosphate dihydrate (DCP). DCP is a low surface area powder showing, from microscopical examination, an irregular shape with a crystalline appearance. It was found that the measured surfaces tended to increase with decreasing porosity: S_K increased and S_M decreased with increasing mean pressures, with a combination of the two terms giving a total specific surface of the packed powder bed.

From Fig. 6.3, where S_V has been plotted against ε, S_V tends towards a constant value as ε is reduced. The low values of S_V at high porosity are almost certainly due to a non-uniformly packed bed which leads to channelling, low resistance, and subsequently low S_V values. For $\varepsilon \leqslant 0.60$, S_V is sensibly constant.

A plot of $\left(\frac{\delta k_o}{k}\right)$ against ε (Fig. 6.4) produces a low "spurious" value for $\left(\frac{\delta k_0}{k}\right)$ at $\varepsilon = 0.60$, for the reasons mentioned above. For $\varepsilon < 0.62$, $\left(\frac{\delta k_0}{k}\right)$ slowly reduces as ε falls, but this small change $0.22 > \left(\frac{\delta k_0}{k}\right) > 0.20$ may not be significant.

Now:

$$\left(\frac{\delta k_0}{k}\right) = N\left(\frac{2-f}{f}\right) \tag{9.1}$$

where [1]

$$N = \left(\frac{3\pi}{8k}\right), \tag{9.2}$$

although the value of this constant depends upon the value of constants used in intermediate equations, that is $f = \dfrac{2N}{\left(\frac{\delta k_0}{k}\right) + N}$. As $\left(\frac{\delta k_0}{k}\right)$ falls f increases. From equations (9.1), (9.2), and the preceding discussion, this implies that the fraction of molecules (f) undergoing diffuse reflection from the powder surface increases as ε decreases, that is, f increases as pore radius decreases, which is in accord with theory.

A clearer relationship is established when the Klinkenberg factor B is plotted against ε. The result (Fig. 6.5) suggests that the Klinkenberg factor B increases linearly (with decreasing ε) if the spurious experimental point is neglected, which is in good agreement with Klinkenberg's experiments with rocks. He did not relate his work to surface area evaluation, hence he failed to state whether the increase in fineness with $\overline{P}$ was linear, exponential, or otherwise.

The relationship between $\left(\frac{\delta k_0}{k}\right)$ and pore radius was further examined by using graded spherical bronze powder. The finer the powder, the smaller the voids, since spheres pack to fairly constant porosity. Plots of $\left(\frac{\delta k_0}{k}\right)$ against S_V, (Fig. 9.1),

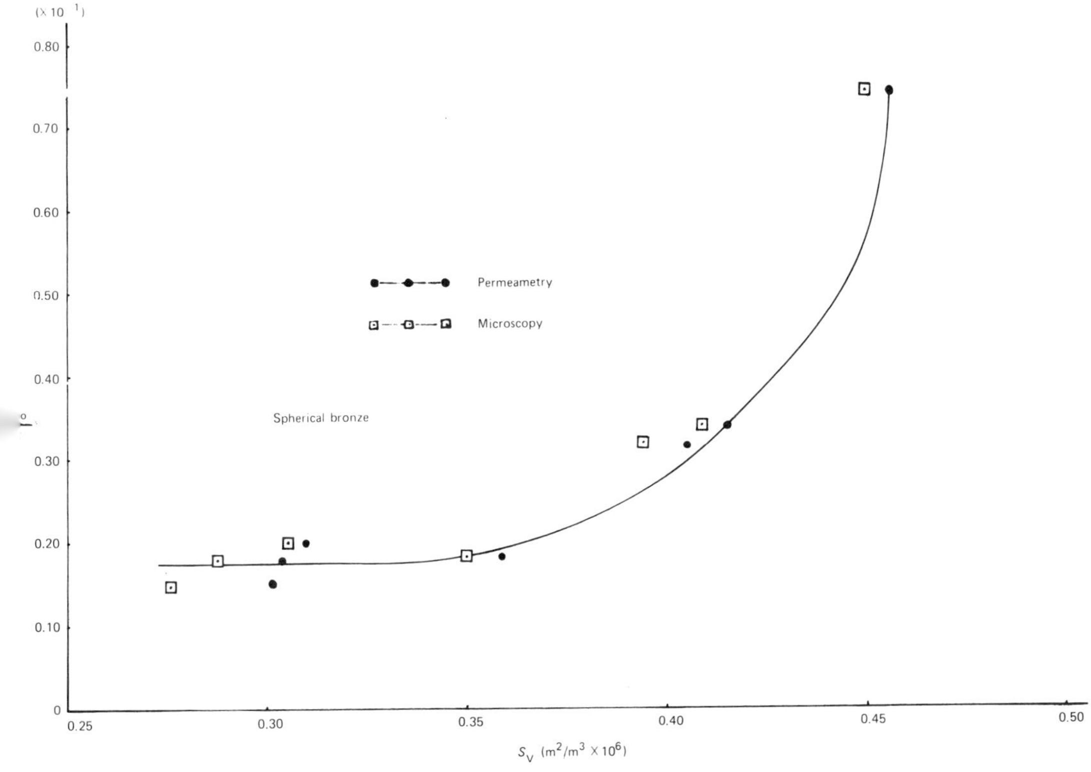

Fig. 9.1 — $\delta k_0/k$ plotted against S_V, where S_V is determined from permeametry and microscopy.

determined both by permeametry and microscopy, show that $\left(\frac{\delta k_0}{k}\right)$ increases rapidly with increasing surface (i.e. decreasing void diameter), which is in accord with the work with DCP, but remains fairly constant at approximately 0.02 for $S_V < 0.36 \times 10^6\,m^2/m^3$. At this value $d_{sv} \simeq 17\,\mu m$, and the error in neglecting the second term in the Carman-Arnell equation is approximately 5%. Therefore the effect of using incorrect values for $\left(\frac{\delta k_0}{k}\right)$ is negligible.

Table 9.1 and Fig. 9.2 are representative of the results from this study. Constantly they may be used to assess the relationship of the coefficients of slip to the mean particle size.

Table 9.1 — Variation of the permeability coefficients of slip with mean particle size (spherical bronze)

Mean particle size (μm)	$\left(\frac{\delta k_0}{k}\right)$ calculated from equation (5.4)	$\left(\frac{\delta k_0}{k}\right)$ calculated from equation (9.5)	Klinkenberg factor B, calculated from equation (5.9) ($Nm^{-2} \times 10^4$)	Klinkenberg factor B, calculated from equation (9.6) ($Nm^{-2} \times 10^4$)
14.6628	0.343×10^{-1}	0.323×10^{-1}	0.1325	0.1210
15.2400	0.322	0.310	0.1186	0.1132
17.1527	0.180	0.254	0.0612	0.0880
19.5759	0.200	0.186	0.0571	0.0560
20.8406	0.183	0.151	0.0507	0.0393
21.7077	0.105	0.126	0.0274	0.0278

From Fig. 9.2 the rate of change in the two variables, $\left(\frac{\delta k_0}{k}\right)$ and B, with the mean particle size, is the same.

Unlike the findings of other investigators (Carman and Arnell proposed a constant factor of 0.45; Klinkenberg suggested simply that the factor B would increase with fineness of the particle without finding out how it would increase), from the results presented here it is apparent that the Carman–Arnell coefficients of slip and the Klinkenberg factor B are inversely related to the mean particle size.

It is proposed here that the coefficients of slip and the Klinkenberg factor, of packed beds, should be expressed by a linear equation of the form:

$$\left(\frac{\delta k_0}{k}\right) = A_{\mathrm{i}} - Cd_{\mathrm{sv}} \tag{9.3}$$

and

$$B = a_i - cd_{\mathrm{sv}} \tag{9.4}$$

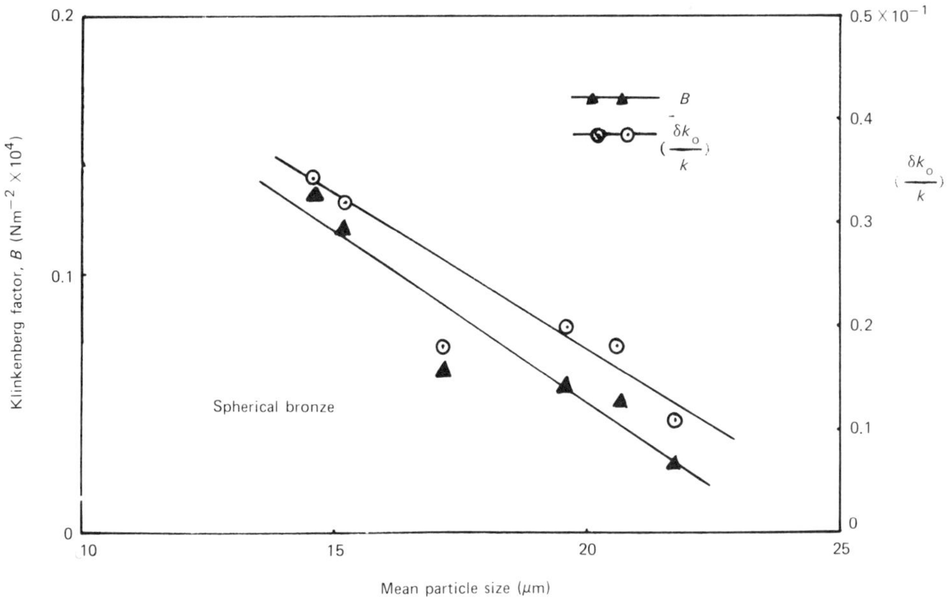

Fig. 9.2 — Variation of the Klinkenberg factor B and Carman–Arnell $\delta k_0/k$, with mean particle size.

where A_i and C depend on factors such as surface roughness of the powder, particle size range, and particle shape. d_{sv} is the surface volume mean diameter of the powder.

The usefulness of equations (9.3) and (9.4) has been correlated, and the linear equation obtained for these relations are:

$$\left(\frac{\delta k_0}{k}\right) = 0.0734 - 0.0028\, d_{sv} \tag{9.5}$$

and

$$B = 0.3144 - 0.0132\, d_{sv} \tag{9.6}$$

These equations are statistically highly significant, with $r = 0.9038$ and 0.9400 respectively, at 95% confidence level. Thus it can be seen that the linear relationships obtained by the use of a simple empirical model is a valid representation of the coefficients of slip and the Klinkenberg factor B.

Cyclone tin powders and copper powders were also examined. They gave results in accord with the earlier work, and validated the point that the numerical values of the coefficient of slip depends to a large extent on the surface characteristics of the powder type and not solely on particle size. This difference in particle fineness has been demonstrated in section 8.1.

Numerical values obtained by comparison of the surface areas by permeametry and microscopy are in fair agreement, as shown in Table 9.2 and graphically in Fig. 9.3.

Table 9.2 — Comparison of surface areas by permeametry and microscopy for spherical bronze powder

Particle size range (μm)	Permeametry S_V (m^2/m^3 $\times 10^6$)	Microscopy S_V (m^2/m^3 $\times 10^6$)	$\left(\frac{\delta k_0}{k}\right)$
13.375	0.4528	0.4486	0.745×10^{-1}
14.663	0.4138	0.4092	0.343
15.240	0.4050	0.3937	0.322
17.153	0.3596	0.3498	0.180
19.576	0.3101	0.3065	0.200
20.841	0.3023	0.2879	0.183
21.708	0.3010	0.2764	0.105

Information obtained from examination of the solid particles are shown in Figs. A1 and A2. Many factors may cause a change in surface areas, for example, larger particles masking smaller particles, or, for some reasons, particles not evenly or tightly packed. The behaviour of the coefficients of slip are of more general interest in this study than S_V, and it is not intended to discuss this further since evaluation of the surface area, although informative, does not give any deep insight into the changes occurring within the solid particles making up the bed.

The above general discussions lead to the following specific conclusions drawn from the evidence presented in this study.

9.2 CONCLUSIONS

An electronic permeameter has been designed to measure the gas flow rate and pressure drops at sub-atmospheric pressures, and the amplified results are reproducible. This is evident from the high correlations found between gas flow rate and mean pressures.

According to the Carman–Kozeny equation the measured surface area should be independent of porosity, and the observed results indicate that this is true, provided that correction is made for slip and that the bed is uniformly packed, i.e. ε is below some critical value.

At reduced pressures, the Carman–Arnell coefficient of slip may be taken as constant for a packed powder bed of irregular shape, but considerably lower than the usually accepted value of 0.45.

In contrast to porous beds, a linear relationship with a high degree of scatter was found between the Carman–Arnell coefficient of slip with the mean particle size, when using spherical bronze particles. The results also showed strong evidence that the Klinkenberg factor B increases linearly with decreasing mean particle size.

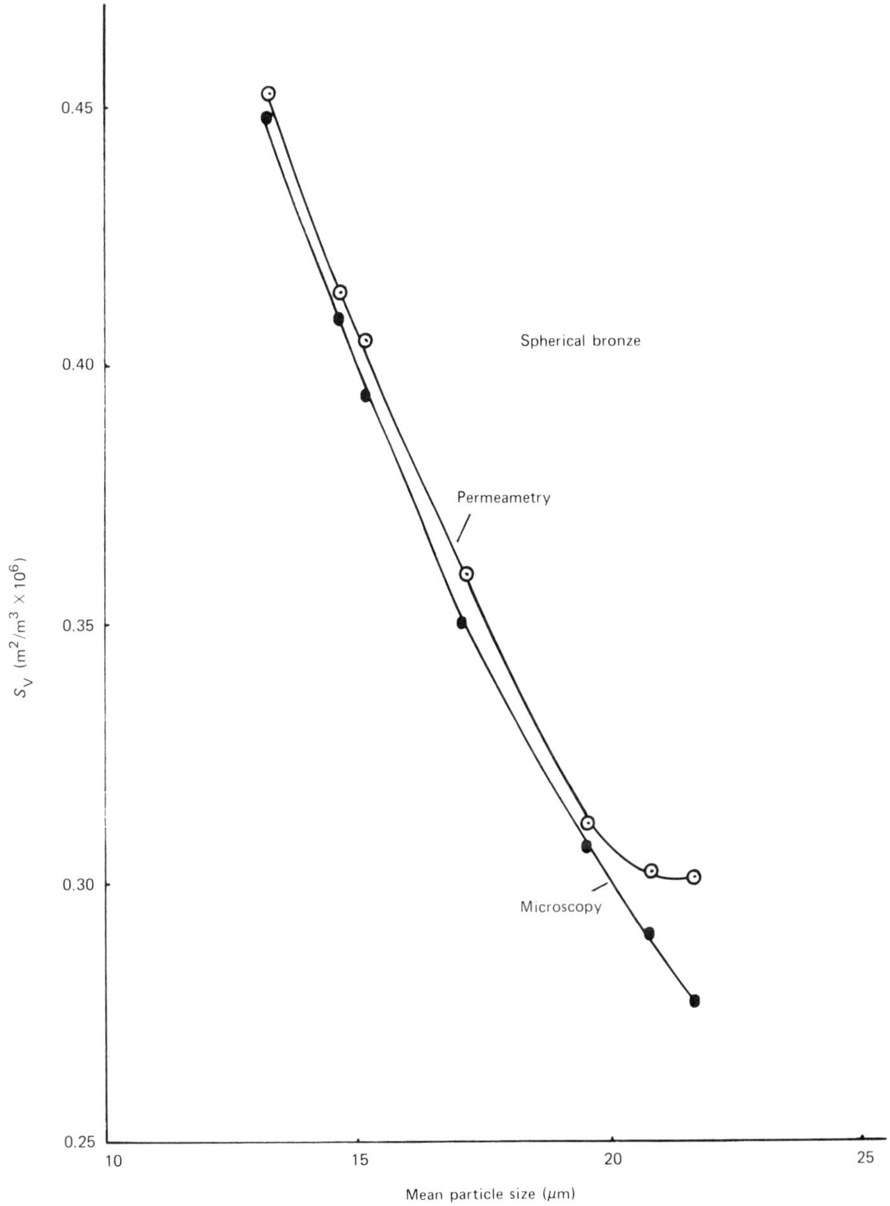

Fig. 9.3 — Comparison of surface area by permeametry and microscopy for spherical bronze.

The particle size, hence the texture of the solid surface characteristics of the bed of particles, determines both the numerical value of the Carman–Arnell coefficient of slip and the Klinkenberg factor B.

It is suggested that for a better understanding of the mechanisms of flow and their application for the evaluation of surface area of powders and the coefficients of slip, further research into measurements of pressure drops across packed beds with powders of both wider and narrower particle size range, is necessary.

REFERENCE

[1] Allen, T., *Particle size measurement*, 3rd edition, Chapman and Hall, (1981).

10

Review of some other available surface area measuring techniques

10.1 INTRODUCTION

In gas permeametry, the envelope surface of particles is deduced from measurements of pressure drops across a bed of powder and flow rates through the bed. This technique gives a useful measure of powder fineness, and it is relatively simple and rapid. However, hidden from view are the internal structure of the pores, their shape and dimensions, their volume and volume distribution, and their contribution to the surface area. By enveloping each particle of a powder in an adsorbed film, the method of gas adsorption can probe the surface irregularities and pore interiors even at the atomic level. The gas adsorption method is therefore a powerful technique for obtaining morphological information about surfaces.

10.2 GAS ADSORPTION

Gas adsorption methods depend upon the determination of the quantity of gas which, when adsorbed on the surface of the solid would completely cover the surface with a single layer of gas molecules. The difference between adsorption and absorption is that whereas in adsorption when a solid is exposed to a gas, the gas molecules impinge upon the solid and may reside upon the surface for a finite time, in absorption we refer to penetration into the solid body [1].

In gas adsorption, experimental results are usually statistical quantities calculated with the aid of mathematical models of adsorption processes. In principle, almost any gas may be used, provided that its behaviour in contact with the solid under investigation conforms with the mathematical model used to analyse the data.

Experimentally, the required data may be obtained by volumetric or gravimetric measurements. The amount adsorbed depends upon the nature of the solid adsorbent and the gas adsorbate and the pressure at which adsorption takes place. The and V_a = volume of gas adsorbed, reduced to s.t.p.†

Note:

Adsorbate = the gas being absorbed
Adsorbed volume (V_a) = the volume of gas (corrected to s.t.p.) adsorbed at the pressure P
Equilibrium adsorption pressure (P) = the pressure of gas in equilibrium with the adsorbed phase.
Hysteresis = non-reversibility of the adsorption isotherm; the desorption isotherm follows a path different from that of the adsorption isotherm.
Isotherm (adsorption isotherm) = a curve obtained by plotting the adsorbed volume against the equilibrium pressure or relative pressure.
Monolayer volume (V_m) = the volume (corrected to s.t.p.) of gaseous adsorbate which forms a monomolecular layer over the surface of the adsorbent.
Relative pressure = the ratio of the equilibrium adsorption pressure P to the saturation vapour pressure P_0.
Saturation vapour pressure (P_0) = the vapour pressure of the bulk liquefied adsorbate at the temperature of adsorption.
Specific surface; specific surface area (S) = the surface area per unit weight (S_W) or volume (S_V) of the solid usually expressed in m^2/g or m^2/m^3).

The following units are usually employed:

Mass, gramme (g)
Length, millimetre (mm)
Molecular area, Å^2
Total surface area, square metres (m^2)
Specific surface area, square metres per gram (m^2/g)
Volume, cubic centimetres (cm^3) or m^3
Temperature, °C or K
Pressure, millimetres of mercury (mmHg) subject to temperature and gravitational corrections, usually negligible, 1 mmHg = 133.3 N/m^2, 760 mmHg = 101 325 N/m^2
Standard temperature and pressure (s.t.p.), 0°C, 760 mmHg, subject to temperature and gravitational corrections, usually negligible, 1 mmHg = 133.3 N/m^2, 760 mmHg = 101 325 N/m^2.

The plot of $P/V_a(P_0 - P)$ against P/P_0 in Fig. 10.2 should be a straight line of slope $C - 1/V_m C$ and intercept $1/V_m C$, from which $V_m C$ and C and hence V_m can be calculated. The range of linearity is usually, but not always, restricted to the relative pressures 0.05 to 0.30.

10.2.2 Flow apparatus

The apparatus consists essentially of the following units:

† Standard temperature and pressure.

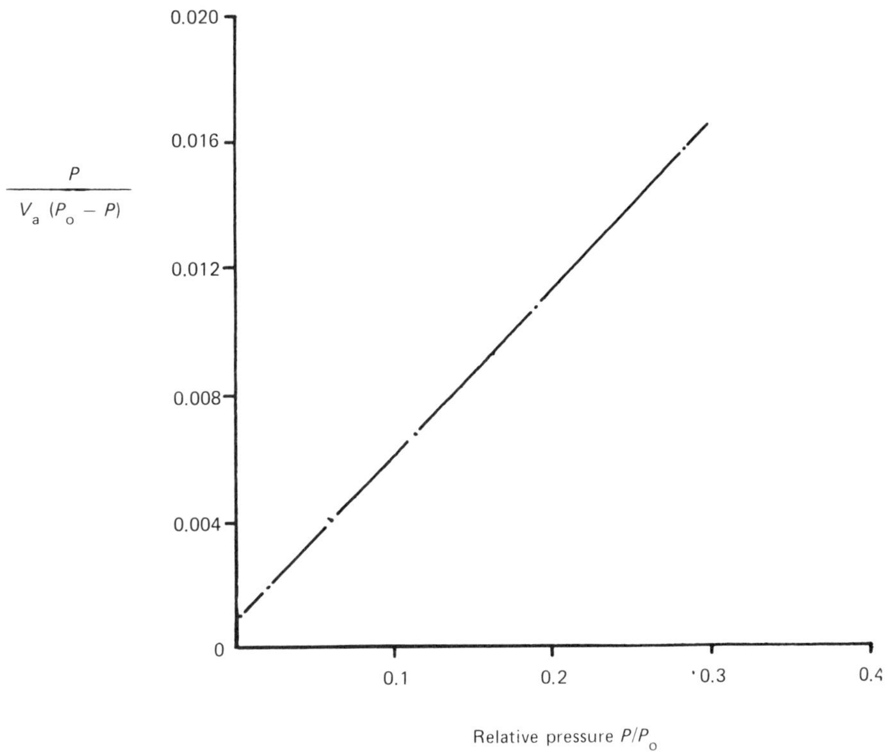

Fig. 10.2 — B.E.T. plot corresponding to adsorption isotherm, Fig. 10.1.

(a) Vacuum manifold
(b) Sample bulb
(c) Gas burette
(d) Manometer, accurate to within 0.1 mmHg
(e) Gas storage vessels.

British Standards [2] recommended the assembly shown in Fig. 10.3, alternative forms of the sample bulbs in Figs 10.4(a) and 10.4(b), and the gas burette in Fig. 10.5. Except for the volumes of the gas burette and sample bulb, the dimensions given in the figures are generally applicable to any apparatus, whatever the purpose or range of specific surface areas. The choice of size of the sample bulb and the volume of the gas burette depends upon the bulk density of the powder and the expected adsorbed volume of gas. Preferably, a size of sample should be selected so that the area available for adsorption is between 20 m^2 and 200 m^2. The volume of the sample bulb should then be chosen so that the selected size of sample occupies two-thirds of the volume of the bulb. The volume of the gas burette bulbs is then chosen so that they can deliver in a few increments the appropriate volume of gas expected to be

10.2.3 Vacuum manifold

The vacuum manifold consists of a rotary pump followed by a two-stage oil or mercury diffusion pump, connected through cold traps cooled with solid carbon dioxide or liquid nitrogen to a main manifold of bore about 2 cm. A McLeod or similar gauge covering the pressure range of 10^{-5} mmHg to 10^{-2} mmHg should be attached to this manifold for initial testing for leaks and outgassing. A suitable layout of apparatus was shown in Fig. 10.3.

10.2.4 Sample bulbs

Sample bulbs of various sizes may be used, depending on the expected specific surface area of the powder. The bulbs in common use have capacities of 5 cm^3, and designs are shown in Fig. 10.4. The narrow tube connecting the bottom of the bulb to the main stem in Fig. 10.4b serves to aid the removal of the gas from the powder during outgassing by reducing the tendency of the powder to become entrained.

10.2.5 Mercury manometer

The mercury manometer is connected at the lower end to a mercury reservoir. In the design shown in Fig. 10.3 the manometer is integral with the gas burette.

10.2.6 Gas burette

The gas burette should be water jacketed, and have several calibrated bulbs so that various volumes may be used. It is connected at the lower end to a mercury reservoir. In the design shown in Fig. 10.5 the gas burette has three bulbs, each of 10–11 cm^3 capacity, and is provided with calibration marks below each of the bulbs. A constriction in the stem prevents rapid movement of the mercury. In the design of apparatus shown in Fig. 10.3 the gas burette and the mercury manometer form the two arms of a constant volume manometer. This arrangement helps to reduce the volume of dead space in the adsorption system. It is generally found that tap water remains at a sufficiently constant temperature (within 1 deg C) during the day to be used in the water jacket of the gas burette. It is recommended [2] that a thermometer should be incorporated in the water feed line adjacent to the burette jacket, and that the temperature of the water should be measured several times during the day.

10.2.7 Rough vacuum line

The rough vacuum line for the mercury reservoirs may consist of a vacuum reservoir of 1000 cm^3 capacity connected through a vacuum tap to a rotary pump, and to the mercury reservoirs.

10.2.8 Ancillary apparatus

Dewar vessels of various sizes are required, and also storage facilities for liquid nitrogen. A Dewar vessel containing hot oil may be used to heat the sample during the vacuum outgassing operation, but for longer periods of outgassing a small electric furnace fitting around the sample bulb may be preferred. A Tesla coil leak detector is required for the detection of pinhole leaks in the vacuum system during construction and use, and other forms of leak detection apparatus are of considerable value [3]. A

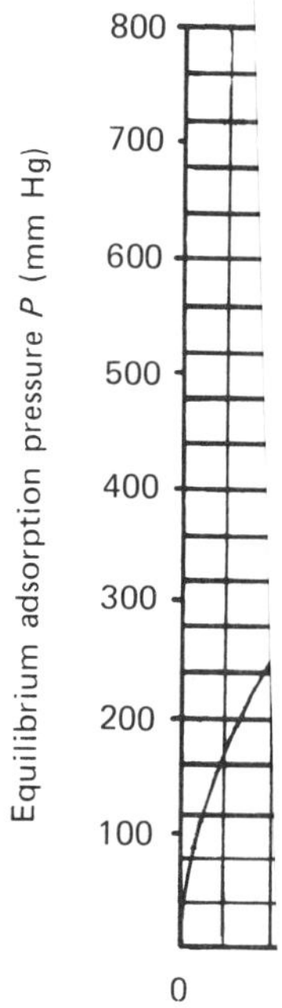

the sample bulb. It
determination; thi
taken to ensure th

10.2.12.3 Outgas
Connect the tube t
in the stem of the
carried out of the b
again while the sam
furnace. Evacuate t
temperature and ti
For measureme
adsorbent is essent
temperature, to re
the surface; a vacuu
continued until the
and 10^{-5} mmHg an
pumping line and co
or liquid nitrogen) s
diffusion of mercury
of water vapour. The

vapour pressure thermometer may be provided to measure the temperature of the liquid nitrogen in the Dewar vessel surrounding the sample bulb (see section 10.2.10).

10.2.9 Construction of the apparatus

Some familiarity with the principles of construction and operation of high-vacuum equipment is assumed [2, 4]. In the construction of the apparatus it is essential to use only high-grade vacuum taps, which must be replaced when any sign of wear is shown. It is impossible to prevent traces of powders from being carried up into the vacuum lines, and it is, therefore, advisable to construct the apparatus so that all parts are easily accessible for cleaning and replacement. The connections within the dashed rectangle in Fig. 10.3 should be made with 2 mm bore capillary tubing, and kept as short as possible to minimize dead space. Other portions of the apparatus should be made with wider bore glass tubing (7–8 mm) to obtain adequate pumping speeds.

The sample bulb, gas burette, and manometer may be duplicated to enable more than one determination to be carried out at the same time.

The volume of the gas burette, from the taps to the various calibration marks, is determined, when the apparatus is being set up, by calibration with mercury.

10.2.10 Vapour pressure thermometer and corrections for deviations from ideal gas laws

To calculate the surface area it is necessary to know the saturation vapour pressure of nitrogen at the temperature of the experiment. The saturation vapour pressure is measured most conveniently with the aid of a vapour pressure probe in the liquid nitrogen bath; this consists merely of a hollow tube containing pure condensed nitrogen or oxygen and connected to a suitable manometer. Values of the saturation pressures of nitrogen and oxygen at various temperatures are given in Table 10.2 (page 190).

Each experimental point in the determination involves calculating the volume of nitrogen unadsorbed at liquid nitrogen temperatures and should allow for deviations from the ideal gas laws. The correction to be applied for nitrogen at −196°C is given in Fig. 10.6.

10.2.11 Preparation

Grease the vacuum taps with a suitable inert lubricant, depending on the ambient temperature. Fill the cold traps with solid carbon dioxide/ethanol or with liquid nitrogen, and evacuate the apparatus by using the rotary pump. If the vacuum as indicated on the McLeod gauge appears to be satisfactory, switch on the diffusion pump. After a period of 10–15 minutes, a vacuum of better than 10^{-5} mmHg should be reached and maintained while pumping, otherwise it is necessary to find and repair the leaks. When the apparatus has not been in use for some weeks it is a wise precaution to regrease all the taps (except those of the gas reservoirs).

If the apparatus is vacuum-tight, fill the nitrogen reservoir with dried nitrogen of high purity (not less than 99.9%) to a pressure of about 600 mmHg. Then seal on the bulb of helium, re-evacuate the connecting lines, and test the seal of leaks. Finally,

Different powders will require periods of outgassing ranging from 15 minutes to several hours. To minimize the time required for outgassing, the temperature chosen should be the maximum compatible with avoidance of sintering, decomposition, or other alteration of the nature of the surface. Judgement as to this temperature is to some extent a matter of experience and of trial and error, guided by general knowledge of the properties of the substance. A solid of high melting point and stable polymorphic modification (e.g. alpha-alumina) could be safely outgassed at 1000°C, but an 'active' oxide prepared by precipitation from solution (e.g. ferric oxide) may sinter appreciably at 100°C. In glass apparatus, the safe working temperature cannot exceed the softening point of glass, viz. 400°C.

For nitrogen adsorption isotherms, an outgassing temperature of 110–130°C is suitable for many oxides, carbonates, sulphates, and other solids such as are encountered amongst pigments, catalysts, and other industrial powders, but lower temperatures may be needed for certain organic compounds. A preliminary drying of the substance in an oven at, say, 110°C, will also accelerate the subsequent outgassing process.

10.2.12.4 Dead space determination with helium

Introduce helium (which is not adsorbed by the powder) into the gas burette, which is levelled to the calibration mark, and record the gas pressure (H_1) to at least 0.1 mmHg. Then allow the helium to expand into the sample bulb, and, after levelling the burette, again record the pressure (H_2). Finally, immerse the sample bulb in liquid nitrogen to a standard depth, indicated by a wire level probe attached to the sample bulb manifold, and after again levelling the burette, record the final pressure (H_3). Then allow the sample bulb to warm up while the helium is being pumped off, until a hard vacuum is again obtained.

10.2.12.5 Nitrogen adsorption

Again cool the sample bulb in liquid nitrogen, the level being kept constant by using the wire gauge, and introduce nitrogen into the gas burette from the reservoir. After levelling the burette and recording the pressure (P_1), admit the nitrogen to the sample bulb. When equilibrium is reached (generally in 10–20 minutes), again level the burette and record the equilibrium pressure (P_2). Repeat several times the cycle of introduction of nitrogen into the burette, measurement, adsorption on the sample, and final measurement of equilibrium pressure, each repetition providing a point on the adsorption isotherm. The time taken to attain equilibrium, and the number of points required, will depend upon the surface area and nature of the sample.

In the selection of the size of the sample bulb, the volume of the gas burette (i.e. which calibration mark to use), and the amount of nitrogen to introduce at each step, it is necessary to use existing knowledge and experience of the particular type of material being examined. In Table 10.1 some guidance was given on the total volume of nitrogen (at s.t.p.) to be introduced into the system with samples of various specific surface areas, during experimental determinations.

It is customary to have at least four points with equilibrium relative pressures (P/P_0) in the range 0.05–0.30.

When the adsorption run is completed, allow the sample to warm up to room

190

with the
plunger i
The a

10.2.12
10.2.12.1
The samp
to 200 m

10.2.12.2
Place the
size that i
to dry the

temperature while the desorbed gas is pumped off. Admit *dry* air, and then weigh the adsorption tube to determine the amount of sample which was used, care being taken to ensure that the joint and tube are free of grease.

It is convenient to record the various measurements on a printed form as they are made. Forms suitable for simplified and fully corrected methods of calculation are shown below, in which the same set of experimental data has been used for the two samples. The experimental measurements made are indicated with an asterisk.

10.2.13 Calculation

10.2.13.1 Method of calculation

Since many of the arithmetical operations involve small differences between larger numbers, it is advisable to use a calculating machine. Not more than four decimal places need to be used in the calculations, and the final result should be suitably rounded off. The use of four significant figures must not be taken to imply accuracy of this order. The figures shown are those actually obtained during calculation, and are retained to avoid errors which might occur if the values of the slope and intercept were small [2].

10.2.13.2 Measurements

- H_1 Initial pressure of helium in gas burette.
- H_2 Pressure of helium in gas burette and bulb at temperature T (not required for simplified calculation).
- H_3 Pressure of helium in gas burette with bulb at temperature T_N.
- T Water jacket temperature.
- T_N Liquid nitrogen temperature.
- P_0 Saturated vapour pressure of nitrogen at T_N.
- P_1 Initial pressure of nitrogen in burette.
- P_2 Equilibrium pressure of nitrogen in burette and bulb.
- v Volume of burette (determined when the apparatus is being set up).
- W Weight of sample.

10.2.13.3 Values calculated

Two methods of calculation are given. In the simplified method, no correction is made for non-ideality of nitrogen at low temperatures, whilst in the refined method the correction is made. For most routine surface area measurements the error introduced by the use of the simplified nethod is negligible; for more accurate work, particularly when measurements are made at high values of P/P_0, as for pore volume measurements, the method with full corrections should be used [2].

When plotting $P/(P_0 - P)V_2$ against P/P_0, it is advisable to use large-scale graph paper. At least three and preferably four points should lie on a straight line, between the relative pressures of 0.05–0.30. If the points do not fall on a straight line the data should be rejected and a fresh determination carried out, since drawing the 'best' line through a number of points can lead to ambiguities. If a straight line is not obtained with any adsorbate, then the B.E.T. method is not applicable.

It has been found that with careful attention to detail, a coefficient of variation of about 2% can be obtained in repeat determinations of the specific surface area of different samples of the same material [2].

Surface area computation form for simplified routine calculations

Date	1 July 1988	Outgassing time	1 h
Sample	Phthalocyanine blue	Outgassing temperature	130°C
Reference	–	Outgassing pressure	2×10^{-5} mmHg

*Weight of bulb	= 6.9248 g	$*T$	= 291.5°K	$v = 23.42\ \text{cm}^3$
*Weight of bulb + sample	= 8.2177 g	$*T_N$	= 77.3°K	$A = \frac{v}{T} \times \frac{273.2}{760} = 0.02888$
Weight of sample, W	= 1.2929 g	$*H_1$	= 136.0 mmHg	
$*P_C = 760$ mmHg		$*H_3$	= 63. 0 mmHg	$u = \frac{H_1 - H_3}{H_3} = 1.1587$
				$Q = A \times u = 0.03346$

Slope (s) = 0.052 18   Intercept (i) = 0.000 95 $S_w = \frac{4.35}{W(s+i)} = 64.4\ \text{m}^2/\text{g}$

D	E	G	H	J	K	M	X	N	Y
		$D - E$	$A.G$	ΣH	$E.Q$	$J - K$	E/P_0	$(1 - X)M$	X/N
$*P_1$	$*P_2$	ΔP	ΔV	$\Sigma \Delta V$	$P_2.Q$	V_a	P_2/P_0	$(1 - P/P_0)V_a$	$\frac{P_3}{(P_0 - P_3)V_a}$
192.3	2.1	190.2	5.4930	5.4930	0.0703	5.4227	0.0028	5.4075	0.00052
219.2	15.3	203.9	5.8886	11.3816	0.5120	10.8696	0.0201	10.6511	0.00189
223.9	43.9	180.0	5.1984	16.5800	1.4690	15.1110	0.0578	14.2376	0.00406
247.9	85.0	162.9	4.7046	21.2846	2.8444	18.4402	0.1118	16.3786	0.00683
323.8	143.2	180.6	5.2157	26.5003	4.7919	21.7084	0.1884	17.6185	0.01069
412.9	213.9	199.0	5.7471	32.2474	7.1577	25.0897	0.2814	18.0295	0.01561

Surface area computation form with full corrections

Date	1 July 1988	Outgassing time	1 h
Sample	Phthalocyanine blue	Outgassing temperature	130°C
Reference	–	Outgassing pressure	2×10^{-5} mmHg

*Weight of bulb	= 6.9248 g	*T	= 291.5°K	$v = 23.42\ cm^3$
*Weight of bulb + sample	= 8.2177 g	*T_n	= 77.3°K	$A = \frac{v}{T} \times \frac{273.2}{760} = 0.02888$
Weight of sample	= 1.2929 g	w	$= \frac{v(H_1 - H_2)}{H_2} = 9.3820$	$B = \frac{z}{T} \times \frac{273.2}{760} = 0.003\,669$
*P_0 = 760 mmHg		y	$= \frac{(v + w)(H_2 - H_3)}{H_3} \frac{T_N}{(T - T_N)}$ $= 6.4071$	$C = \frac{v}{T_N} \times \frac{273.2}{760} = 0.029\,80$
*H_1 = 136.0 mmHg *H_2 = 97.1 mmHg *H_3 = 63.0 mmHg		z	$w - y = 2.9749$	

Slope (s) = 0.052 37	Intercept (i) = 0.000 93	$S_w = \frac{4.35}{W(s + i)} = 64.3\ m^2/g$

D	E	F	G $D - E$	H $A.G$	J ΣH	K $B.E$	L $C.F$	M $J - (K + L)$	X E/P_0	N $(1 - X)M$	Y X/N
*P_1	*P_2	P'_3	ΔP	ΔV	$P_3\left(\frac{z}{T}\right)$ $\Sigma \Delta V$	$P'_2\left(\frac{y}{T_N}\right)$ $\times \frac{273.2}{760}$	$\times \frac{273.2}{760}$	V_a	P_1/P_0	$(1 - P_2/P_0)V_a$	$\frac{P^2}{(P_0 - P_2)V_a}$
192.3	2.1	2.1	190.2	5.4930	5.4930	0.0077	0.0626	5.4227	0.0028	5.4075	0.000 52
219.2	15.3	15.4	203.9	5.8886	11.3816	0.0561	0.4589	10.8666	0.0201	10.6482	0.001 89
223.9	43.9	44.2	180.0	5.1984	16.5800	0.1611	1.3172	15.1017	0.0578	14.2288	0.004 06
247.9	85.0	85.7	162.9	4.7046	21.2846	0.3119	2.5539	18.4188	0.1118	16.3596	0.006 83
323.8	143.2	144.6	180.6	5.2157	26.5003	0.5254	4.3091	21.6658	0.1884	17.5840	0.010 71
412.9	213.9	216.9	199.0	5.7471	32.2474	0.7848	6.4636	24.9990	0.2314	17.9643	0.015 66

10.2.13.4 Simplified method

Basic factors

$$A = \frac{v}{T} \times \frac{273.2}{760}$$

$$u = \frac{H_1 - H_2}{H_3}$$

$$Q = A \times u$$

Calculated values	Operation (symbols refer to columns on form)
$\Delta P = P_1 - P_2$	$G = D - E$
$\Delta V = A \times \Delta P$	$H = A \times G$
$\Sigma\Delta V$ = cumulative ΔV	$J = \Sigma H$
$P_2 Q$ = dead space correction	$K = E \times Q$
V_a = volume of nitrogen adsorbed	$M = J - K$
P/P_0 = relative pressure, X	$X = E/P_0$
$(1 - P/P_0)V_a$ = part of B.E.T. function	$N = (1 - X)M$
	$Y = X/N$

$$\frac{P}{(P_0 - P)V_a} = \text{B.E.T. function}$$

$$= \frac{X}{(1 - X)V_a}$$

10.2.13.5 Method with full corrections

Basic factors	Calculated values	Operation (symbols refer to columns on form)
$w = \frac{v(H_1 - H_2)}{H_2}$	P_2' = pressure corrected for non-ideality	$F = E$ + correction from Fig. 10.6
$y = \frac{(v + w)(H_2 - H_3)}{H_3} \times \frac{T_N}{(T - T_N)}$	$\Delta P = P_1 - P_2$	$G = D - E$
	$\Delta V = A \times \Delta P$	$H = A \times G$
$z = w - y$	$\Sigma\Delta V$	$J = \Sigma H$
$A = \frac{v}{T} \times \frac{273.2}{760}$	$P_2\left(\frac{z}{T}\right)\frac{273.2}{760}$ = Dead space corrections	$K = B \times E$
$B = \frac{z}{T} \times \frac{273.2}{760}$	$P_2'\left(\frac{y}{T_N}\right)\frac{273.2}{760}$	$L = C \times F$
$C = \frac{y}{T_N} \times \frac{273.2}{760}$	V_a = volume of nitrogen adsorbed	$M = J - (K + L)$
	P/P_0 = Relative pressure	$X = E/P_0$
	$(1 - P/P_0)\ V_a$ = Part of B.E.T. function	$N = (1 - X)M$

$$\frac{p}{(P_0 - P)V_a} = \text{B.E.T. function} \qquad Y = X/N$$

$$= \frac{X}{(1-X)V_a}$$

The two factors mentioned in section 10.2.10 make significant differences when working with pressures approaching saturation, that is, for pore size distribution measurements, but in the limiting range of relative pressures used for surface area measurement, the differences are not as significant.

10.2.13.6 Values of the cross-sectional area for various adsorbates

Table 10.3 lists values of the cross-sectional area, σ, of various adsorbates for use

Table 10.3 — Values of the cross-sectional area for various adsorbates

Adsorbate	Temperature of adsorption	Cross-sectional area σ	
		Customary value	Quoted values
	°C	Å^2	Å^2
Nitrogen	−196	16.2	13.6 − 16.2
Argon	−196	13.7	13.7 − 16.7
Krypton	−196	19.5	17.7 − 22
Xenon	−196	—	18 − 27
Oxygen	−183	14.1	13 − 17
Nitrous oxide	−78.5	16.8	16.8 − 20
Benzene	25	40	30 − 50

with the B.E.T. method of specific surface area measurement. In addition to the customary value, a range of values is also quoted for each adsorbate. In certain cases, particularly where the point of inflection of the adsorption isotherm is not well defined, it may be necessary to use a figure other than the customary value in order to bring measurements made when using different adsorbates into agreement.

10.2.13.7 Comparative tests of the B.E.T. method

To assess reproducibility, tests were made on samples of powders by a number of laboratories. Results are shown in Table 10.4.

10.2.13.8 Worked example [1]

The predetermined cumulative volumes of the gas burettes by filling with mercury and weighing are:

$$V_1 = 9.19 \text{ ml};\ V_2 = 17.56 \text{ ml};\ V_3 = 26.85 \text{ ml}.$$

Table 10.4 — Specific surface of powders determined by six laboratories using the B.E.T. method
All results in square metres per gram

Laboratory	Alumina R52		Titanium dioxide R53		Non-porous silica R54		Porous silica R55	
A	1.5							
B	1.63	1.63	8.59	8.80	293	286	293	304
	1.61	1.63	8.76	8.40	280		296	298
	1.60	1.58	8.79	8.79			291	307
							296	
C			9.60	9.40	294	299		
D	1.67	1.64	8.36	8.20	305	305		
E	1.40		8.00	8.50	296		316	
F	1.50	1.78	8.30	8.50	290		310	
	1.80		8.72	8.54				

The sample is weighed in the sample tube and fitted to the apparatus, and the system is then evacuated. The heating furnace is then placed round the sample tube and degassing proceeds under vacuum. The tube is then immersed in liquid nitrogen, and helium is drawn into the burette, and the pressure is noted. The helium is then allowed to expand into the sample tube, and the pressure is again noted. The volume of gas in the burette at s.t.p. is calculated as follows, using the experimental data presented in Table 10.5.

$$V_0 = \frac{P}{P_0}\frac{T_0}{T}V = \frac{163.5}{760}\frac{273}{285}9.19 = 1.892 \text{ ml}$$

After expansion:

$$V_0 = \frac{46.0}{760}\frac{273}{285}9.19 = 0.533 \text{ ml}$$

Thus, the sample tube contains (1.892–0.533) ml of gas under a pressure of 46.0 mmHg.

$$\text{The free space factor} = \frac{1.892 - 0.533}{46.0} \text{ ml mm}^{-1} \text{ pressure}$$

$$= 0.029\,56 \text{ ml mm}^{-1}$$

Table 10.5 — Observed and calculated data for sample [1]

Sample	Gas	Burette volume	Pressure (mm)	Volume of the gas in the burette (m l)	Volume of the gas in the sample tube (ml)	Volume of the gas in the free space (ml)	Volume of the total gas (ml)	V_{ad}	$V_{ad/s}$
Off	He	1	163.5	1.892		1.359	Free space factor		
On	He	1	46.0	0.533		1.359	$\frac{1.359}{46} = 0.02956 \frac{ml}{mm}$		
Off	N_2	3	142.5	4.830			4.830		
On	N_2	1	27.0	0.312	4.518				
Off	N_2	3	159.0	5.380			9.898		
On	N_2	3	86.5	2.930	6.968	2.575		4.393	0.431
On	N_2	2	103.5	2.290	7.608	3.090		4.518	0.443
On	N_2	1	125.0	1.448	8.450	3.722		4.728	0.464
Off	N_2	3	174.5	5.920			14.370		
On	N_2	3	147.0	4.980	9.390	4.390		5.000	0.490
On	N_2	1	211.5	2.445	11.925	6.350		5.575	0.546

Sample: Cement, Hoogovern cement (cemy) Klasse — B; *Weight*: 10.2028 g; *Degassing*: 300°C, 3 h; *Temperature of the water jacket*: 12°C; *Burette used*: B.

After the free-space factor determination, the helium is pumped out of the system and nitrogen is admitted into the burette. Time is allowed for the gas to reach equilibrium temperature, and the pressure is then noted. The gas is then admitted to the sample tube and the pressure is again noted. All volumes are reduced to s.t.p.

(a) Volume of nitrogen admitted $= \frac{142.5}{760} \cdot \frac{273}{285} \cdot 26.85 = 4.830$ ml

Volume remaining in burette after expansion

$$\frac{27.0}{760} \cdot \frac{273}{285} \cdot 9.19 = 0.312 \text{ ml}$$

Volume in free space $= 0.02956 \times 27.0 = 0.789$ ml
Volume adsorbed $= 4.830 - (0.312 + 0.798) = 3.720$ ml.

(b) More nitrogen admitted to the burette $= 5.380$ ml
Nitrogen already in the sample tube $= 4.518$ ml
Total nitrogen in the system $= 9.898$ ml.

The calculations are carried out cumulatively in steps in the same manner. The observed and calculated data are given in Table 10.6.

Table 10.6 — B.E.T. and HJr data

P	V ml g^{-1}	P/P_0	$P/V(P_0 - P)$	$1/V^2P$ per g
86.5	0.431	0.114	0.298	5.38
103.5	0.443	0.136	0.356	5.09
125.0	0.464	0.165	0.424	4.63
137.0	0.490	0.194	0.489	4.15
211.5	0.546	0.278	0.706	3.36

B.E.T. plot
Slope $= 2.47$
Intercept $= 0.02$
$1/V_m = 2.49$
$V_m' = 0.405$ ml
$S_w = 1.76$ m^2 g^{-1}

HJr plot
Slope $= 0.196$
$S_w = 1.79$ m^2 g^{-1}

A B.E.T. plot of $P/V(P_0 - P)$ against P/P_0 gives a value of V_m from which the surface is determined from the equation:

$$S_w = 4.35\ V_m\ \text{m}^2\ \text{g}^{-1}$$

A Harkins and Jura plot of P/P_0 against $1/V^2$ on semi-log paper has a negative slope A from which the surface area is determined using:

$$S_w = 4.06\,A \text{ m}^2 \text{ g}^{-1}$$

Typical B.E.T. and HJr plots are shown in Figs 10.7 and 10.8, as well as the tabulated data in Table 10.5.

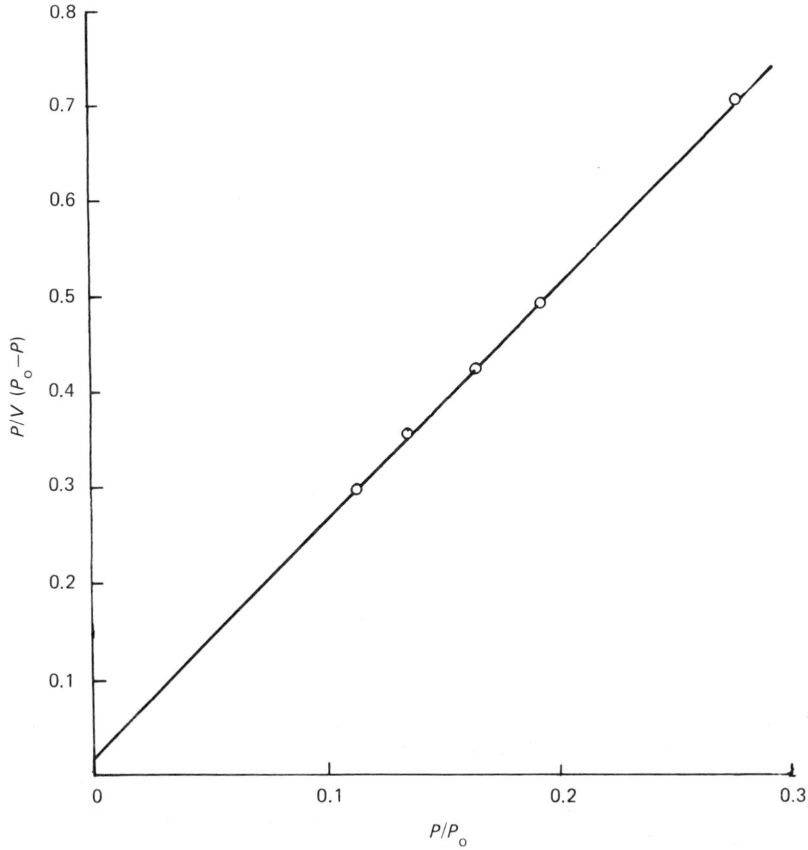

Fig. 10.7 — B.E.T. plot.

10.2.14 Theoretical background of adsorption

Depending on the nature of the relevant forces, adsorption processes may be classified as physical or chemical. Physical adsorption, also known as van der Waals adsorption, is caused by molecular interaction forces. Here, the formation of a

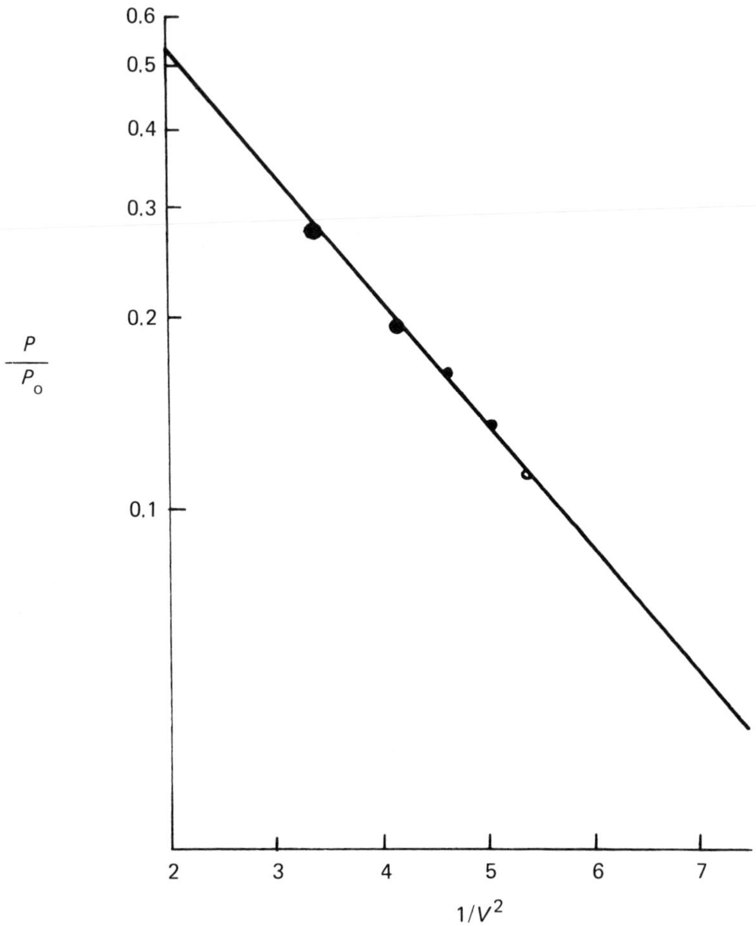

Fig. 10.8 — HJr plot.

physically adsorbed layer may be likened to the condensation of a vapour to form a liquid. Physical adsorption is therefore of importance only at temperatures below the critical temperature for the gas. The heat of adsorption is not only of the same order of magnitude as that of liquefaction, but physically adsorbed layers behave in many respects like two-dimensional liquids [1, 2, 3].

Chemical adsorption, also termed chemisorption, involves some degree of specific chemical interaction between the adsorbate and the adsorbent, hence the energies of adsorption may be quite large and comparable to those of chemical bond formation.

As physical adsorption is the result of relatively weak interaction between solids and gases, almost all the gas adsorbed can be removed by evacuation at the same temperature at which it was adsorbed. The quantity of physically adsorbed gas at a given pressure increases with decreasing temperature. Consequently, most adsorption measurements for the purpose of determining surface areas are made at low temperatures. Gas that is chemisorbed may be difficult to remove merely by reducing the pressure, and when it does occur, it may be accompanied by chemical changes [1, 2, 3].

10.2.14.1 The Langmuir isotherm

Langmuir [5] proposed the first equation relating the quantity of adsorbed gas to the equilbrium pressure of the gas. In his model, adsorption is limited to a monolayer, and his equation has limited applicability to physical adsorption with wider application to chemical adsorption and the adsorption of solute, including dye molecules, from solution.

Langmuir's method was to equate the number of molecules evaporating from the surface with the number condensing on the surface. Since surface forces are short-range, only molecules striking a bare surface are adsorbed; molecules striking a previously adsorbed molecule are elastically reflected back into the gas phase.

The number of molecules striking a unit area in unit time is given from the kinetic theory as:

$$Z = \frac{P}{\sqrt{(2\pi mkT)}} \tag{10.4}$$

where k is Boltzmann's constant,
m is the mass of a molecule,
P is the pressure, and
T is the absolute temperature.

The number evaporating n depends upon the energy binding the molecules to the surface. If Q is the energy evolved when a molecule is adsorbed and τ_0 the molecular vibration time, the residence time is given by:

$$\tau = \tau_0 \exp\left(+\frac{Q}{RT}\right)$$

where τ_0 is of the order of 10^{-13} s and, for physical adsorption, Q has a value [6] between about 6 and 40 J mol^{-1}. The number of molecules evaporating from unit area per second is therefore given by $(1/\tau)$.

If the fraction of surface covered with adsorbed molecules at pressure P is θ, then the rate of adsorption on an area $(1-\theta)$ equals the rate of desorption from an area $(1-\theta)$.

$$\frac{p}{\sqrt{(2\pi mkT)}} \cdot \alpha_0(1-\theta) = \left[\frac{1}{\tau_0}\exp\left(-\frac{Q}{RT}\right)\right]\theta$$

where α_0, the condensation coefficient, is the ratio of elastic collisions with the bare surface (α_0 tends to unity under conditions of dynamic equilibrium).

If the gas adsorbed at pressure P is V, and the volume required to form a monolayer is V_m, then:

$$\theta = \frac{V}{V_m} = \frac{bP}{1 + bP}$$

where

$$b = \frac{\alpha_0 \tau_0}{\surd(2\pi mkT)} \exp\left[\frac{Q}{RT}\right] . \tag{10.5}$$

The equation is usually written in the form:

$$\frac{P}{V} = \frac{1}{bV_m} + \frac{P}{V_m} . \tag{10.6}$$

A plot of P/V against P yields the monolayer capacity, V_m. To relate this to surface area it is necessary to know the area occupied by one molecule, σ. Surface area from the monolayer capacity can then be calculated using:

$$S_w = \frac{N\sigma V_m}{M_v} \tag{10.7}$$

where S_w = specific surface area in $m^2\ g^{-1}$
N = Avagadro number, 6.023×10^{23} molecules/g mol
σ = area occupied by one adsorbate molecule, usually taken as $16.2 \times 10^{-20}\ m^2$, for nitrogen
V_m = monolayer capacity cm^3 per gram of solid
M_v = gram molecular volume = 22 410 cm^3/g mol

$$S_w = \frac{(6.023 \times 10^{23})\ (16.2 \times 10^{-20})}{(22\ 410)} V_m$$
$$= 4.35\ V_m \text{ for nitrogen at liquid nitrogen temperature.} \tag{10.8}$$

One basic assumption in deriving the Langmuir equation is that the energy of adsorption Q is constant, thus making b constant. This, in turn, implies that the surface is entirely uniform, although this is not supported by experimental evidence.

It is usually assumed that molecules are adsorbed as whole (discrete entities) on to definite points of attachment on the surface, and each point can accommodate only one adsorbed molecule. If adsorption takes place first on high-energy-level sites, this must be compensated for by lateral interaction increasing the energy of adsorption of the molecules adsorbed later. Alternatively, if there are no high-energy-level sites, the energies of the adsorbed molecules are independent of the presence or absence of other adsorbed molecules on neighbouring points of attachment.

From equation (10.6), at low pressures $(1 + bP)$ tends to unity, bP may be neglected and Henry's law is obeyed [7]:

$$V = bPV_m \tag{10.9}$$

If, in deriving the Langmuir equation, it is assumed that adsorption is not localized, the rate of condensation is proportional to the total surface and not the bare surface, hence:

$$\frac{P}{\sqrt{(2\pi mkT)}}\alpha_0 = \left[\frac{1}{\tau_0}\exp\left(-\frac{Q}{RT}\right)\right]\theta\ ,$$

i.e. Henry's Law is obeyed at all pressures.

At high pressures bP is large compared with unity and $V = V_m$, therefore the isotherm approaches saturation. If these requirements break down and Q is a linear function of θ, the following equation develops [8]:

$$mRT \ln P/P_0 = 1 - V/V_m \tag{10.10}$$

V is plotted against log P, intersecting the ordinate at $V = V_m$. This equation has been applied to the adsorption of carbon dioxide on to alumina at 22°C, $(80 < P < 400$ mmHg) and $P_0 = 4500$ mmHg. V was found to equal V_m when $x = 0.10$, the same relative pressure as found for adsorption of carbon dioxide at -78°C and the same value for V_m.

If Q is a logarithmic function of θ, the Freundlich equation develops:

$$mRT \ln P/P_0 = \ln \theta\ . \tag{10.11}$$

This has been applied to the adsorption of hydrogen on metallic tungsten [9]. Sips [10] considered a combination of the Langmuir and Freundlich isotherms:

$$\theta = \frac{AP^{1/n}}{1 + AP^{1/n}} \tag{10.12}$$

which has the proper limits for monolayer adsorption but reduces to equation (10.11) at low pressures.

In a later paper, Sips [11] revised his theory and arrived at:

$$\theta = \left(\frac{P}{a + P}\right)^c \quad \text{where } a \text{ and } c \text{ are constants.}$$

The Langmuir equation has also been derived from thermodynamic [12] and statistical [13] considerations.

10.2.14.2 B.E.T. isotherm

The most important step in the study of adsorption came with a derivation by Brunauer, Emmett, & Teller [14, 15] for the multilayer adsorption of gases on solid surfaces. This multilayer adsorption theory, known generally as the B.E.T. theory, has occupied a central position in gas adsorption studies and surface area measurements ever since.

Brunauer, Emmett, & Teller extended Langmuir's kinetic theory to multilayer adsorption. The B.E.T. theory assumes that the uppermost molecules in adsorbed stacks are in dynamic equilibrium with the vapour. This means that where the surface is covered with only one layer of adsorbate, an equilibrium exists between that layer and the vapour, and where two layers are adsorbed, the upper layer is in equilibrium with the vapour, and so forth. Since the equilibrium is dynamic, the actual location of the surface sites covered by one, two, or more layers may vary, but the number of molecules in each layer will remain constant. In other words, B.E.T. is based on the assumption that the forces that produce condensation are chiefly responsible for the binding energy of multimolecular adsorption, hence the concept of localization prevails at all layers, and the forces of mutual interaction are neglected.

The commonly known form of the B.E.T. equation may be written as:

$$\frac{P}{V(P_0 - P)} = \frac{1}{V_m c} + \frac{c - 1}{V_m c}\frac{P}{P_0} . \tag{10.13}$$

A plot of $P/V(P_0 - P)$ against P/P_0 should yield a straight line having a slope $(c - 1)/V_m c$ and an intercept $1/V_m c$.

Equation (10.13) is capable of describing type 1, type 2, and type 3 isotherms,† depending on the constant c. In general it has been found that only type 2 isotherms (those with high values of c) have well-defined 'knee-bends', which are essential for accurate V_m values. The preference for using nitrogen at liquid nitrogen temperature

† See section 10.2.14.5.

is due to the fact that at liquid nitrogen temperatures, with all solids so far reported, this gas exhibits higher c values than the c values for alternative gases.

For type 2 isotherms, the B.E.T. equation has been found to be valid generally between 0.05 and 0.3 relative pressure, but examples have also been reported where this range has been extended or shortened [16].

The internal consistency of the B.E.T. method has been demonstrated by many authors [17, 18] by their measurements on several solids. The degree of correspondence between the specified surfaces obtained with several adsorbates allows confidence to be placed in the method.

10.2.14.3 Criticism of the B.E.T. theory

The B.E.T. model has been criticized [19, 20] on the grounds that although c is assumed to be constant, there is a great deal of evidence that this is not so. For constant c it is necessary that the surface be energetically homogeneous, i.e. there are no high-energy-level sites. Experimental measurements of variations of heats of adsorption with coverage show that the first molecules to be adsorbed generate more energy than subsequent molecules. Braunauer, Emmet, & Teller suggest that this is why their theory breaks down at low surface coverage [21]. Further criticisms can be found elsewhere [22–25].

10.2.14.4 The B.E.T. equation

The B.E.T. equation is equivalent to the difference between the upper branches of two rectangular hyperbolas [26] which may be written as:

$$\frac{V}{V_m} = \frac{1}{1-x} - \frac{1}{1+(c-1)x} \,. \qquad (10.14)$$

For c less than 2 and positive, a type 3 isotherm is generated. Values of c less than unity imply that the cohesive force between the adsorbing molecules is greater than the adhesive force between the adsorbate and the adsorbent. The point of inflection of the isotherm was determined by differentiating twice with respect to x and equating to zero. Other explanations of the mathematical nature of the B.E.T. equation have been discussed by many workers [1, 202].

10.2.14.5 The shapes of isotherms

Generally, the majority of adsorption isotherms may be classified [30, 31] in the six types shown in Fig. 10.9a,b,c.

Type 1 isotherm is characterized by a rapid initial rise in amount adsorbed at low pressures followed by a flat region. In some cases the curve is reversible and the amount adsorbed approaches a limiting value. In others the curve approaches the line $P/P_0 = 1$ asymptotically, and the desorption curve may lie above the adsorption curve shown down to very low pressures. For many years it was thought that the shape of this isotherm was due to adsorption being limited to a monolayer, and the isotherm was interpreted on the basis of the Langmuir theory [14, 32]: This type of isotherm is still known as a Langmuir isotherm. It is now generally accepted that the shape is characteristic of micropore filling, and the limiting amount adsorbed is a

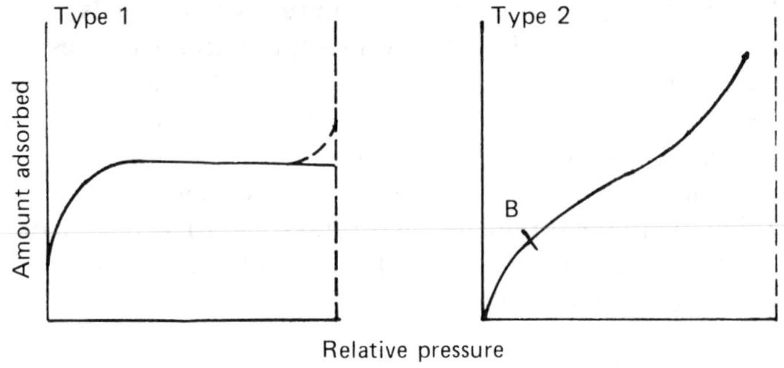

Fig. 10.9a — Types 1 and 2 isotherms.

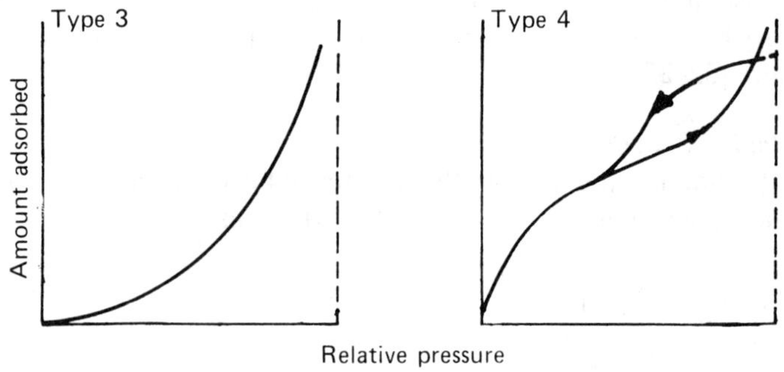

Fig. 10.9b — Types 3 and 4 isotherms.

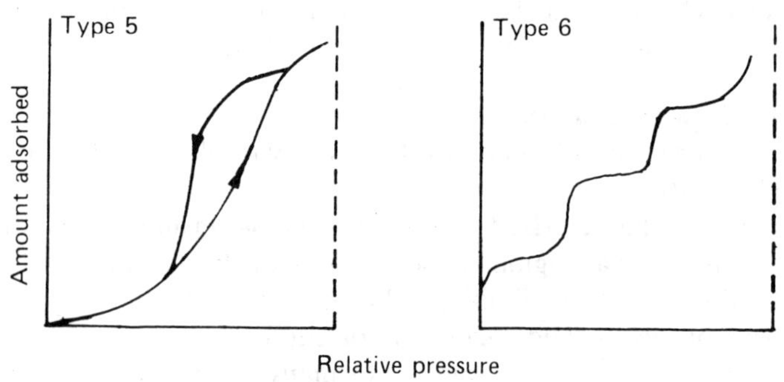

Fig. 10.9c — Types 5 and 6 isotherms.

measure of micropore volume rather than monolayer surface. Type 1 isotherms may also occur for adsorption on high-energy-level surfaces [33].

The reversible type 2 isotherm is obtained by adsorption on many non-porous or mesoporous powders, and it represents unrestricted multilayer adsorption on a heterogeneous substrate. Although layers at different levels may exist simultaneously, monolayer completion occurs at the point of inflection of the isotherm. This is known as point B, and was first identified by Emmett & Brunauer [34]. They subsequently developed a theory containing a constant c, to locate this point. Type 2 isotherms occur for high c values, and the 'knee' at the point of inflection becomes more pronounced as the c value increases. Increasing c values indicate increasing affinity between the adsorbate and the adsorbent.

Type 3 isotherms arise when adsorbate–adsorbent interaction is weak and c is less than 2.

A characteristic feature of type 4 isotherms is the hysteresis loop. This has been explained as being due to capillary phenomena [6, 17], and this part of the isotherm is used for pore size distribution evaluation. As the pressure is reduced from the saturation value, the gas molecules condensed in the capillary cracks of adsorbents do not evaporate as readily as they would from the bulk liquid, owing to the lowering of the vapour pressure over the concave meniscus formed by the condensed liquid in the pores.

Type 5 isotherms are similar to type 4 except that the adsorbate–adsorbent interaction is weak.

Type 6 isotherms arise with stepwise multilayer adsorption or noble gas molecules on a uniform substrate [35–38].

10.3 OTHER MODIFICATIONS OF THE B.E.T. EQUATION AND METHOD

Pickett [39] modified the n-layer B.E.T. treatment to take into account the decrease in probability of escape from an elemental area covered with n layers as adjacent-elemental areas also become covered with n layers. This leads to the result that there can be no evaporation from such an area. Other modifications can be found elsewhere [40–44].

10.3.1 The Hüttig equation

Several modifications have been derived to extend the scope of the B.E.T. equation. Hüttig [45, 46], derived a formulation of the localized multilayer adsorption on a uniform surface. The difference between the B.E.T. and the Hüttig theories is in the assumption by the latter that the evaporation of the ith layer molecule is entirely unimpeded by the presence of molecules in the $(i + 1)$th layer, whereas the B.E.T. contention is that $(i + 1)$th layer molecules are completely effective in preventing the evaporation of underlying molecules. For the majority of gas-solid systems the V_m values calculated from the Hüttig equation exceed the B.E.T. values by 2 to 20%, depending [7] upon the value of c. Compromise equations between Hüttig and B.E.T. have also been attempted [7].

10.3.2 The relative method of Harkins & Jura (HJr)

Harkins & Jura [47] derived an isotherm, by analogy with condensed liquid layers, independent of V_m, hence avoiding any explicit assumption of the value of the molecular area of the adsorbate in the calculation of the surface area. Fundamentally the HJr method is empirical. The B.E.T. method yields a value for V_m from which the surface area is to be calculated, whereas the HJr method yields the surface area directly without giving the value of V_m.

Livingstone [48, 49] and Emmett [50] have found that in the linear B.E.T. region ($P/P_0 = 0.05$ to 0.3) a linear HJr plot is obtained only when $50 < c < 250$. For $c = 10$, 5, or 2, and $P/P_0 < 0.4$, there is no linear relationship, while for $c = 100$ the range of mutual validity of the two equations is limited to the region 0.01 to about 0.13 relative pressure. Smith & Bell [51] extended this enquiry to the n-layer B.E.T. equation.

Both B.E.T. and HJr methods are open to criticism in that they involve the arbitrary selection of constants (k and σ) which undoubtedly depend on the nature of the solid surface [2].

Of the two, the HJr method is inferior for the following reasons [7]:

(1) The quantity $1/V^2$ is sensitive to slight errors in V.
(2) The range of relative pressure over which a linear HJr plot obtains is variable, depending upon the value of c in the B.E.T. equation. For each new solid, a large number of experimental points may be needed to locate this linear region.
(3) Some systems yield HJr plots with more than one linear section.

However, the adsorption of nitrogen at $-195°C$ on the majority of solids is characterized by c values in the range 50–240, so that the surface areas obtained from the two equations are in agreement.

10.3.3 The Frenkel–Halsey–Hill equation (FHH)

Hill [52] took into account the decrease in interaction energy for molecules adsorbed in second and subsequent layers, and he derived the equation

$$\ln x = -b \left(\frac{V}{V_m} \right)^s \tag{10.15}$$

where b and s are constants. This equation has been discussed by Sing [35], and an analysis of the two constants has been carried out [53].

10.3.4 The Dubinin–Radushkevich equation (D–R)

Dubinin [54–57] derived an equation, based on the Polanyi potential theory, for the determination of micropore volume. In this theory it is postulated that the force of attraction at any given point in the adsorbed film is given by the adsorption potential (ε), defined as the work done by the adsorption forces in bringing a molecule from the gas phase to that point. Polanyi expressed the adsorption potential as the work required to isothermally compress one mole of gas from the equilibrium pressure P to the vapour pressure P_0:

$$\varepsilon = RT \ln \frac{P_0}{P} \tag{10.16}$$

Dubinin considered that the adsorption potential varies according to the nature of the adsorbate and adsorbent. Thus the characteristic curves for different absorbates on the same adsorbent are homologous and can be superimposed by use of an affinity coefficient (β) where:

$$\beta = \frac{\varepsilon_1}{\varepsilon_2}\ , \tag{10.17}$$

where the suffixes refer to the respective vapours. If adsorbate 2 is taken as an arbitrary standard then $(\varepsilon/\varepsilon_0) = \beta$, where the suffix zero refers to the standard vapour and that without suffix to the other vapour. Experimental extension of the Dubinin equation can be found elsewhere [58–77].

10.3.5 The V_A-t method

For a number of non-porous solids, the ratio between the adsorbed volume V and the volume of the unimolecular layer V_m, if plotted as a function of x, could be represented approximately by a single curve. Schüll *et al.* [78] showed that with the aid of this function the thickness of the adsorbed layer as a function of x could be calculated.

The film thickness for nitrogen is given by the equation

$$t = 0.354 \left(\frac{V}{V_m}\right) \text{nm} \tag{10.18}$$

(the constant is the thickness of one layer of nitrogen molecules). This equation may be written as

$$S_t = 15.47 \left(\frac{V}{t}\right) \tag{10.19}$$

with V $cm^3\ g^{-1}$ (vapour at s.t.p.), t is the film thickness in nm and S_t is the specific surface in $m^2\ g^{-1}$.

Lippens *et al.* [79] have published tables of t against x, for the construction of the $V-t$ curve, from $t = 0.996$ nm at $x = 0.76$ to $t = 0.351$ nm at $x = 0.08$. Brockhoff [80] extended the curve to $x = 0.92$.

These values cannot be used for all substances, and other t-curves are available. One of the most widely used is de Boers' common t-curve [81, 82] shown in Fig.

10.10. The experimental points, for a variety of substances, deviate by 10% or more from the average curve. When the t-curve is being used for pore size analysis these errors are small enough to be neglected. For surface area evaluation t-curves with small deviations are required, and these should be common for groups of materials such as halides, metals, graphite [83]. Further discussions of the V_A-t methods can be found elsewhere [84–89]. It may be necessary to mention that experimentally, values of V measured as a function of relative pressure are transformed to functions of t. This gives a straight line passing through the origin, and the specific surface may be obtained from the slope.

The t-method is based on the B.E.T. conception, but yields additional information. For non-porous solids a graph of V against t yields a straight line as shown in Fig. 10.11. Deviations from the straight line are interpreted as either a decrease in accessible surface area due to blocking of micropores, or the onset of capillary condensation in intermediate pores. A second linear portion gives the surface area of what is left, that is, the surface area of the wider capillaries or of the outside area of the granules [2].

10.3.6 Volumetric apparatus for high surface areas

Allen [12] has reviewed volumetric apparatuses hereunder, the earlier ones of which have been reviewed by Joy [90]. In all the volumetric methods, the principle underlying the determination is the same. The pressure, volume, and temperature of a quantity of adsorbate are measured and the amount of gas present is calculated. The material is then brought into contact with the adsorbate, and when the constant pressure, volume, temperature conditions show the system to have attained equilibrium, the amount of gas is again calculated. The difference between the amount of gas present initially and finally represents the adsorbate 'lost' from the gas phase to the adsorbed phase. The accurate determination of the amount of gas unadsorbed at equilibrium depends upon the precise knowledge of the 'dead-space' or the space surrounding the adsorbent particles. The dead-space volume is usually determined by expansion measurements using helium, whose adsorption can be assumed to be negligible. Estimation of the quantity of unadsorbed gas is often complicated by the fact that part of the dead-space is at room temperature and part at the temperature of the adsorbent.

Since the amount adsorbed represents the difference between the amount admitted to the dead-space and the amount remaining in the dead-space at equilibrium, it can be evaluated with confidence only if these two quantities are of unlike magnitude. To achieve this, the apparatus is so designed as to minimize the dead-space volume. In practice, it is convenient to fix the volume and temperature and measure the changes in pressure.

Regardless of the particular design, the basic apparatus must provide means for removing gases and vapours which all materials pick up when exposed to the atmosphere. The apparatus must also provide means for permitting readsorption of known quantities of the gas on to the material. It should also have evacuating systems, gauges to measure the vacuum, a gas storage part, and the analytical part.

Conventional types of nitrogen adsorption apparatus invariably follow the assembly originally described by Emmett [91] shown in Fig. 10.12. Adsorbate gas is taken into the burette and its pressure measured on a manometer. The stopcock

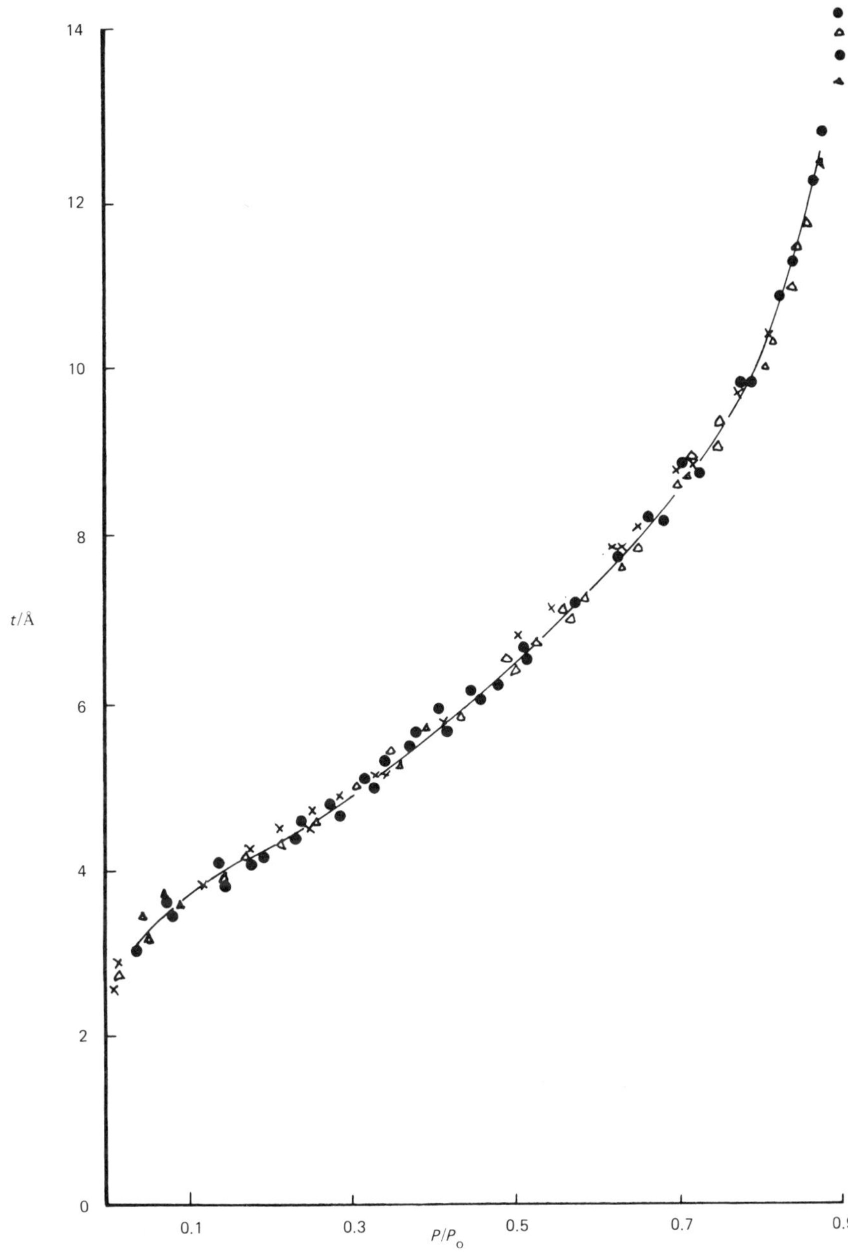

Fig. 10.10 — The common *t*-curve of multimolecular nitrogen adsorption on oxidic and hydroxidic adsorbents.

between the sample and the burette is then opened and the new pressure, after allowing time for the equilibrium to be established, is read on the manometer. The volume of the gas admitted to the sample bulb is proportional to the difference in the

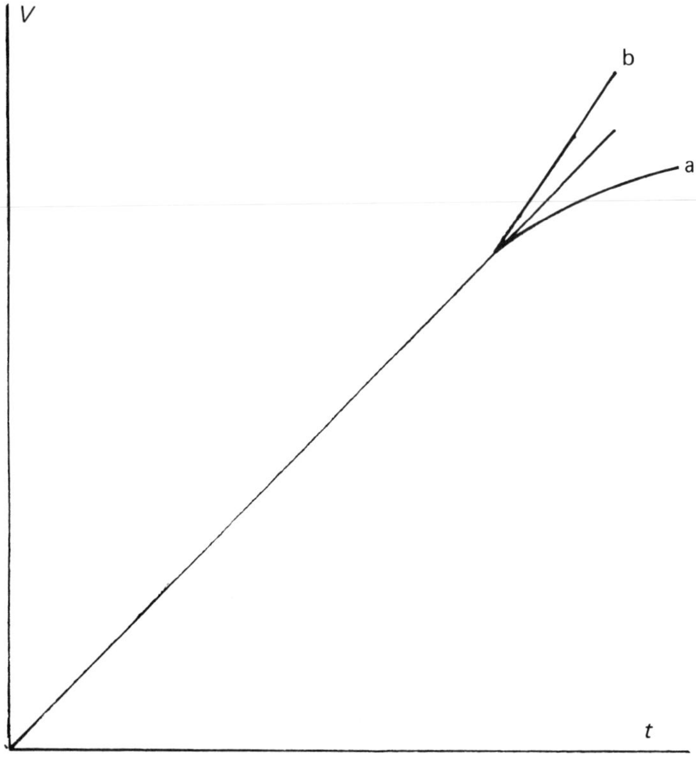

Fig. 10.11 — The *V-t* curve.

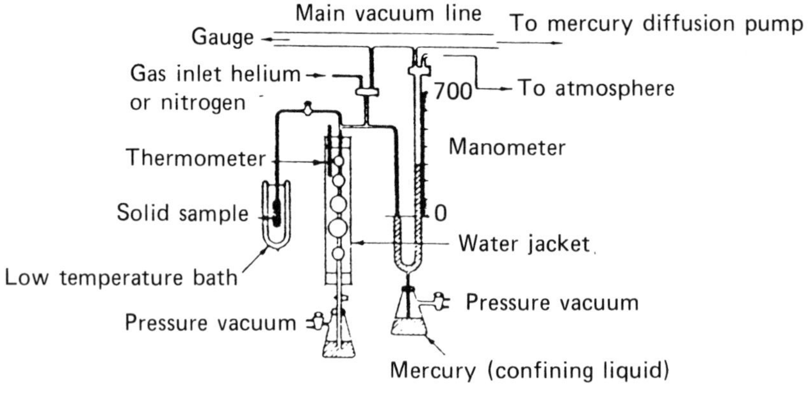

Fig. 10.12 — Emmett's apparatus for surface area determination by gas adsorption.

pressures before and after opening the stopcock. This latter pressure is also the equilibrium adsorption pressure. The volume adsorbed is equal to the volume admitted less the volume of gas required to fill the dead-space in the sample bulb and the burette connections. To obtain more adsorption points, the mercury level is raised to the next volume mark and a new pressure is established. Helium is used to calibrate the dead-space. In all adsorption calculations a correction for the non-ideal behaviour of nitrogen at liquid nitrogen temperature is included. In a method like this, the total volume admitted is found by summing the separate doses.

The main disadvantage of the original design [92–96], suggestions for refinements [96–103], and accuracy and automation [104, 105], have been published by many researchers.

10.3.7 Volumetric apparatus for low surface area

A semi-micro unit has been described [103] for surface area evaluation down to 2 m^2 g^{-1}, and a capillary differential manometer to keep the dead-space low [79]. Surface areas down to 1 m^2 g^{-1} can also be determined by using nitrogen adsorption, provided that great care is taken. With decreasing surface area, the error in the dead-space factor makes nitrogen unsuitable. Since the amount of gas in the dead-space is proportional to the absolute pressure it is preferable to use gases with low saturation vapour pressures. Krypton gas with a saturated vapour pressure of 1.76 mmHg at − 195°C and ethylene gas with 0.1 mmHg saturated vapour pressure at liquid air temperatures lend themselves readily to low-pressure measurement.

Researchers [106–116] in this field have reviewed various gases used for the adsorption measurements on low surface areas, and it seems that krypton at liquid nitrogen temperature is the popular choice. In measurements of such low pressures, two obvious risks must be considered, namely, the desorption of water and other vapours from the glass walls of the apparatus and thermal transpiration. By heating the entire system for a short time, or by permanently keeping the system under vacuum, most of the adsorbed vapours from the glass walls should be removed. Otherwise, the slowly desorbing vapours will increase the pressure in the system during adsorption measurements, leading to erroneous results.

10.3.8 Continuous-flow gas chromatographic methods

A continuous-flow method, based on the gas chromatographic technique, has been introduced for the measurement of surface area of fine powders by gas adsorption. A scheme was first proposed by Loebenstein & Deitz [100] for reducing the vacuum requirements by using a flowing mixture of the adsorbate and an inert gas, such as helium, to obtain the low adsorbate gas pressure required.

The method, a modification of gas-adsorption chromatography in which the column packing is the sample itself and the mobile gas phase is a mixture of a suitable adsorbate and an inert gas, was developed by Nelsen & Eggertsen [117]. They used nitrogen as the adsorbate and helium as the carrier gas in the following manner.

A known mixture of nitrogen and helium is passed through the sample and then through a thermal conductivity cell connected to a recording potentiometer. When the sample is cooled in liquid nitrogen, the sample adsorbs the nitrogen from the mobile phase; this is indicated by a peak on the recorder chart, and after equilibrium is established, the recorder-pen resumes its original position. Removing the coolant gives a desorption peak equal in area and in the opposite direction to the adsorption peak, and either peak may be used to measure the nitrogen adsorbed.

Calibration for such a system may be either absolute (by injecting a known amount of nitrogen into the mobile phase at the point normally occupied by the sample and obtaining a factor for the amount of nitrogen per unit peak area on the resulting recorded curve), or by comparison with a sample of known surface area.

A schematic diagram of their apparatus is shown in Fig. 10.13. Nitrogen flow

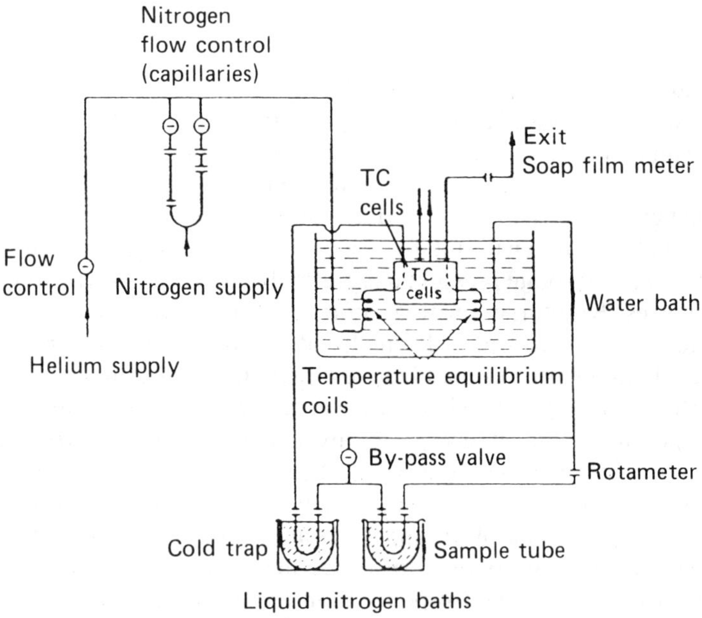

Fig. 10.13 — Nelsen & Eggertsen's apparatus.

control was achieved by two capillary tubes in parallel, 0.25 mm inner diameter, and 150 and 300 mm long. The capillaries were used independently or together to give three nitrogen flow rates in the range of 5–20 ml min^{-1} with a pressure head of 2 lb in^{-2} g. The helium flow was controlled by needle valves. The flow measurements were made by rotameter and soap-film meter.

The mobile phase was first passed through the reference arm of the thermal conductivity cell, on to the sample and then, again, to the thermal conductivity cell, which was immersed in a temperature-controlled bath.

Nelsen & Eggertsen measured adsorption at three flow rates, i.e. at three partial pressures. The sample was outgassed at the desired temperature (up to 500°C) while being purged with He at 20 ml min^{-1} for 1 h. Nitrogen relative pressures in the range of 0.05–0.3 and a total nitrogen and helium flow of 50 ml min^{-1} were used. The desorption peaks were used for measurement because they were relatively free from tailing effects.

Calculation is essentially the same as for the pressure–volume method, but is much simpler since no dead-space corrections are needed. The authors assumed complete linearity of the thermal conductivity cell over the concentration range employed. They analysed samples ranging in surface area from 3 to 450 $m^2\,g^{-1}$, and the results obtained on continuous-flow and pressure–volume methods were in good agreement.

The main advantages of the continuous method over the conventional B.E.T. method are [1]:

(a) Elimination of fragile and complicated glassware.
(b) Elimination of a high-vacuum system.
(c) Permanent records obtained automatically.
(d) Speed and simplicity.
(e) Elimination of dead-space correction.

Ellis, Forrest, & Howe [118] made some modification and improvements to the original technique for their specific applications. The schematic diagram of their apparatus is shown in Fig. 10.14. All flow controls were done by needle valves and all flow measurements were made by rotameters. They used a helium flow rate of 150 ml min^{-1} and a nitrogen flow rate of 3, 5 or 10 ml min^{-1}.

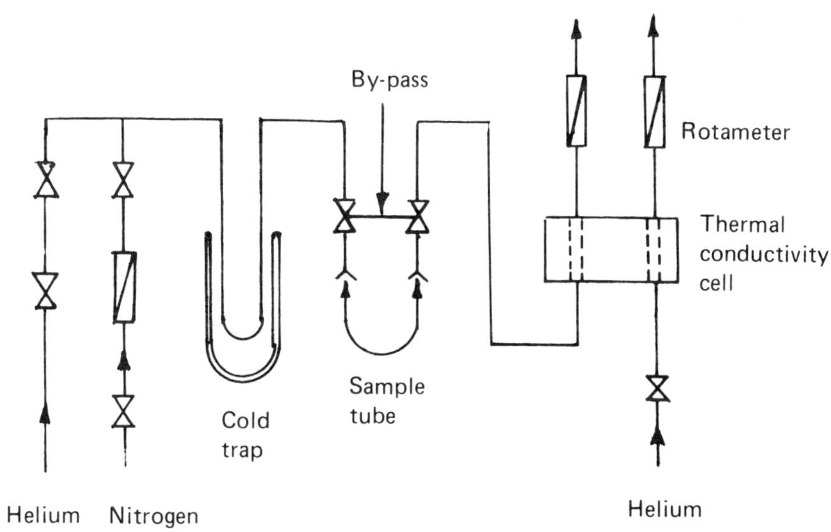

Fig. 10.14 — Ellis, Forrester & Howe's apparatus.

By taking more care (when chilling the sample to avoid shock effects in the gas stream) they could extend the method to surface areas in the region of 0.01 $m^2\,g^{-1}$.

They obtained good linearity in the B.E.T. plot, even with surface areas as low as 0.02 $m^2\ g^{-1}$. They analysed samples with surface areas of 14.2 to 0.005 $m^2\ g^{-1}$.

Below 0.01 $m^2\ g^{-1}$ conventional adsorption methods with nitrogen are not practical, hence no results using the PVT methods were given by them. Above that range agreement between the continuous flow and PVT methods was good, but in their results the surface-area value by the new method was slightly higher than that given by the PVT method.

Ellis *et al.* [118] also developed a shortened method using only one flow rate, i.e. a single-point method. Since, usually, the B.E.T. intercept is very small, the intercept term can be ignored and $S_w \propto V(P_0 - P)/P$. For constant P values, i.e. N_2/He flow-rate ratio constant in the flow stream, $S_w \propto V$. Since $V \propto$ area of recorder chart, a plot of S_w against area of fixed sample weight and fixed N_2/He ratio should be linear. They analysed a number of samples (300 mg each) and determined the adsorption peak areas for 10 ml min^{-1} flow of nitrogen and 50 ml min^{-1} helium, and they obtained a linear graph from which the surface areas of subsequent samples were obtained. Results by this method were also comparable to those obtained by the normal B.E.T. method based on PVT measurements. This method was especially good compared with multipoint methods for routine measurements, as one sample can be analysed per hour [2].

Atkins [119] developed a precision instrument for use in the carbon black industry. For precision measurements he stated that it is necessary to consider the effect of changes in ambient temperature, barometric pressure, liquid nitrogen temperature, and nitrogen concentration in the gas mixture. Correction for the non-linearity of the katharometer was also necessary, and was achieved either by adjusting the sample size so that the desorption and calibration peaks were of the same height, or by using a correction factor; this was determined as a function of relative peak areas, i.e. desorption peak area divided by the calibration peak area. Atkins used heat exchanger coils in the detector circuit in addition to the temperature control of the detector. Two different premixed gases were connected to the apparatus so that either of them could be used or both together. The apparatus also had special calibration valves and calibration loops to select the volume of gas nearest to the desorption volume for calibration, so that similarly shaped calibration peaks and heights of similar magnitude were obtained. He has also tabulated the variables like percentage nitrogen in the mixture, calibration loop temperature, barometric and saturated vapour pressure of the nitrogen, and the corresponding percentage changes in calculated surface area due to their variation, and has discussed the errors due to the non-linearity of the katharometer.

Haley [120] extended the continuous-flow measurement to include the size distribution in the 10 to 300 Å radius range. He used 10% nitrogen in helium (as mobile phase) at various pressures up to 150 lb in^{-2}, causing the nitrogen partial pressure to reach its liquefaction point, consequently varying the nitrogen relative pressure in the sample tube in the range 0.16 to 1. Nitrogen adsorbed or desorbed, by increasing or decreasing the pressure in the tube, was measured continuously. He also measured surface areas and obtained a variation of approximately ±2.5% in the range of 40 to 1250 $m^2\ g^{-1}$.

Since helium is extremely expensive [1, 2], it may be replaced by other gases which are not adsorbed under the experimental conditions, e.g. hydrogen. Whitehead [121] used hydrogen as a carrier gas at a flow rate of 50 ml min^{-1} and found it was very satisfactory.

Several problems arise when the Nelsen & Eggertsen-type apparatus is used for measuring surface areas less than 500 $cm^2\ g^{-1}$, the adsorption and desorption peaks having the shapes shown in Fig. 10.15.

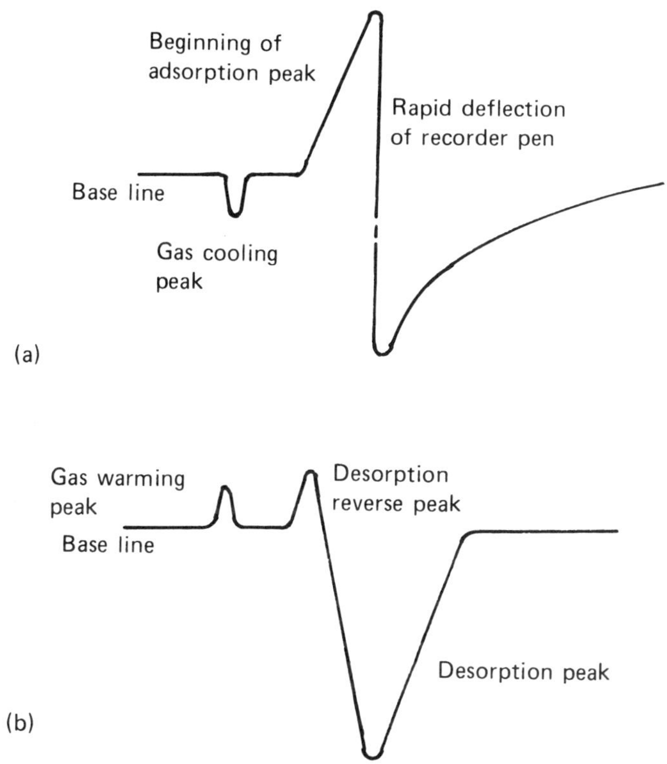

Fig. 10.15 — Shapes of peaks obtained with Nelsen & Eggertsen type equipment of low surface area: (a) absorption reverse peak, and (b) description reverse peak.

During adsorption a peak is produced when the sample tube is immersed in liquid nitrogen. This immersion causes a contraction of the gas inside the sample tube and a reduction in the gas flow through the katharometer; the thermistor on the measurement side warms up, causing a change in its resistance and a peak on the recorder chart. The adsorption peak results from a cooling of the thermistor due to removal of nitrogen from the stream, which causes an increase in the thermal conductivity of the gas mixture.

A reverse peak occurs midway through the adsorption. During desorption a gas-warming peak occurs, and immediately before desorption, a desorption-reverse peak. Since the desorption peak is used for measurement purposes, effort has been mainly directed at finding an explanation for the desorption reverse peak and correcting for it.

Lovell [122] considered that the reverse peak on desorption was due to transverse thermal diffusion. When the sample tube is removed from the coolant, pre-cooling of the gas stream in the inlet section quickly ceases and gas enters the sample-catching section at very nearly room temperature. Since the sample is still cold, partial separation of the gases takes place owing to transverse thermal diffusion, nitrogen moving to the walls of the container and helium to the centre. Gas flow is more rapid at the centre, and helium-rich gas is carried to the katharometer, giving rise to a thermistor cooling peak. Immediately afterwards, when the temperature of the sample tube has risen considerably, the separated nitrogen diffuses into the gas stream and produces a peak in the opposite direction and necessarily of the same area. It can be seen that, when a sample is present in the container, the area of desorption peak will be increased by the area of the reverse peak. Thus, the true area of the desorption trace may be obtained by subtracting the area of the reverse peak from that of the observed peak.

Lovell's experimental refinement was to allow the desorbed gases to expand into a vessel whose volume could be adjusted by altering the amount of mercury in it (see Fig. 10.16).

As the gas expands, the mercury level is forced down and the pressure is equalized by lowering the mercury level in a second container. When the desorption and expansion process is complete and enough time has elapsed for mixing to take place, the adsorbed gas is swept through the thermal-conductivity cell and the area of the desorption peak is measured with no interference from a reverse peak. Tucker [123] removed the anomalous peaks by using an interrupted flow technique. Bhat & Krishnamoorthy [124] describe a simple continuous gas flow adsorption apparatus, and Yen & Chang [125] introduced a modified system.

The Ströhline areameter in Fig. 10.17 consists of an adsorption vessel with the sample and a reference vessel on either side of a mercury manometer [126, 127]. Degassing is carried out by purging with dry measuring gas with the sample tube in a heating block. Both vessels are then filled with nitrogen at ambient temperature and atmospheric pressure, after which they are immersed in a liquid-nitrogen bath. Adsorption of nitrogen on to the sample creates a pressure difference across the manometer from which the surface area of the sample may be determined, correction being applied for the sample volume and the effect of the change in height of the manometer level. The surface may be determined by a simple nomogram or by calculation. This is essentially a one-point method of determining specific surface [2] and may be used for surfaces in the range 0.2 to 1000 $m^2 g^{-1}$. The reproducibility is of the order of 1%, and the error introduced by assuming that the B.E.T.-plot passes through the origin produces specific surfaces about 10% lower than those obtained when using the conventional multipoint method. Using one-point determinations, up to 16 analyses per day may be performed. Alternatively, the nitrogen may be fed in under pressure and a full isotherm determined [128].

The Gravimat is a beam-balance for surface-area and pore size distribution measurement. A full description of this instrumnent is given by Robens [129], and a description of the measuring technique for nitrogen adsorption by Roth [128, 131]. It

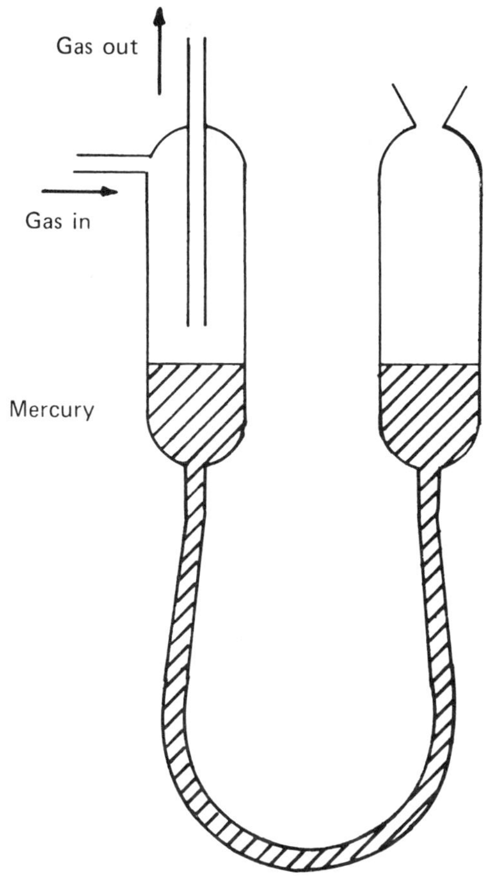

Fig. 10.16 — The expansion chamber.

is possible to measure and control pressures from atmospheric down to 10^{-3} mmHg, hence the instrument can be used with a range of adsorbates.

The balance has a sensitivity of 0.1 μg, but thermal effects and buoyancy reduce the measuring accuracy; a monolayer of 34.8 cm^2 of nitrogen weighs 1.0 μg, giving a lower limit of surface-area measurement of 100 cm^2. Conditions are more favourable with krypton of which 18.7 cm^2 weighs 1.0 μg.

10.4 GRAVIMETRIC METHODS

The gravimetric techniques have the great advantage over the volumetric methods in that the volume of the adsorption system is quite immaterial, and the amount of gas adsorbed is observed directly by measuring the increase in the weight of the solid sample upon exposure to a gas or vapour. The tedious volume calibration and dead-space determinations are thus eliminated [1, 2, 3].

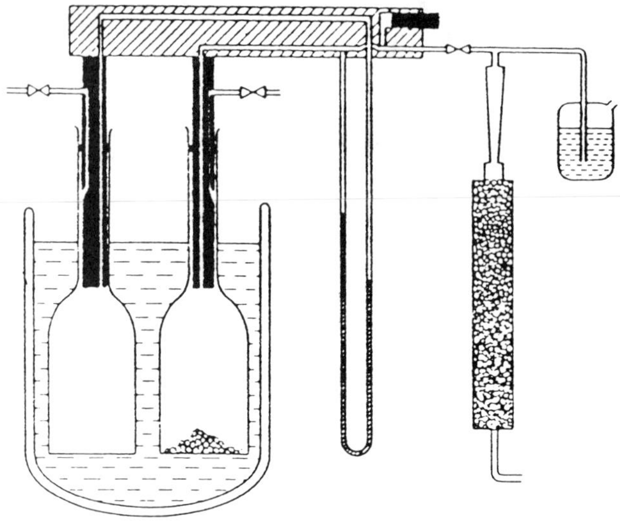

Fig. 10.17 — The Ströhline areameter.

The main disadvantages of the method are [1]:

(a) The apparatus is much less robust and correspondingly more difficult to construct and maintain than volumetric apparatus.
(b) The apparatus has to be calibrated by placing known weights in the adsorbent-pan, and the method is hence subject to the errors always attached to determination which are dependent on the constancy of calibrations of easily fatigued and strained mechanical systems.
(c) Buoyancy corrections have to be made.

10.4.1 Single-spring balances

McBain & Bakr [132] introduced a sorption balance, the essential features of which are a quartz helical spring supporting a small gold or platinum bucket in which the sample is placed. The spring is calibrated by adding small known weights to the bucket and measuring the increase in length of the spring with a reading microscope. These calibrations must be done over the entire range of temperatures at which adsorption measurements are made. The liquid adsorbate, free from dissolved gases, is sealed in a small glass bulb and placed with a magnetic hammer in a glass or quartz envelope ('balance case'). The balance is heated in vacuum for outgassing. Finally, the case is sealed off, the lower end cooled, and the bulb broken. Adsorption measurements are carried out with the lower end in one thermostat bath to control the temperature of the liquid and the upper end in another to control that of the solid. The equilibrium pressure can be calculated from the temperature of the liquid, provided that vapour pressure data are available. The amount adsorbed is proportional to the spring extension, and the correction for buoyancy is significant at higher pressures.

Other techniques attempted may be found elsewhere [133–140].

10.4.2 Multiple-spring balances

Gravimetric methods have the additional advantage that several determinations can be carried out simultaneously by connecting several balance cases to the same gas or vapour manifold and observing the individual spring extension [1].

Seborg & Stamm [141] connected five of six simple spring units in series. Pidgeon & Maass [142], Mulligan *et al.* [143] and Stamm & Woodruff [144] have all described similar multiple-spring balances. Mulligan *et al.* connected as many as fifteen springs to the same apparatus.

10.4.3 Beam balances

Beam-type vacuum balances have greater sensitivity than the helical-spring balances, and also the troublesome buoyancy correction at higher pressures is eliminated, at least partly if not completely [1].

Beam balances can be of either high sensitivity at very low total loads or of medium sensitivity at large total loads, which is in contrast to the normal short-spring balances which have a medium sensitivity at low total loads.

The majority of the high-sensitivity low-load balances are based on those originally designed by Barrett, Birnie, & Cohen [145] and by Gulbransen [146]. Barrett *et al.* used a glass beam 40 cm long supported on a tungsten torsion wire, and they enclosed the whole assembly in a tubular glass casing connected to the vapour and vacuum manifolds. Calibration was effected by moving a small soft-iron rider along the beam by means of a magnet outside the case. Gulbransen's balance was constructed from glass rod, quartz fibres, and metal wires on the same principles as an ordinary chemical balance.

Rhodin's microbalance [147–149] is essentially a modification of these, in which some stability is sacrificed for increased sensitivity by the use of thinner and lighter wires. This balance has been adopted by Bowers & Long [150] for adsorption at liquid helium temperatures. Rhodin's balance was made as symmetrical as possible, to eliminate buoyancy corrections and to minimize thermal eddy currents. The adsorbent and counter weights were matched to within 10^{-5} g and immersed to the same depth in identical thermostatic baths, and the outgassing was done at 400°C in a vacuum of 10^{-7} mmHg. With this balance, it was possible to observe a vertical displacement of 10 mm to better than 0.01 mm, and with loads up to 1 g, it was possible to observe weight changes of 10^{-7} g $\pm$ 20% in a reproducible manner.

Beam balances have also been operated as null-point instruments. The beam is acted upon by a solenoid outside the balance housing, the current through the solenoid being adjusted to restore the beam to its horizontal position. One such balance by Gregg [151] uses two concentric solenoids, the inner one suspended from the beam and the outer one fixed to the envelope. The original balance had a sensitivity of 0.3 mg, the range of load being as high as 10 to 20 mg. In an automatic version of this instrument described by Gregg & Wintle [152] a photoelectrically operated relay adjusts a potentiometer slide-wire contact which is connected to the solenoid on the balance.

Although so many different types of gravimetric apparatus have been reported, they have not become popular owing to their delicate nature and the difficulty of accurately compensating for buoyancy effects.

Fig. 10.18 is a rough drawing of the Cahn [153] vacuum microbalance. The

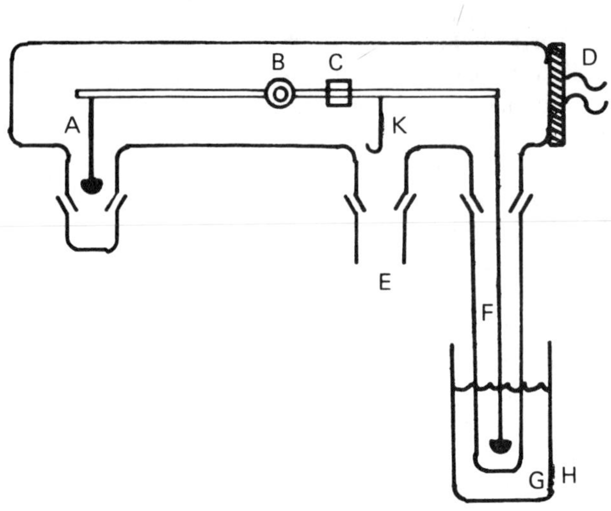

Fig. 10.18 — Cahn vacuum microbalance.

sensitivity is 1 μg with a load of 0.5 gram. Counterweight A is used to balance the adsorbent contained in bucket G, which is immersed in a liquid nitrogen filled Dewar flask H. Photocell C senses movement of the beam and provides current to the coil B to restore the beam. A metal plate D is fastened against the ground glass end of the vacuum bottle with electrical connections for bringing the signal to a recorder to continuously monitor the weight changes. Hang-down wire F is made of fine stainless steel. Ground joint E leads to pumps and gauges. For larger weights the inner hook K can be used.

10.4.3.1 Buoyancy correction

When using beam microbalances the effects due to buoyancy and thermal transpiration must be considered [154]. The gas displaced by the sample produces a buoyancy force that reduces the weight by exactly the mass of displaced gas. Buoyancy effects both the mass being weighed and the counterweight. Also, the buoyancy changes with pressure. The mass of gas displaced is given by

$$W = \frac{\overline{M}PV}{RT} \quad . \tag{10.20}$$

Allowing V_w and T_w to be the volume and absolute temperature of the counterweight, V_c and T_c to be the same for the sample and its container, the buoyancy correction becomes

$$\Delta W = \frac{P\overline{M}}{R}\left(\frac{V_c}{T_c} - \frac{V_w}{T_w}\right) \quad . \tag{10.21}$$

To estimate the error introduced by buoyancy, using nitrogen as the adsorbate, allow T_w and T_c to equal 295 and 77°K, respectively, and $V_w = V_c = V$; equation (10.21) then becomes

$$\Delta W = 0.00327PV \tag{10.22}$$

with P in atmospheres and V in cubic centimetres. In a completed monolayer 1 μg of nitrogen will occupy 34.8 cm^2 of surface. Therefore, if the volumes V_c and V_w are each 2 cm^3, an error of 0.1 torr in pressure measurement will produce a weighing error of 0.86 μg or a surface area error of 29.9 cm. For 1% accuracy it would then be necessary for the sample to exhibit an area not less than 2990 cm^2 (0.299 m^2).

The error caused by pressure measurement is not nearly as significant as that produced by measurements of the sample and counterweight volumes [154]. By evacuating the container and admitting helium to a known pressure, say 300 torr, the term in parenthesis of equation (10.21) can be calculated. With this information, ΔW for nitrogen can be calculated at any desired pressure. If the buoyancy effect with helium can be evaluated to within 1%, then the error, ΔW for nitrogen, assuming V in equation (10.22) is 2.5 cm^3 and P is 200 torr near the region of monolayer coverage, would be

$$\Delta W = 0.00327 \left(\frac{200}{760}\right) (2.5)(0.01) = 21.5\ \mu\text{g} \tag{10.23}$$

The corresponding surface area error would be 748 cm^2. Thus, to ensure accuracy of 1% or better it would be necessary to measure [154] no less than about 7.5 m^2.

Here,
W = mass of gas displaced
$\overline{M}$ = molecular weight of adsorbate
P = adsorbate vapour pressure (mm Hg)
R = gas constant, 82.1 ml atm. deg.$^{-1}$ mole^{-1}
T = absolute temperature (K)

10.5 PORE SIZE DISTRIBUTION AND GAS ADSORPTION

It is not possible to cover all the techniques used in determining surface areas, but the most widely used method is the low-temperature gas adsorption, particularly nitrogen and krypton at liquid–nitrogen temperature. Permeametry and, to a lesser extent, gas diffusion are used for comparison purposes owing to their ease of operation and simplicity. Surface areas may also be calculated from size distribution data.

Other adsorption techniques include adsorption from solution [155–158], and here the problem is one of determining the amount adsorbed since this is usually very small [1]. Adsorption studies have been described with fatty acids [159–161], polymers, ions, dyestuffs [162–167], and electrolytes [168–169], using a range of analytical techniques. The most usual experimental method of determining a single point on the adsorption isotherm of a binary solution is to bring a known amount of solution of known composition into contact with a known weight of adsorbent in a vessel at the required temperature and to stir for several hours. After equilibrium an aliquot part of the bulk liquid is separated and the concentration change is determined by some suitable method [1]. The amount adsorbed will then be some function of the final concentration.

Surface areas may also be determined from heats of adsorption, and this technique has been greatly simplified with the introduction of the flow microcalorimeter [170–174]. This instrument can be used with gas or liquid mixtures to determine heats of adsorption and amount adsorbed; it thus provides information on molecular areas as well as energies of adsorption [1, 2].

Pores can be classified [175, 176] into three groups:

(a) Macropores having widths in excess of one or two thousand Å; capillary condensation does not take place in these pores, which are essentially an avenue of transport to smaller pores;
(b) Intermediate, also known as transition or mesopores, having widths of between 15 and 16 Å up to 1000 to 2000 Å; these mark the limit of applicability of the Kelvin equation, which deals with total increase in free energy in three parts, i.e. evaporation, expansion, and condensation of moles of vapour to liquid at given pressures.
(c) Micropores, from 5 to 6 Å up to 13 to 14 Å. Since the concept of surface of a solid body is a macroscopic notion [1], surface area loses its significance when micropores are present. However, pore volume is still an applicable concept.

Pore size and size distribution have significant effects over a wide range of phenomena from the adsorbency of fine powders in chemical catalysis to the frost resistance of bricks. Because of this, pore size measurements have been described which use a wide range of techniques and apparatus. Pore surface area is generally accepted as being the difference between the area of the surface envelope of the particle and its total surface area. The pores may be made up of fissures and cavities in the particle: they may be V-shaped, i.e. wide-necked; or 'ink-bottle' pores, i.e. narrow-necked. In order that their volume distribution may be determined, it is necessary that they are not totally enclosed and that the molecules used for measurement purposes may enter through the neck. The extent and presence of small open pores may be determined by finding the volume of a powder by immersing it in mercury and finding the volume displacement, then finding the volume in a gas pyknometer, using helium as the gas [177]. Since mercury does not wet most solids it leaves the pores unfilled, and the difference between the two volumes is the pore volume. Closed pores may be evaluated by grinding the powder which opens out some of the pores, thus decreasing the apparent solid volume [178].

Total pore volume may be determined by boiling the powder in a liquid, decanting the liquid, and determining the volume of liquid taken up by the solid after it has been superficially dried. Pore size distribution may be found by using a range of liquids of different molecular sizes [178]. Direct visual examination of particle sections under the optical and electron microscopes has also been used with a range of powders [179–182]. Porosity may also be measured from the absorption of gamma rays [183].

Pore volume and pore surface distributions may be determined from gas adsorption isotherms. If the amount of gas adsorbed on the external surface is small compared with the amount adsorbed in the pores, the total volume is the condensed volume adsorbed at the saturation pressure.

With many adsorbents a hysteresis loop occurs between the adsorption and desorption branches of the isotherm as shown in Fig. 10.19. This has been explained

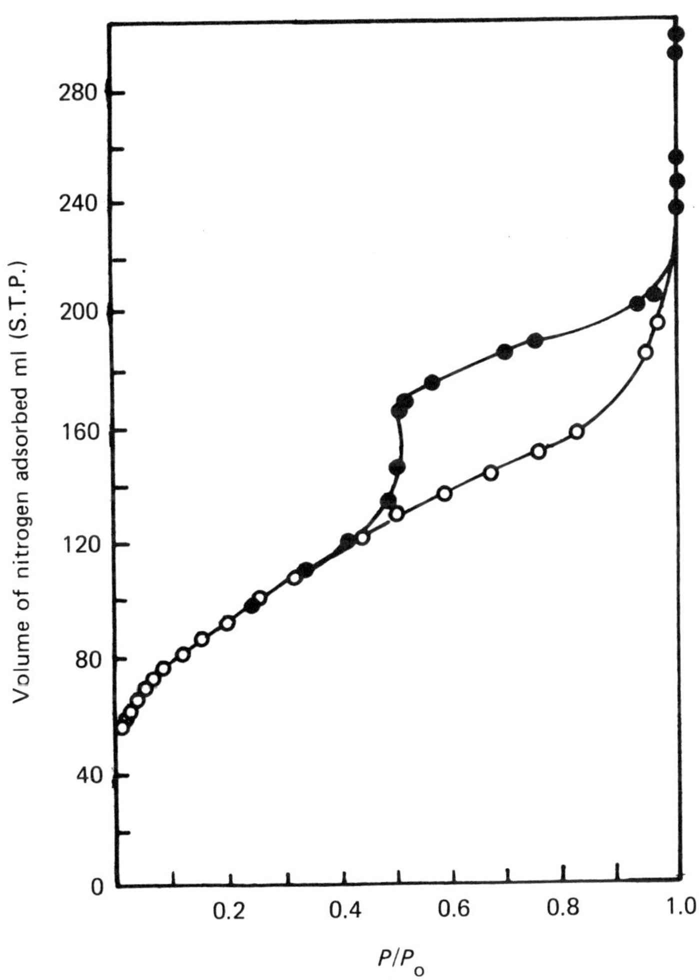

Fig. 10.19 — Adsorption of nitrogen on activated clay catalyst. Open circles indicate adsorption; solid circles desorption. Total (B.E.T.) area $S_w = 339$ mm^2 g^{-1}, monomolecular volume $V_m = 78.0$ cm^3 gas at s.t.p.

as being due to capillary condensation augmenting multilayer adsorption at the pressures at which hysteresis is present, the radii of curvature being different during adsorption from the radii during desorption [177].

de Boer [178] analysed fifteen shape groups of capillaries from a consideration of five types of hysteresis loop, as shown diagrammatically in Fig. 10.20a.

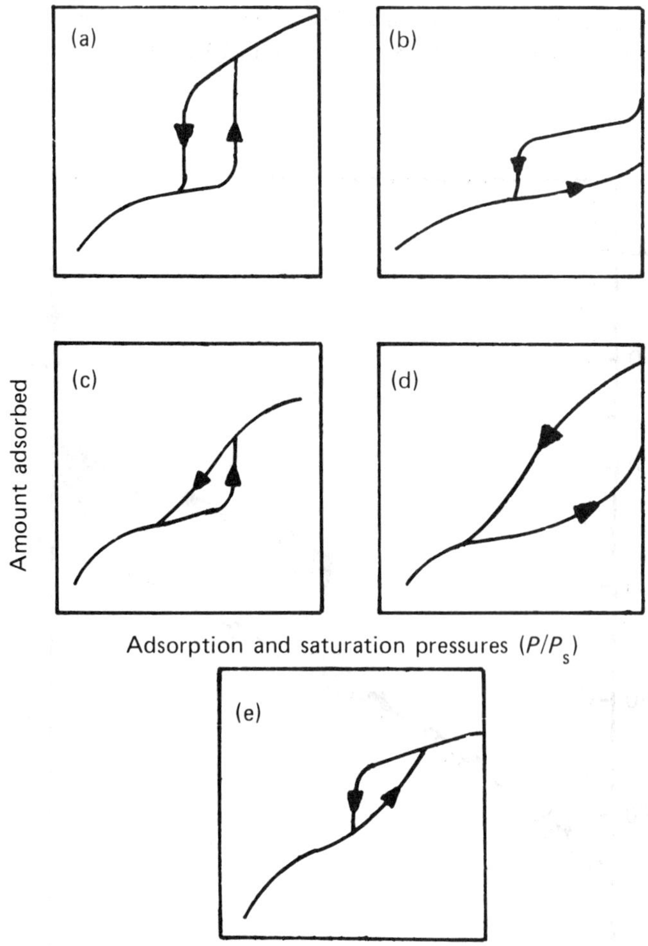

Fig. 10.20a — de Boer's five types of hysteresis loop.

Type A Both adsorption and desorption branches are steep at intermediate relative pressures.

Type B The adsorption branch is steep at saturation pressure; the desorption branch at intermediate relative pressures.

Type C The adsorption branch is steep; at intermediate relative pressures the desorption branch is sloping.

Type D The adsorption branch is steep at saturation pressure; the desorption branch is sloping.

Type E The adsorption branch has a sloping character; the desorption branch is steep at intermediate relative pressures.

Type A includes tubular capillaries open at both ends; tubular capillaries with slightly widened parts; tubular capillaries of two different main dimensions, wide-necked 'ink-bottle' pores provided that $r_n < r_w < 2r_n$; tubular capillaries with one narrowed part, narrow-necked 'ink-bottle' pores; capillaries with wide parts and narrow short necks open at both ends; trough-shaped capillaries (r_n, r_w are the radii of the narrow and wide parts respectively).

Type B includes open split-shaped capillaries with parallel walls, capillaries with very wide bodies and narrow short necks.

Type C typifies a heterogeneous distribution of pores of some of the following shape groups; tapered or doubly tapered capillaries and wedge-formed capillaries with closed sides and open ends.

Type D loops occur for a heterogeneous assembly of capillaries with wide bodies of sufficiently wide dimension r_w and having a greatly varying range of narrow necks; wedge-shaped capillaries open at both ends.

Type E loops can be formed by assemblies of capillaries of one of the shape groups for type A when the dimensions responsible for the adsorption branch are heterogeneously distributed and the dimensions responsible for the desorption branch are of equal size.

10.5.1 Pore shapes

de Boer & Broekhoff [201] classified the relationship between the form of the hysteresis loop and capillary pore shapes. This is shown in Fig. 10.20b.

From the five types of hysteresis loops in Fig. 10.20a, de Boer classified in Table 10.7, fifteen pore shapes and two shape groups from the space between particles.

In type A of Fig. 10.20a, the adsorption and desorption curves are steep at intermediate relative pressure. From pore types 1 to 8 and 16 in Fig. 10.20b, this occurs with tubular capillaries; tubular capillaries with two different dimensions; wide and narrow necked 'ink bottle' pores; rectangular cross-section pore; and trough shaped pores.

In type B, the adsorption curve is steep at saturation and the desorption curve steep at intermediate pressure. Pore types 6, 7, 9 and 10 in Fig. 10.20b shows that this occurs with open slit shaped capillaries with parallel walls and capillaries with wide bodies and narrow necks.

With type C, the adsorption branch is steep at intermediate relative pressures with the desorption branch sloping at intermediate relative pressure. Pore types 4 to 8 and 11 and 12 in Fig. 10.20b shows that this occurs with heterogeneous distribution of pores; tapered and wedge shaped capillaries.

The adsorption curve of type D is steep at saturation pressure, while the desorption curve is sloping. Pores type 9, 10, and 13 in Fig. 10.20b shows that this

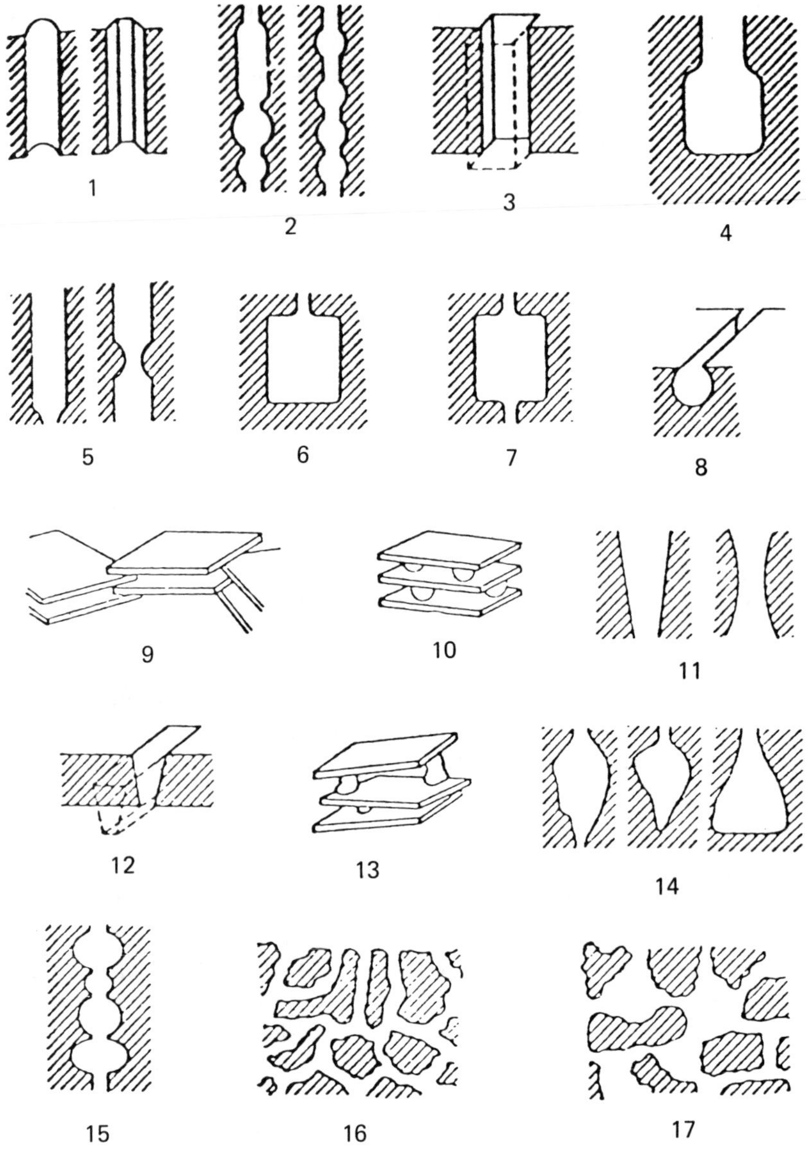

Fig. 10.20b — The fifteen pore types and the two interparticle voids as classified by de Boer.

occurs with a heterogeneous distribution of pores; open slip shaped capillaries with parallel walls.

The adsorption curve of type E is sloping, and the desorption curve is steep at intermediate relative pressure. Pores types 4 to 8, 14, 15, and 17 of Fig. 10.20b suggests that this occurs with similar shapes to Type A and with multidimensional and varying size along the capillary length.

Table 10.7 — Hysteresis loop correlation

Pore shape (Fig. 10.20b)	Hysteresis loop type (Fig. 10.20a)				
	A	*B*	*C*	*D*	*E*
1	×	—	—	—	—
2	× [a]	—	—	—	—
3	× [b]	—	—	—	—
4	× [c]	—	×	—	×
5	× [c]	—	×	—	×
6	× [d]	×	×	—	×
7	× [d]	×	×	—	×
8	×	—	×	—	×
9	—	×	—	×	—
10	—	×	—	×	—
11	—	—	× [e]	—	—
12	—	—	×	—	—
13	—	—	—	×	—
14	—	—	—	—	× [f]
15	—	—	—	—	×
16	×	—	—	—	—
17	—	—	—	—	×

[a]When $r_w \leq 2r_n$. [b]When $1 > a$. [c]When $r_n < r_w < 2r_0$. [d]When $r_w < 2r_n$. [e]When $R \leq 2r$. [f]When r_w, r'_w and r''_w are large relative to r_n, r_w signifies the widest dimesnion and r_n the narrowest.

Pore shape is a difficult parameter to characterize [202], but can be explained by using the geometric ratio of surface area to pore volume:

$$\beta = \frac{1}{r} \times \frac{2V}{A}$$

where: β = pore shape factor
A = surface area
V = volume of pores
r = average pore size.

The last method of surface area determination to be considered is mercury porosimetry.

10.6 MERCURY POROSIMETRY

When a small drop of mercury is placed on the surface of a smooth bed of powder it assumes a spherical shape except for the portion in contact with the surface [154, 184]. As additional mercury is added to the drop, the height increases until it reaches a maximum value. As shown in Fig. 10.21, further additions of mercury increase the drop diameter with no additional increase in the height.

The relationship between angle θ and the maximum drop height h_{max} can be written as:

$$\cos \theta = 1 - \frac{\rho g h_{max}^2}{2\gamma} \tag{10.24}$$

where ρ and γ are the density and surface tension of mercury respectively, and *g* is the acceleration due to gravity.

Mercury prosimetry is widely used for determining the pore size distribution of porous material and the void size distribution of tablets and compacts. The method is based on the capillary rise phenomenon whereby an excess pressure is required to cause a non-wetting liquid to climb up a narrow capillary. The pressure difference across the interface may be written by using the Young & Laplace [185] equation; its sign is such that the pressure is less in the liquid than in the gas phase if θ is greater then 90°, and more if θ is less than 90°.

$$\Delta P = \gamma \left[\frac{1}{r_1} + \frac{1}{r_2}\right] \cos \theta \tag{10.25}$$

Here, γ is the surface tension of the liquid, r_1, r_2 are mutually perpendicular radii, and θ is the angle of contact between the liquid and the capillary walls measured within the liquid.

10.6.1 Operating procedure

The principle of the technique [1, 186–190] may be understood by considering Fig. 10.22. The sample cell, which is constructed of glass, is shown in Fig. 10.23. It consists of two pieces, (a) the portion which contains the sample and (b) the cap with a precision bore tubing constituting the penetrometer (or dilatometer). The two parts are fitted together by means of ground surfaces. A weighed sample of powder is placed in the sample space, part (b) is then fitted on to part (a), and the sample holder is inserted in the pressure chamber. The pressure chamber is a heavy-walled steel vessel since it has to withstand very high pressures.

The sample is next subjected to vacuum to remove adsorbed gases. The time required for degassing depends upon the material to be tested. If the material is thoroughly dried beforehand it will usually suffice to reduce the pressure in the sample chamber at 20 μm of mercury. For high surface area materials which cannot withstand high temperatures it may be necessary to degas for two or three days to obtain reproducible results. This time can be greatly reduced if the sample can be heat treated under vacuum before analysis.

Mercury is next introduced into the sample chamber until it completely covers the sample so that the sample container is completely full of mercury. Excess mercury is then drained off.

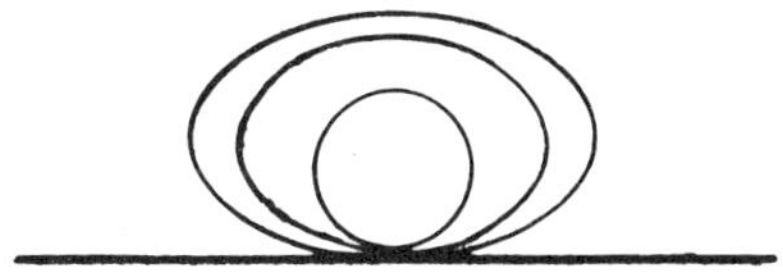

Fig. 10.21 — Change of mercury drop shape with size.

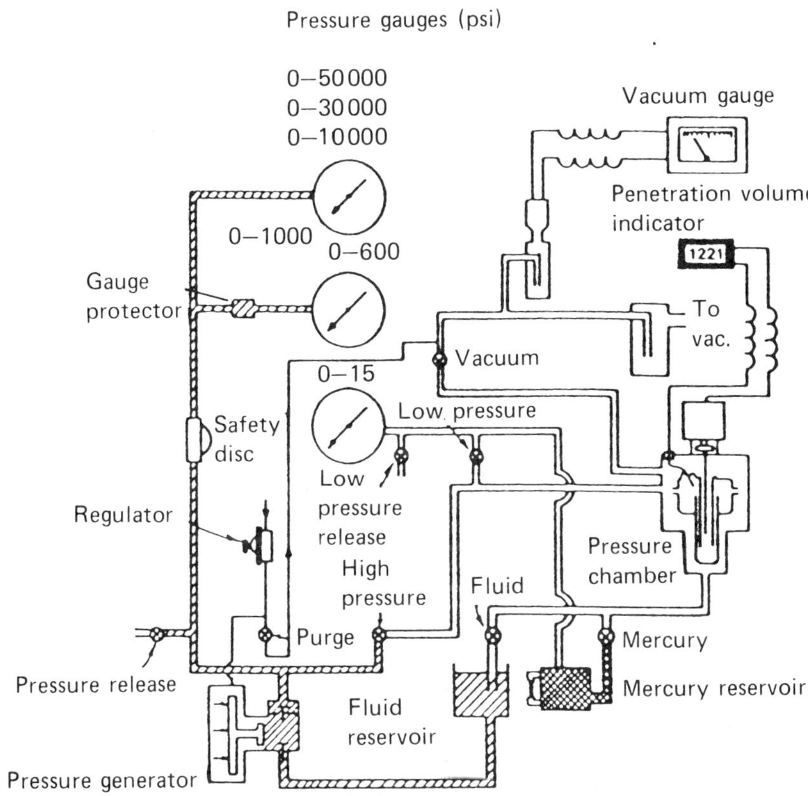

Fig. 10.22 — Micrometres' mercury porosimeter.

The pressure in the sample chamber is then raised to 0.5 psi from which point the analysis has begun. The pressure is then raised in increments of 1 psi to atmospheric pressure. After each increment the reading on the digital counter corresponding to the void or pore volume penetrated is recorded along with the pressure.

Next, the chamber around the sample cell is filled with hydraulic fluid, and the pressure is increased in small increments to the upper pressure level. The applied

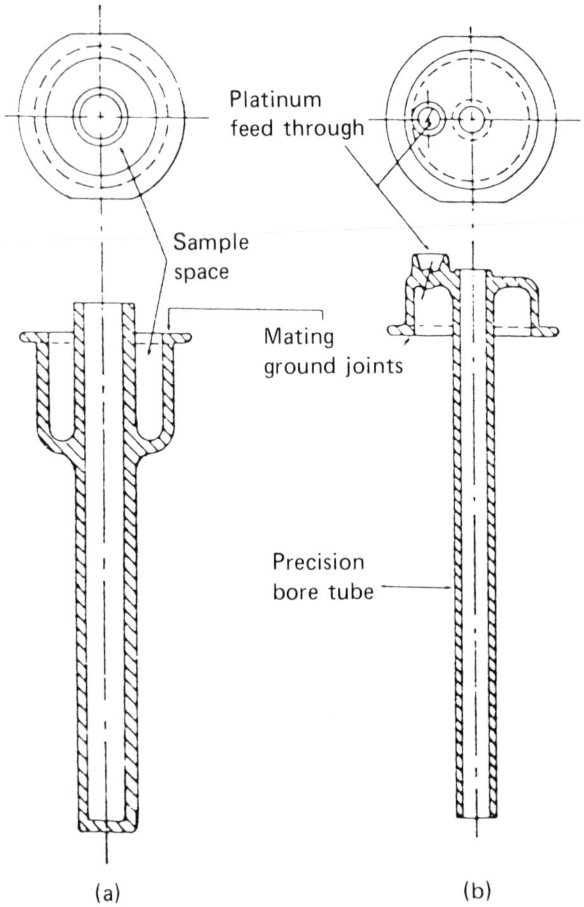

Fig. 10.23 — Sample cell.

pressure forces the mercury into sample voids and pores. Readings are taken at the various pressure levels to determine the volume of mercury penetrating the pores at each pressure. A typical test will require fifty or so separate points. One problem that arises with this particular equipment is mercury contamination. This is eliminated with the Carlo Erba and American Instrument Company's porosimeters in which the sample cell is filled with clean mercury under vacuum before inserting it in the pressure chamber.

The Micromeretics porosimeters have been specially designed to enable the sample to be evacuated to a pressure of approximately 2×10^{-2} torr inside the pressure chamber of the instrument. An alternative filling technique has been described [191] to allow evacuation, outside the pressure chamber, to a pressure of 10^{-5} torr. The advantages of this technique over the conventional filling method are:

(1) The process is under visual control and the fluidizing motion that can occur with very fine powders can be minimized.

(2) Under poorly controlled filling conditions air bubbles can be trapped in the cell, leading to spurious results.
(3) Mercury is not contaminated by the hydraulic fluid during the filling operation, and clean mercury is used for each sample.
(4) Samples can be evacuated for long periods at low pressures.

This filling process is at a disadvantage where pores are to be examined at less than atmospheric pressure. The errors introduced by adsorbed gases are less important for these large pores, and such samples should be examined in the usual manner, [1].

Table 10.8 shows typical experimental data [192] in the form of P against volume intruded. Using Washburn's equation [193]

$$\Delta P = -\frac{2\gamma}{r}\cos\theta \tag{10.26}$$

P is converted to pore radius, and assuming a surface tension for mercury of 0.480 N m^{-1} and a contact angle of 130°C, with pressure P in psi and radius r in μm, equation (10.26) becomes

$$r = \frac{89.5}{P} \tag{10.27}$$

When the pore size range is narrow, (Fig. 10.24) it is possible to plot the cumulative pore size distribution by volume on linear paper and to differentiate the curve to obtain the relative pore size distribution by volume. For a wide distribution, where the radii may vary from 0.0015 to 180 μm, it is necessary to plot the data on log–linear paper with V on the linear axis as shown in Fig. 10.25. This curve of V against P is called the pressurizing curve, and from it the various distributions may be found [1]. Fig. 10.25 may also be used for narrow pore size distributions with V plotted against r to give the cumulative volume distribution and the curve differentiated to give the relative volume distribution. Alternatively, a tabular method may be used. Total surface area may be found from a graphical integration of the PV curve with the application of the equation developed by Rootare & Prenslow [194, 195], in its final form:

$$S_w = -\frac{0.0223}{m}\int_{V_0}^{V} P\,dV\ m^2 g^{-1}\ , \tag{10.28}$$

where m, the sample mass, is in grams and P in psi.

This equation may also be used to determine the cumulative distribution in steps, as shown in Table 10.9. Further, these data may be plotted to give the cumulative

Table 10.8 — Table of experimental pressurizing data

Pressure P ($lb_f\ in^{-2}$)	Volume intruded V ($cm^3\ g^{-1}$)	Pore radius r (μm)
0		
400	0	0.224
500	0.0105	0.179
600	0.0181	0.149
750	0.0251	0.119
1000	0.0321	0.0895
1250	0.0495	0.0716
1600	0.0670	0.0559
1750	0.0809	0.0511
1900	0.120	0.0471
1950	0.148	0.0459
2000	0.191	0.0448
2050	0.218	0.0437
2100	0.240	0.0426
2150	0.254	0.0416
2200	0.275	0.0407
2250	0.289	0.0398
2350	0.318	0.0381
2500	0.352	0.0358
2650	0.374	0.0338
2750	0.388	0.0325
3000	0.423	0.0298
3250	0.445	0.0275
3500	0.458	0.0256
3750	0.472	0.0239
4000	0.479	0.0224
4500	0.493	0.0198
5000	0.500	0.0179
6000	0.508	0.0149
7000	0.512	0.0128
8000	0.515	0.0112
10000	0.515	0.00895

surface distribution shown in Fig. 10.26, and the curve can be differentiated to give the relative surface distribution. Other studies on mercury porosimetry may be found elsewhere [196–200].

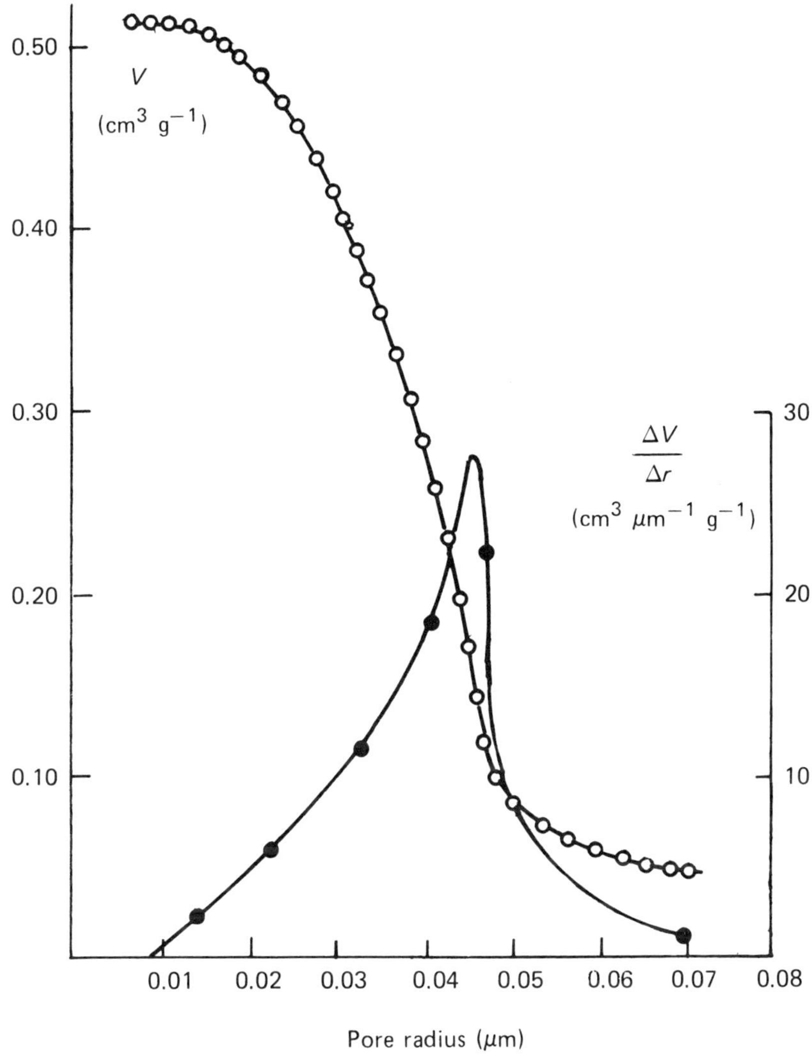

Fig. 10.24 — Cumulative and relative pore size distribution by volume. The relative distribution is obtained by differentiating the cumulative curve to smooth out experimental errors. Tubular points are also included on the relative distribution curve.

REFERENCES

[1] Allen, T., *Particle size measurement*. 3rd Edition, Chapman and Hall, London, New York (1981).

[2] British Standard Institutions: 3636. *Methods for proving the gas tightness of vacuum or pressurized plants*.

[3] British Standard Institutions: 4359, 1969, Part 1: *Nitrogen absorption BET method*.

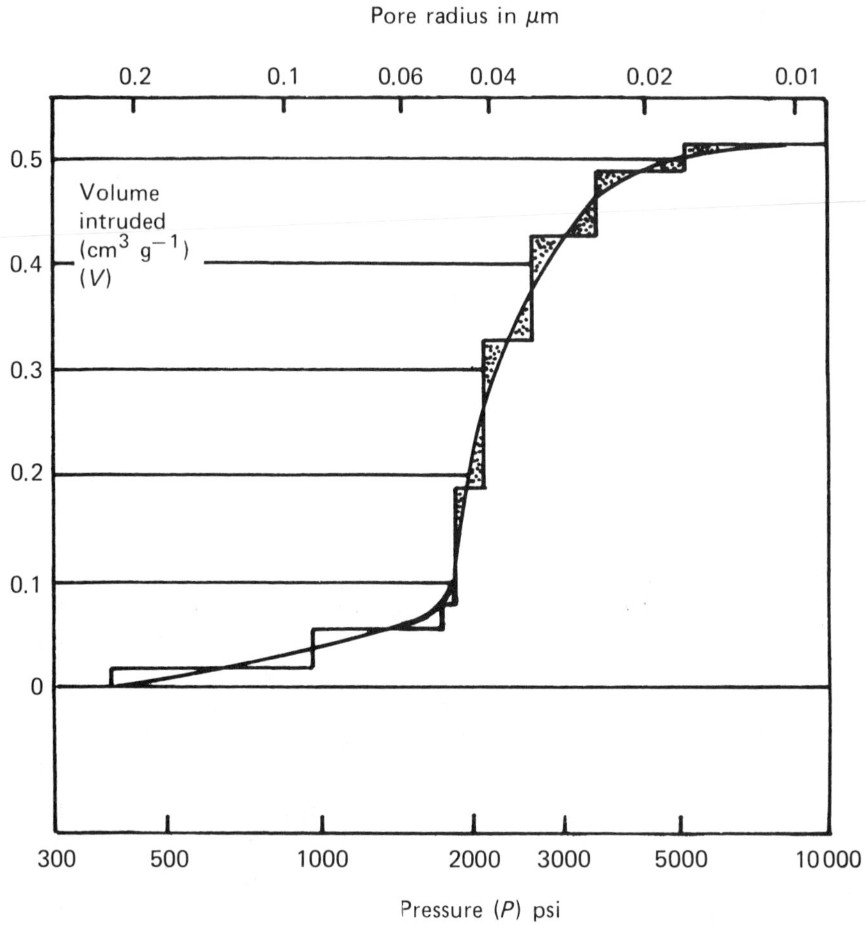

Fig. 10.25 — Pressurizing curve.

[4] Barr, W. E. & Anhorn, V. J., *Scientific and industrial glass blowing and laboratory techniques* (1949).

[5] Langmuir, I., *J. Am. Chem. Soc.*, **40**, 1361, 1368 (1918).

[6] Adamson, A. W., *Physical chemistry of surfaces*, Interscience, NY (1960).

[7] Young, D. M. & Crowell, A. D., *Physical adsorption of gases*, Butterworths (1962).

[8] Halsey, G. & Taylor, H. S., *J. Chem. Phys.*, **15**, 9, 624–30 (1947).

[9] Burevski, D., PhD thesis, University of Bradford (1975).

[10] Sips, R., *J. Chem. Phys.*, **16**, 490 (1948).

[11] Sips, R., *J. Chem. Phys.*, **18**, 1024 (1950).

[12] Volmer, M., *Z. Phys. Chem.*, **115**, 253 (1925).

[13] Fowler, R. H., *Proc. Camb. Phil. Soc.*, **31**, 260 (1935).

[14] Brunauer, S. & Emmett, P. H., *J. Am. Chem. Soc.*, **59**, 2682 (1937).

Table 10.9 — Calculation of cumulative and relative frequency distributions

Pressure P ($lb_f\ in^{-2}$)	Volume V ($cm^3\ g^{-1}$)	Pore radius r (μm)	Mean pressure P ($lb_f\ in^{-2}$)	Volume increment ΔV ($cm^3\ g^{-1}$)	$\overline{P}\Delta V$	$\Sigma\overline{P}\Delta V$	Surface S_w ($m^2\ g^{-1}$)	Δr (μm)	$\Delta V/\Delta r$ ($cm^3\ \mu m^{-1}\ g^{-1}$)	$\Delta S/\Delta r$ ($m^3\ \mu m^{-1}\ g^{-1}$)
400	0	0.224				0	0			
			700	0.0321	22			0.1345	0.24	4
1000	0.0321	0.0895				22	0.5			
			1400	0.0494	69			0.0398	1.24	38
1800	0.0815	0.0497				91	2.0			
			1900	0.1095	208			0.0049	22.3	959
2000	0.191	0.0448				299	6.7			
			2200	0.1390	306			0.0075	18.5	907
2400	0.330	0.0373				605	13.5			
			2710	0.1000	271			0.0086	11.1	698
3120	0.430	0.0287				876	19.5			
			4260	0.0730	311			0.0121	6.0	573
5400	0.503	0.0166				1187	26.5			
			6700	0.0120	80			0.0044	2.2	409
8000	0.515	0.0122				1267	28.3			

From equation (10.28) $S_w = 0.0223 \times 1267$
$= 28.3\ m^2\ g^{-1}$

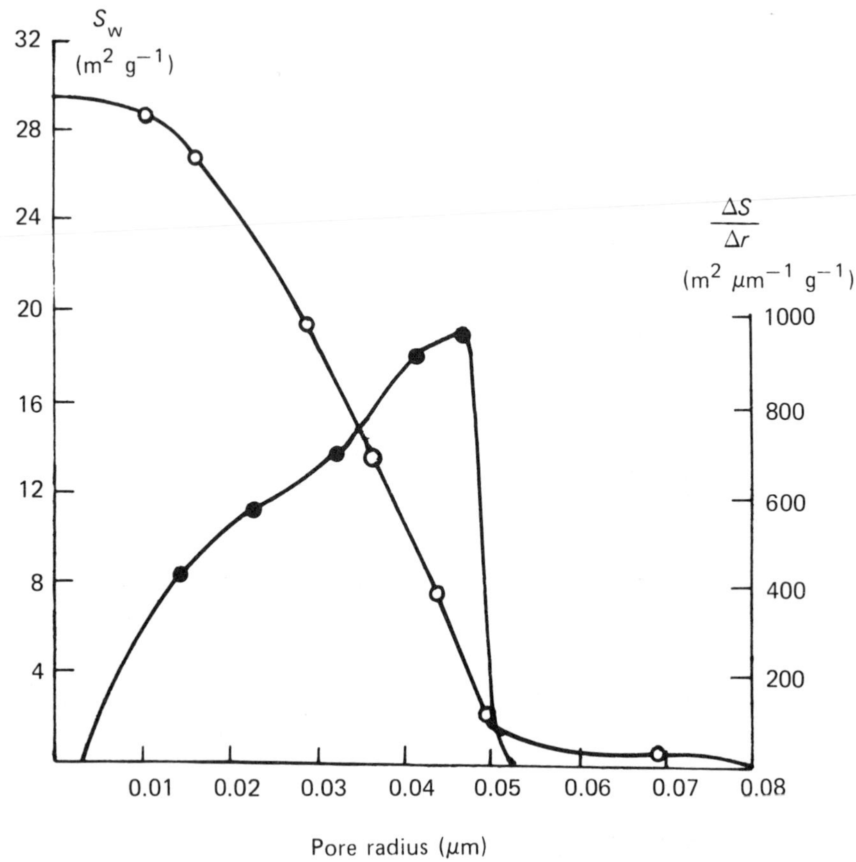

Fig. 10.26—Cumulative and relative pore size distribution by surface. The relative distribution is obtained by differeentiating the cumulative curve to smooth out experimental errors. Tabular points are also indicated on the relative distribution curve.

[15] Brunauer, S., Emmett, P. H., & Teller, E., *J. Am. Chem. Soc.*, **60**, 309 (1938).

[16] Corrin, M. L., *J. Am. Chem. Soc.*, **75**, 4623 (1953).

[17] Orr, C. & Dallavale, J. M., *Fine particle measurement,* Macmillan, N.Y. (1959).

[18] Sandstede, G & Robins, E., *Chem. Ing. Tech.*, **32**, 413 (1960).

[19] Cassel, H. M., *J. Chem. Phys.*, **12**, 115 (1944).

[20] Cassel, H. M., *J. Phys. Chem.*, **48**, 195 (1944).

[21] Brunauer, S., *Solid surfaces and the gas solid interface. Advances in Chemistry Series,* No. 33, American Chemical Society, Washington, D.C. (1961).

[22] Halsey, G. D., *Adv. Catalysis,* **4**, 259 (1952).

[23] Gregg, S. J. & Jacobs, J., *Trans. Faraday Soc.*, **44**, 574 (1948).

[24] Halsey, G. D., *J. Chem. Phys.*, **16**, 931 (1948).

[25] Dollimore, D., Spooner, P., & Turner, A., *Surface Technol.*, **4**, 2, 121–160 (1976).
[26] Gregg, S. J. & Jacobs, J., *Trans. Faraday Soc.*, **44**, 574 (1948).
[27] Genot, B., *J. Colloid Interfac. Sci.*, **50**, 3, 413–418 (1975).
[28] Hill, T. L., *J. Chem. Phys.*, **14**, 268 (1946).
[29] Lowell, S., *Powder Technol.*, **12**, 291–293 (1975).
[30] Brunauer, S., *et al.*, *J. Am. Chem. Soc.*, **62**, 1723 (1940).
[31] Joyner, L. G., Weinberger, E. B., & Montgomery, C. W., *J. Am. Chem. Soc.*, **67**, 2182–2188 (1945).
[32] Emmett, P. H. & Dewitt, T. W., *J. Am. Chem. Soc.*, **59**, 2682 (1937).
[33] Wade, W. H. & Blake, T. D. *J. Phys. Chem.*, **75**, 1887 (1971).
[34] Emmett, P. H. & Brunauer, S., *J. Am. Chem. Soc.*, **59**, 1553 (1937).
[35] Parfitt, G. D. & Sing, K. S. W., (eds.) *Characterization of powder surfaces*. Academic Press (1971).
[36] de Boer, J. H., Kaspersma, J. H., & Van Dongen, R. H., *J. Colloid Interfac. Sci.*, **38**, 1, 97–100 (1972).
[37] Singleton, J. H. & Halsey, G. D., *J. Phys. Chem.*, **58**, 1011 (1954).
[38] de Boer, J. H., *et al.*, *J. Catalysis*, **7**, 135–139 (1967).
[39] Pickett, G., *J. Am. Chem. Soc.*, **67**, 1958 (1945).
[40] Anderson, R. B., *J. Am. Chem. Soc.*, **68**, 686 (1946).
[41] Anderson, R. B. & Hall, W. K., *J. Am. Chem. Soc.*, **70**, 1727 (1948).
[42] Keenan, A. G., *J. Am. Chem. Soc.*, **70**, 3947 (1948).
[43] Brunauer, S., Skalney, J., & Bodor, J. *Colloid Interfac. Sci.*, **30**, 546 (1969).
[44] Barrer, R. M., MacKenzie, N., & McLeod, D., *J. Chem. Soc.*, 1736 (1952).
[45] Hüttig, G. F., *Mschr. Chem.*, **78**, 177 (1948).
[46] Ross, S., *J. Phys. Chem.*, **53**, 383 (1953).
[47] Harkins, W. D. & Jura, G., *J. Chem. Phys.*, **11**, 430, 431 (1943).
[48] Livingstone, H. K., *J. Chem. Phys.*, **15**, 617 (1947).
[49] Livingstone, H. K., *J. Chem. Phys.*, **12**, 466 (1944).
[50] Emmett, P. H., *J. Am. Chem. Soc.*, **68**, 1784 (1946).
[51] Smith, T. D. & Bell, R., *Nature*, **162**, 109 (1948).
[52] Hill, T. L., *Adv. Catalysis*, **4**, 211 (1952).
[53] Alzamora, L. & Cortes, J., *J. Colloid Interfac. Sci.*, **56**, 2, 347–349 (1976).
[54] Dubinin, M. M., *Adsorption-desorption Phenomena*, Academic Press (1972).
[55] Dubinin, M. M., *Q. Rev.*, **9**, 101, London (1955).
[56] Dubinin, M. M., *J. Colloid Interface. Sci.*, **23**, 487–499 (1967).
[57] Dubinin, M. M., *Chemistry and physics of carbon*, (ed. Walker, P. L.), 2, Arnold, London (1966).
[58] Lamond, T. G. & Marsh, H., *Carbon*, **1**, 293 (1963).
[59] Walker, P. L. & Shelef, M., *Carbon*, **5**, 7 (1967).
[60] Freeman, E. M. & Marsh, H., *Carbon*, **8**, 19 (1970).
[61] Toda, Y., *Carbon*, **8**, 565 (1970).
[62] Freeman, E. M., *Carbon*, **8**, 7 (1970).
[63] Dovaston, N. G., McEnaney, B., & Weedon, C. J., *Carbon*, **10**, 277 (1972).
[64] Bering, B. P., Dubinin, M. M., & Serpinsky, V. V., *J. Colloid Interfac. Sci.*, **21**, 378 (1966).
[65] Bering, B. P., *Izv. Akad. Nauk SSSR, Ser. Khim.* (English Trans.), **17** (1971).

[66] Shram, A., *Nuovo Cimento, Suppl.* **5**, 309 (1965).
[67] Ricca, F., Medana, R., & Bellarda, A., *Z. Phys. Chem.*, **52**, 276 (1967).
[68] Marsh, H. & Rand, B., *J. Colloid Interfac. Sci.*, **33**, 101–116 (1970).
[69] Marsh, H. & Rand, B., *3rd Conf. Industrial Carbon and Graphite Soc. Chemy. Ind.*, 212 (1971).
[70] Rand, B., *J. Colloid Interfac. Sci.*, **56**, 2, 337–346 (1976).
[71] Marsh, H. & Siemiensiewska, T., *Fuel,* **44**, 335 (1965).
[72] Kaganer, M. G., *Russ. J. Phys. Chem.*, **33**, 352 (1959).
[73] Jovanovic, D. S., *Kolloid Z. Z. Poly,* **235**, 1203 (1969).
[74] Jovanovic, D. S., *Kolloid Z. Z. Poly,* **235**, 1214 (1969).
[75] Ramsay, J. D. F. & Avery, R. G., *J. Colloid Interfac. Sci.*, **51**, 1, 205–206 (1975).
[76] Rosai, L. & Giorgi, T. A., *J. Colloid Interfac. Sci.*, **51**, 2, 217–224 (1975).
[77] Stoeckli, H. F., *J. Colloid Interfac. Sci.*, **59**, 1, 184–185 (1977).
[78] Schüll, C. G., Elkin, P. B., & Roess, L. C., *J. Am. Chem. Soc.*, **70**, 1405 (1948).
[79] Lippens, B. C., Linsen, B. G., & de Boer, J. H., *J. Catalysis,* **3**, 32 (1964).
[80] Brockhoff, J. C. P., Thesis, Delft Univ., (cit. by 86) (1969).
[81] de Boer, J. H., Linsen, B. G., & Osinga, J., *J. Catalysis,* **4**, 643 (1965).
[82] Lippens, B. C. & de Boer, J. H., *J. Catalysis,* **4**, 319 (1965).
[83] Parfitt, G. D., Sing, K. S. W., & Urwin, D., *J. Colloid Interfac. Sci.*, **53**, 2, 187–193 (1975).
[84] Leclous, A., *J. Catalysis,* **81**, 22 (1970).
[85] Mikhail, R. Sh., Guindy, N. M., & Ali, I. T., *J. Colloid Interfac. Sci.*, **55**, 2, 402–408 (1976).
[86] Sing. K. S. W., *Proc. Int. Symp. Surface Area Determination* (eds. Everett, D. H. & Ottewill, R. M.), Butterworths, London (1970).
[87] Lamond, T. G. & Marsh, H., *Carbon,* **1**, 281, 293 (1964).
[88] Marsh, H. & Rand, B., *J. Colloid Interfac. Sci.*, **33**, 3, 478–479 (1970).
[89] Dubinin, M. M., *Izv. Akad. Nauk SSSR, Ser. Khim.*, **6**, 1232–1239 (1975).
[90] Joy, A. S., *Vacuum,* **3**, 254 (1953).
[91] Emmett, P. H., *12th Report of the Committee on Catalysis, Physical Adsorption in the Study of the Catalysis Syrface,* Wiley, NY (1940).
[92] Joyner, L. G., *Scientific and industrial glass blowing and laboratory techniques,* Instruments Publ. Co., Pittsburgh, Pennsylvania, USA (1949).
[93] Emmett, P. H., *Colloid chem.*, V. Reinhold, New York (1944).
[94] Harvey, E. N., *ASTM Symp. Paint and Paint Material* (1947).
[95] Shubert, Y. & Kopelman, B., *Powder Metall. Bull.*, **6**, 105 (1952).
[96] Vance, R. F. & Pattison, J. N., *Special Report on Apparatus for Surface Area Determination and Other Adsorption Studies on Solids*. Battelle Memorial Institute, Ohio, USA (1954).
[97] Harkins, W. D. & Jura, G., *J. Am. Chem. Soc.*, **66**, 1366 (1944).
[98] Thompson, J. B., Washburn, E. R., & Guildner, L. A., *J. Phys. Chem.*, **56**, 979 (1952).
[99] Bensen, S. W. & Ellis, D. A., *J. Amer. Chem. Soc.*, **70**, 3563 (1948).
[100] Loebenstein, W. V. & Deitz, V. R., *J. Res. Nat. Bur. Stand.*, **46**, 51 (1951).
[101] Dollimore, D., Rickett, G., & Robinson, R., *J. Phys. E.*, **6**, 94 (1973).

[102] Bugge, P. E. & Kerlogue, R. H., *J. Soc. Chem. Ind., Lond.*, **66**, 377 (1947).
[103] Harris, M. R. & Sing, K. S. W., *J. Appl. Chem.*, **5**, 223 (1955).
[104] Lippens, B. C., Linsen, B. G., & de Boer, J. H., *J. Catalysis,* **3**, 32–37 (1964).
[105] Rasneur, B. & Charpin, J., *Fine Particle Int. Conf.*, (ed. Kuhn, W. E.) Electrochem. Soc. Inc., Princeton, NJ (0000).
[106] Wooten, L. A. & Brown, C., *J. Am. Chem. Soc.,* **65**, 113 (1943).
[107] Lister, B. A. J. & MacDonald, L. A., *UK AERE Report C/R 915* (1952).
[108] Pickering, H. L. & Eckstrom, H. C., *J. Am. Chem. Soc.,* **74**, 4775 (1952).
[109] Tomlinson, L. *UKAEA Report 1, G.R.-TN/S — 1032* (0000).
[110] Haul, R. A. W., *Angew. Chem.*, **68**, 238 (1956).
[111] Carden, J. L. & Pierotti, R. A., *J. Colloid Interfac. Sci.,* **47**, 2, 379–394 (1974).
[112] Rosenberg, A. J., *J. Am. Chem. Soc.*, **78**, 2929 (1956).
[113] Dollimore, J., *Chemy. Ind.,* **18**, 742 (1963).
[114] Leipziger, F. D. & Altamari, L. A., *Nucl. Sci. Engng,* **8**, 312 (1961).
[115] Aylmore, D. W. & Jepson, W. B., *J. Scient. Instrum.*, **38**, 4, 156 (1961).
[116] Choudhar, V. R., *J. Sci. Ind. Res.,* **33**, 12, 634–641 (1974).
[117] Nelson, F. M. & Eggertsen, F. T., *Analyt. Chem.,* **30**, 1387 (1958).
[118] Ellis, J. F., Forrest, C. W., & Howe, D. D., *UK AERE Report D-E, G.R. 229 (CA)* (1960).
[119] Atkins, J. H., *Analyt. Chem.,* **36**, 579 (1964).
[120] Haley, A. J., *J. Appl. Chem.,* **13**, 392 (1963).
[121] Whitehead, R. C., Private communication with Ref. 1 (1966).
[122] Lovell, G. H. B., Surface area determination. *Proc. Conf. Soc. Chem. Ind., Bristol, 1969*, Butterworths (1970).
[123] Tucker, B. G., *Analyt. Chem.,* **47**, 4, 78–79 (1975).
[124] Bhat, R. K. & Krishnamoorthy, T. S., *Indian J. Technol.,* **14**, 4, 170–171 (1976).
[125] Yen Chi-Min, Chang, Chi-Yuan, *Hua Hsueh Hsueh Pao,* **35**, 3–4, 131–140 (1977).
[126] Haul, R. & Dumbgen, G., *Chem. Ingr. Tech.*, Part 1, **32,** 349; (1960); Part 2, **35**, 586 (1963).
[127] Gall, L., *Angew. Mess. Regeltech.,* **4**, 12, 107–111 (1964).
[128] Roth, R., *Staub-Reinhalt Luft,* **31**, 8, 320–322 (1971).
[129] Robens, E., *Lab. Pract.,* **18**, 3, 292 (1969).
[130] Robens, E. & Sandstede, G., *Z. Instrum.,* **75**, 167–178 (1967).
[131] Robens, E. & Sandstede, G., *J. Phys. E.,* **2**, 4, 365–368 (1969).
[132] McBain, J. W. & Baker, A. M., *J. Am. Chem. Soc.,* **48**, 690 (1926).
[133] Morris, H. E. & Maass, P., *Can. J. Res.,* **9**, 240 (1933).
[134] Dunn, R. C. & Pomeroy, H. H., *J. Phys. Colloid Chem.,* **52**, 981 (1947).
[135] McBain, J. W. & Britton, H. T. S., *J. Am. Chem. Soc.,* **52**, 2198 (1930).
[136] McBain, J. W. & Sessions, R. F., *J. Colloid Sci.,* **3**, 213 (1948).
[137] Boyd, G. E. & Livingstone, H. K., *J. Am. Chem. Soc.,* **64**, 2838 (1942).
[138] Seborg, C. O., Simmons, F. A., & Baird, P. K., *Ind. Engng. Chem. Ind. Edn.,* **28**, 1245 (1936).
[139] Dubinin, M. M. & Timofeev, D. P., *Zh. Fiz. Khim.,* **21**, 1213 (1947).
[140] Lemcke, W. & Hofmann, U., *Angew. Chem.,* **47**, 37 (1934).
[141] Seborg, C. O. & Stamm, A. J., *Ind. Engng. Chem. Ind. Edn.,* **23**, 1271 (1931).

[142] Pidgeon, L. M. & Maass, O., *J. Am. Chem. Soc.*, **52**, 1053 (1950).
[143] Mulligan, W. O., *Analyt. Chem.*, **23**, 739 (1951).
[144] Stamm, A. J. & Woodruff, S. A., *Ind. Engng. Chem. Analyst. Edn.*, **13**, 386 (1941).
[145] Barrett, H. M., Birnie, A. W., & Cohen, M., *J. Am. Chem. Soc.*, **62**, 2839 (1940).
[146] Gulbransen, E. A., *Rev. Scient. Instrum.*, **15**, 201 (1944).
[147] Rhodin, T. N., *J. Am. Chem. Soc.*, **72**, 4343 (1950).
[148] Rhodin, T. N., *J. Am. Chem. Soc.*, **72**, 5691 (1950).
[149] Rhodin, T. N., *Adv. Catalysis*, **5**, 39 (1953).
[150] Bowers, R. & Long, E. A., *Rev. Scient. Instrum.*, **26**, 337 (1955).
[151] Gregg, S. J., *J. Chem. Soc.*, **561**, 564 (1946).
[152] Gregg, S. J. & Wintle, M. F., *J. Scient. Instrum.*, **23**, 259 (1946).
[153] Cahn, L. & Schutz, H. R., *Vac. Microbalance Tech.*, **3**, 29 (1962).
[154] Lowell, S., *Introduction to powder surface area*. John Wiley & Sons, New York, Chichester, Toronto (1979).
[155] Adam, N. K., *The physics and chemistry of surfaces*, Oxford University Press (1941).
[156] Allen, T. & Patel, R. M., *J. Colloid Interfac. Sc.*, **35**, 4, 647–655 (1971).
[157] Allen, T. & Patel, R. M., *J. Appl. Chem.*, **20**, 165–171, (1970) and *Particle Size Analysis, Soc. Analyt. Chem.*, London (1971).
[158] Orr, C. & Dallavalle, J. M., *Fine particle measurement*, Macmillan, NY (1959).
[159] Adamson, A. W., *Physical chemistry of surfaces*, Interscience, NY (1963).
[160] Weissberger, A., *Organic solvents*, Interscience, NY (1955).
[161] de Boer, J. H., *The dynamical characteristics of adsorption*, Princeton University Press (1953).
[162] Gregg, S. J. & Sing, K. S. W., *Adsorption surface area and porosity*. Academic Press, London (1967).
[163] Allen, T. & Burevski, D., *Powder Technology*, **18**, 2, 139–148 (1977).
[164] Allen, T. & Burevski, D., *Powder Technology*, **17**, 3, 265–272 (1977).
[165] Allen, T. & Burevski, D., *Powder Technology*, **21**, 1, 91–96 (1978).
[166] Giles, G. H., Silva, A. P. D., & Trivedi, A. S., *Surface area determination* (eds. Everett, D. H. & Ottewill, R. H.), Butterworths (1970).
[167] Eltekov, Yu. A., *Surface area determination*, Butterworths (1970).
[168] Lyklema, J. & Van der Hul, H. J., *Surface area determination* (eds. Everett, D. H. & Ottewill, R. H.), Butterworths, 341–354 (1970).
[169] Adamson, A. W., *Physical chemistry of surfaces*, Interscience, NY (1963).
[170] Bickerman, J. J., *Physical surfaces*, Academic Press, NY (1970).
[171] Wade, W. H. & Hackerman, N., *J. Phys. Chem.*, **64**, 1196 (1960).
[172] Partyka, S., *4th Int. Conf. Thermodynamic Chemistry (CR)*, **7**,46–55 (1975).
[173] Harkins, W. D. & Jura, G., *J. Am. Chem. Soc.*, **66**, 1362 (1944).
[174] Gregg, S. J. & Sing, K. S. W., *Adsorption surface area and porosity*. Academic Press, London (1967).
[175] Dubinin, M. M., *J. Colloid Interfac. Sci.*, **23**, 487–499 (1967).
[176] Bering, B. P., Dubinin, M. M., & Serpinsky, V. V., *J. Colloid Interfac. Sci.*, **38**, 1, 184–194 (1972).

[177] Gregg, G. J. & Sing, K. S. W., *Adsorption surface area and porosity.* Academic Press, London (1967).
[178] de Boer, J. H., *The structure and properties of porous materials*, Butterworths (1958).
[179] Oulton, T. D., *J. Phys. Colloid Chem.*, **52**, 1296 (1948).
[180] Wheeler, A., *Catalysis,* **2**, Reinhold, NY (1955).
[181] Schüll, C. G., *J. Am. Chem. Soc.*, **70**, 1405 (1948).
[182] Barrett, E. P., Joyner, L. G., & Halenda, P. O., *J. Am. Chem. Soc.*, **73**, 373–380 (1951).
[183] Kurz, H. P., *Powder technology,* **6**, 167–170 (1972).
[184] Osipow, L. I., *Surface chemistry,* Reinhold, Nre York (1964).
[185] ASdamson, A. W., *Physical chemistry of surfaces,* 2nd Edn., Interscience, NY (1967).
[186] Rootare, H. M. & Spencer, J., *Powder Technology,* **6**, 1, 17–24 (1972).
[187] Rootare, H. M., *Advanced experimental techniques in powder metallurgy, VS: Perspectives in Powder Metallurgy,* Plenum Press, NY, 225–254 (1970).
[188] Scholten, J. J. F., *Porous carbon solids* (ed. Bond, R. L.), Academic Press, 225–249 (1967).
[189] Svata, M. & Zabransky, Z., *Powder Technology,* **2**, 159–161, (1968/69).
[190] Henderson, L. M., Ridgeway, C. M. & Ross, W. B., *Refin. Natur. Gas Mfr.*, **19**, 6, 69–74 (1940).
[191] Palmer, H. K. & Rowe, R. C., *Powder Technology,* **11**, 195–196 (1975).
[192] Rootare, H. M. & Spencer, J., As Ref. [186] and [187].
[193] Washburn, E. W., *Proc. Natn. Acad. Sci.*, **7**, 115–116 (1921).
[194] Rootare, H. M. & Prenzlow, C. F., *J. Phys. Chem.*, **71**, 2734–2736 (1967).
[195] Rootare, H. M., *Advanced experimental techniques in powder metallurgy, VS: Perspectives in Powder Metallurgy*, Plenum Press, NY, 225–254 (1970).
[196] Spencer, D. J. M., *British Coal Utilization Research Assoc. Bull.,* **33**, 10, 228–239 (1969).
[197] Juhola, A. J. & Wiig, E., O., *J. Am. Chem. Soc.*, **71**, 2078–2080 (1949).
[198] Roberts, N. K., *J. Chem. Soc.*, 1907–1915 (1964).
[199] Szeitering, P., *The structure and properties of porous solids*. Butterworths, 287 (1958).
[200] Stanleywood, N. G., *Analyst,* **104**, 1235, 97–105 (1979).
[201] de Boer, J. H. & Briekhoff, J. P. C., *J. Catal.*, **9**, 8, 15 (1967); **10**, 153 (1968); **10**, 368, 377, 391 (1968).
[202] Stanley-Wood, N. G., (Ed.) *Enlargement and compaction of particulate solids*. Butterworths Monographs in Chemical Engineering, London, Butterworths & Co. (Publishers) Ltd. (1983).

11

Structural characterization during gas–solid reactions

11.1 INTRODUCTION

The effect of pore structure and the influence of product layer diffusion on gas–solid reactions are of fundamental importance in the development of local raw materials and processes based on gas–solid reactions. For such reactions, a knowledge of the rate of reaction is necessary for a proper analysis of the process. Structural changes are known to occur in gas–solid reactions which may result in swelling, shrinking, cracking, or sintering during the course of the reaction [1]. Such structural changes affect the overall reaction rate as the diffusion of reactant or product gases through the product layer are altered by the concomitant structural changes as reported for the reduction of nickel oxide with hydrogen [2]. These structural changes are consequently incorporated in the establishment of the reaction mechanism and in the mathematical modelling of the gas–solid reaction [3,4,5,6]. Porosity and internal surface area are conventionally used to characterize the structural changes of the solid material. These two parameters are determined by using a mercury porosimeter and a B.E.T. set-up, respectively.

A method is presented in this study [21] for the derivation of structural characterization parameters using only the B.E.T. data and thereby obviating the need for a porosimeter. The method is applied to the gaseous reduction of the iron oxide, wustite, which is of industrial significance in the direct reduction process for the production of sponge iron and is of topical interest to emerging iron — and steel-makers.

11.2 APPARATUS AND EXPERIMENTAL PROCEDURES

The wustite particles used [21] as the starting material for the reduction experiments were prepared by the thermogravimetric reduction of Carol Lake hematite particles with 60% CO–40% CO_2 at the temperatures 850, 950, and 1000°C at which the particles were subsequently reduced respectively. The apparatus used for the reduction comprised the standard gravimetric arrangement that allowed about 200 mg of the wustite material to be spread evenly in a (Pt + 10% Rh) crucible of diameter 1.7 cm which was hung from the balance arm of a semi-microforce balance and enclosed in a vycor reaction tube. The balance arm provided a continuous monitoring of the weight sample, and this was coupled to a servoscribe for the provision of a permanent record. The reacting sample was heated by a kanthal wound furnace under an arrangement that allowed quenching of the sample to be carried out at any extent of the reaction. A total flow rate of 1 L/min (at s.t.p.) of the reacting gas mixture was used for the experiments. Surface area measurements were carried out by the B.E.T. technique [7].

11.3 RESULTS

Figs 11.1–11.3 show the plot of fractional reduction of the wustite particles against time, and they reveal that at all the reaction temperatures the highest initial rates of reduction occurred during the reduction with pure H_2. The following sections show the methods used to derive the structural characterization parameters.

11.3.1 Calculation of volume of wustite in the sample

Two assumptions are made in this calculation:

(1) The reacting particles do not undergo any dimensional changes, that is, there is no swelling or shrinking of the particles during the reaction. This assumption is consistent with the relatively high resistance of Carol Lake iron oxide to swelling and cracking during reduction [8].
(2) The gangue materials in the oxide do not participate in the reaction, on the basis that their melting point was not attained during the reaction.

Let Z = weight in grams of the wustite material which had a weight change of w% because of the reduction reaction:

$$Fe_yO = yFe + \tfrac{1}{2}O_2 \tag{11.1}$$

Reaction (1) shows that

$$\frac{Zw}{100} \text{ grams of } O_2 \text{ will be lost by} \tag{11.1}$$

$$\frac{Zw}{100} \times \frac{(56y + 16)}{16} \text{ grams of } Fe_y0 \tag{11.2}$$

The volume of the remaining wustite material is then given as

$$\frac{Z}{5.73} \times (1 - w/100(56y + 16)\text{cm}^3 \tag{11.3}$$

where 5.73 g cm^{-3} is the density of wustite [9].

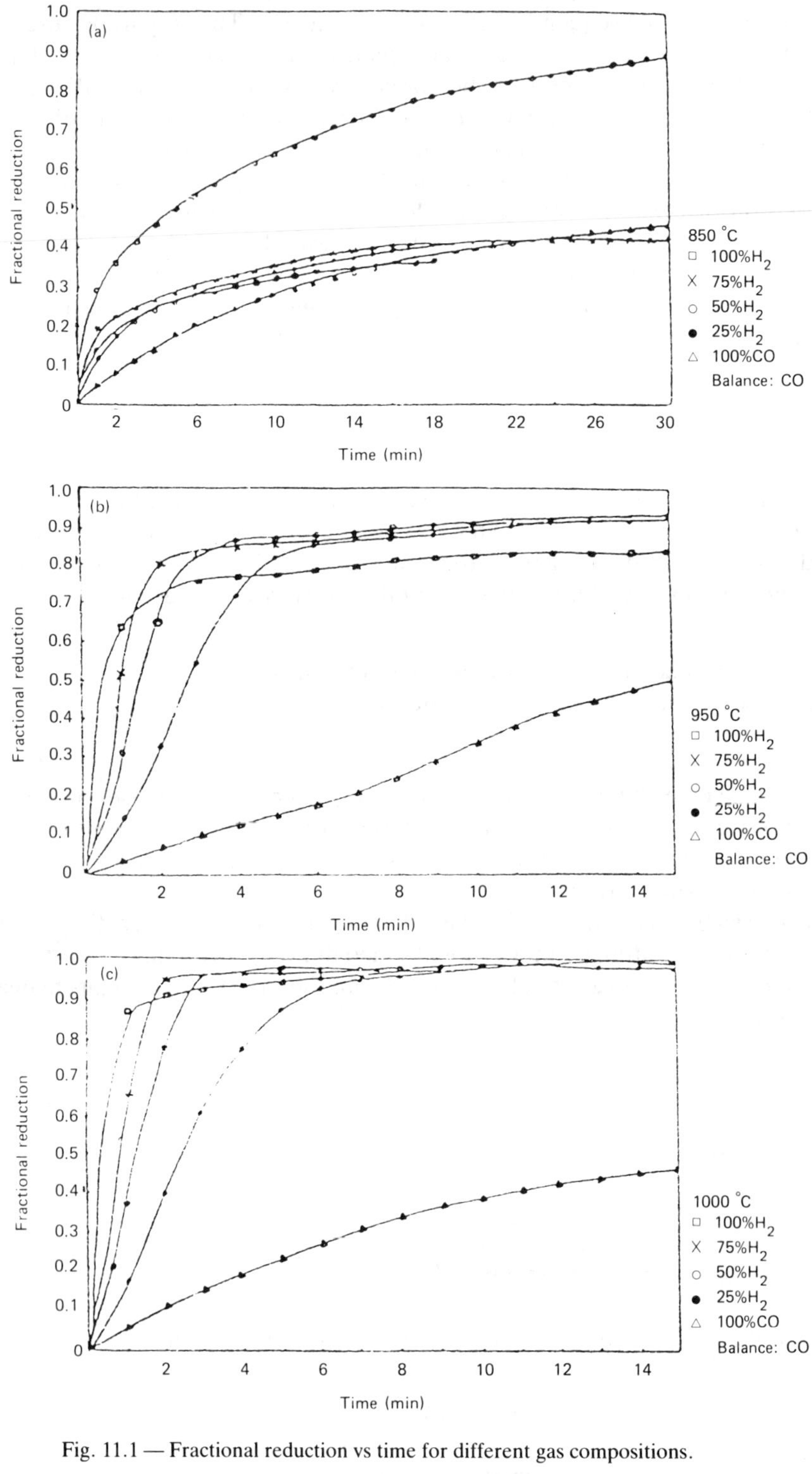

Fig. 11.1 — Fractional reduction vs time for different gas compositions.

11.3.2 Calculation of volume of iron in the sample

From equation (11.1), a weight change of 16 grams of oxygen produces $56y$ grams of iron, thus the volume of the iron produced by a weight loss of $\frac{zw}{100}$ grams is given as

$$\left(\frac{Zw}{100}\times\frac{56y}{16}\right)\Big/7.87\,\mathrm{cm}^3 \tag{11.4}$$

where 7.87 gram^{-3} is taken as the density of the iron [10].

11.3.3 Estimation of effective pore volume

Representing volume as V, the specimen volume at any extent of the reaction is made up of:

$$V_{\mathrm{oxide}} + V_{\mathrm{iron}}\,V_{\mathrm{pores}} = \text{total specimen volume} \tag{11.5}$$

from which

$$V_{\mathrm{pores}} = \text{total volume} - V_{\mathrm{oxide}} - V_{\mathrm{iron}} \tag{11.6}$$

The mean pore radius, r_{mp} is then obtained by using [11].

$$r_{\mathrm{mp}} = (2V_{\mathrm{pores}})/S \tag{11.7}$$

where S, cm^2, is the surface area of the sample determined by the B.E.T. technique.

Computed values of V_{pores} and r_{mp} for the various experimental conditions are shown in Table 11.1.

11.3.4 Apparent particle size from surface area measurements

The mean surface diameter is useful in studying the flow of fluids through porous media where it is used as the diameter of the particle. It is defined as that diameter the square of which when multiplied by the number of particles and also by a suitable constant (π for spheres, 6 for cubes) gives the total surface of the aggregate number of particles. It is, however, easier to use the specific surface area to characterize a particulate system because of the difficulties associated with the determination of the number of particles in such a system [12]. If the particles approximate the spheres, the surface area is πD^2 where D is the diameter of the spheres and the mass is $(\rho\pi D^3)/6$ where ρ is density, so that the specific surface is given as $6/\rho D$. Table 11.2 shows the particle radius obtained from the specific surface areas of the initial wustite samples. Density of wustite was taken as 5.73 g cm^{-3}.

11.3.5 Estimation of porosity

Porosity, θ, was defined as [13].

$$\theta = \frac{(\text{total volume of micropores})}{(\text{total volume of solid} + \text{total volume of micropores})}$$

which simplifies to

Table 11.1 — Structural data of the reduced wustite sample

Reducing gas	Temperature, °C											
	850				950				1000			
	S	V ($\times 10^4$)	r	θ ($\times 10^3$)	S	V (x10^4)	r	θ ($\times 10^3$)	S	V ($\times 10^4$)	r	θ ($\times 10^3$)
100% H_2	401.4	94	0.47	300	411.9	118	0.57	369	435.9	140	0.64	439
75% H_2	535.6	51	0.19	165	418.2	127	0.61	410	348.4	134	0.77	441
50% H_2	470.1	47	0.20	152	289.8	128	0.88	416	268.6	135	1.01	442
25% H_2	367.8	49	0.27	161	292.7	126	0.86	407	258.2	137	1.06	442
0% H_2	524.8	52	0.15	131	178.9	48	0.54	154	123.6	115	1.86	383

Balance: CO
S = surface area, cm^2
V = pore volume, cm^3
r = mean pore radius, μm
θ = porosity.

Table 11.2 — Effective particle radius, r, (μm)

Temperature (°C)	S (m²/g)	r
850	0.21	2.49
950	0.16	3.27
1000	0.12	4.36

$$\theta = \frac{m_p V_g}{m_p V_g + \left(\frac{m_p}{\rho}\right)} \tag{11.8}$$

where V_g = void volume per gram of catalyst
ρ = true density of the bulk solid.

Calculated values of θ from equation, (11.8) are shown in Table 11.1, while the effective particle radius, r, can be seen in Table 11.2.

11.3.6 Calculation of pore structure parameter

Bhatia and Perlmutter [14] have related internal pore structure to area and porosity by use of the pore structure parameter, ϕ, defined by

$$\theta = \frac{4\pi L_0(1-\theta)}{S_0^2} \tag{11.9}$$

where S_0 = initial surface area/unit volume
L_0 = initial total length of pore system per unit volume, which are determined from the relations:

$$S_0 = 2(1-\theta)\int_0^\infty \frac{dV_0}{r(1-V_o)} \tag{11.10}$$

and

$$L_0 = (1-\theta)\int_0^\infty \frac{dV_0}{\pi^2(1-V_o)} \tag{11.11}$$

where V_o is pore volume and r is pore radius. The pore structure parameters, ϕ, for the solid material investigated in this analysis is shown in Table 11.3.

11.4 DISCUSSION

The structural change that accompanied the reduction process was exhibited by the development of pores. The role of pores during the reaction is that of promoting the reduction as the reducing gas penetrates and reacts with the oxide along the pores. These pores become more numerous with increase in the extent of the reaction.

Table 11.3 — Pore structure parameter for reduced wustite

Temperature (°C)	Surface area, S_0 (cm^2/cm^3)	Pore structure parameter, ϕ
850	14.46	242
950	18.6	141
1000	4.9	403

Their origin can be explained thus: In ionic compounds, such as iron oxide, diffusion coefficients are a function of composition, small deviations from stoichiometry then invariably give rise to large changes in the diffusion coefficients. The departure of wustite from stoichiometry results in an increase in the number of cation vacancies, so that it is the cation that is the mobile species, while the oxygen ions, in comparison, are immobile [15]. It is the difference in the rate of diffusion of one kind of atom and another that Greenwood [16] suggested could lead to the creation of vacant lattice sites in metals and alloys. The coalescence of these vacant sites can then lead to the development of porosity in the sample.

The lower porosity development observed during the wustite reduction at 850°C with pure H_2 as compared to that at 1000°C can be explained in accordance with the view of Brill-Edwards *et al.* [17] that pores are created when the rate at which vacancies are generated by the reaction exceeds the rate at which they are diffused out of the grain. At the lower temperature of 850°C, it is reasonable to expect that the pore diffusion process is relatively faster than the interfacial reaction that results in the generation of vacancies. The result is that the rate at which vacancies are generated does not exceed the rate at which they can diffuse out of the grain.

Furthermore, at lower temperatures, it has been reported [18] that the vacancies in wustite tend to cluster, with the result that only few vacancies are available for migration. With only few vacancies available for mobility, the concentration gradient is less steep and so the diffusion is slower. Consequently, iron ion diffusion through the lattice will be retarded since the diffusion of an iron in one direction is always accompanied by the movement of a vacancy in the opposite direction [18]. Moreover, transformation of body centred cubic (bcc) structure of γ-Fe to face centred cubic (fcc) α-Fe above 911°C results in a distributed lattice in which the diffusion of iron ions is accelerated and a large concentration of vacancies can arise where mass (oxygen) is lost.

The higher porosity development observed during the reduction with H_2 as compared to that of CO can also be attributed to the evidence that H_2, unlike CO, reduces the mechanical strength of iron through segregation at boundaries and at interfaces and causes decohesion which could result in a relative ease of void nucleation at the iron–wustite interface. In effect, the pores could have been voids that were produced preferentially along grain boundaries during the reaction. Such preferential void formation at grain boundaries has been observed by Greenwood [16] and Brill-Edwards [17].

In view of these explanations, and with the observed increased porosity development during the rapid rate of wustite reduction with pure H_2, this study is in contrast to the prediction by Turkdogan [19] and Nakiboglu [20] that less porous iron is apt to form when the rate of iron oxide reduction is high.

It may therefore be summarized that wustite reduction with H_2, CO, and H_2-CO gas mixtures can be seen as a typical industrial gas–solid reaction. Structural characterization of the solid structure was based on the parameters of surface area, pore volume, mean pore radius, porosity, and pore structure, which were all generated from B.E.T. data only, obviating therefore the need for a mercury porosimeter. Variation in the porosity of the reaction sample was dependent on both the temperature of the reaction and on the reducing gas composition.

11.5 CONCLUSION

Structural changes that accompany the gaseous reduction of wustite particles at 850, 950, and 1000°C have been characterized with parameters derived from only the B.E.T. data without the use of a porosimeter. The procedure for the structural characterization can be applied to similar gas–solid reactions. In the reduction of wustite, a lower pore development occurred at the lower reduction temperature and also during the reduction with pure CO. The origin of the pores has been attributed to the coalescence of vacant lattice sites. It is postulated that more porous iron is apt to form when the rate of reduction of iron oxide is high.

REFERENCES

[1] Sakai, S., Chida, T., & Tadaki, T., *J. Chem. Eng. Japan*, **14**, 450 (1981).
[2] Szekely, J. & Evans, J. W., *Met. Trans.*, **2**, 1699 (1971).
[3] Bhatia, S. K. & Perlmutter, D. D. *AIChE J.*, **26**, 379 (1980).
[4] Park, J. Y. & Levenspiel, O., *Chem. Eng. Sci.*, **30**, 1207 (1975).
[5] Wen, C. Y. & Ishida, M., *Environ. Sci. Technol.*, **7**, 703 (1973).
[6] Kimura, S., Takagi, Y., Tone, S., & Otake, T., *J. Chem. Eng. Japan*, **14**, 456 (1981).
[7] Flood, E. A., *The solid–gas Interface*, Edward Arnold, London (1967).
[8] Surtees, N., Stone, H. E. N., & Daniel, B. L., *J. I. S. I.*, **208**, 361 (1970).
[9] Quets, J. M., Wadsworth, M. E., and Lewis, J. R., *Trans. AIME*, **218**, 543 (1960).
[10] Bogdandy, L. V. & Engell, H. J., *The reduction of iron ores*, Springer Verlag, Berlin (1971).
[11] Satterfield, C. N., *Mass transfer in heterogeneous catalysis*, M.I.T. Press, Mass., Cambridge (1970).
[12] Brown, G. G., (ed) *Unit operations*, John Wiley and Sons, New York (1966).
[13] Bradshaw, A. V. & Matyas, A. G., *Met. Trans.*, **B**, **7B**, 81 (1976).
[14] Bhatia, S. K. & Perlmutter, D. D., *AIChE J.*, **27**, 247 (1981).
[15] Himmel, L., Mehl, R. F., & Birchenall, C. E., *Trans. AIME*, **197**, 827 (1953).
[16] Greenwood, J. N., *J.I.S.I.*, **171**, 380 (1952).
[17] Brill-Edwards, H., Daniel, B. L., & Samuel, R. L., *J.I.S.I.*, 203, 361 (1965).
[18] Stangway, P. K. & Ross, H. U., *Trans. AIME*, **239**, 1394 (1967).
[19] Turkdogen, E. T., *Met. Trans.*, *B*, **9B**, 163 (1978).

[20] Nakiboglu, F., *Australia-Japan Symp. Extraction Metallurgy, Sidney, Australia* (1980).

[21] Obaro, D. M., Development of Local Raw Materials and Processes. *UNESCO/ ANSTI Chemical Engineering Sub-Network Disciplinary Area Seminar, University of Lagos, Nigeria* (April, 1986).

12

General concept of powder characterization and compaction

12.1 INTRODUCTION

When powders are at rest, they are solids, when aerated they may behave as liquids, and when suspended in gases take on some of the properties of gases. The industrial applications of this new science are far reaching. For example, the size of fine particles affects the properties of a powder in many ways: it determines the setting time of cement, the hiding power of pigments, and the activity of chemical catalysts; the taste of food, the potency of drugs, and the sintering shrinkage of metallurgical powders are also strongly affected by the size of the particles of which the powder is made up [1].

12.2 PROBLEMS IN THE CHARACTERIZATION OF POWDERS

A powder is a complicated and complex form of a solid material, and a powder particle may best be described as a solid of very large surface area compared to its rather small volume [2]. Table 12.1 illustrates some of the characteristics and complexities of a single powder. If we consider a mass of particles, i.e. the powder, then there are as many or more characteristics as can be seen in Table 12.2. The behaviour of powder particles to each other and the effects of the environment are still areas of which very little is known. In fact, if we consider a container filled with a powder, we shall observe that this consists of a relatively small volume of powder particles and usually a larger volume of voids or pores between the particles, and that these voids have their own characteristics, as shown in Table 12.3. In many powder processes these voids determine the behaviour of the powder in processing.

12.2.1 Particle size

Hausner *et al.* [2,3,4], have stated that it is probably a great mistake to express the size of a complicated shaped particle with one figure: the diameter, and to character-

Table 12.1 — Characteristics of a powder particle

A.	MATERIAL CHARACTERISTICS	
	1.	Structure
	2.	Theoretical density
	3.	Melting point
	4.	Plasticity
	5.	Elasticity
	6.	Purity (Impurities)
B.	CHARACTERISTICS DUE TO THE PROCESS OF FABRICATION	
	1.	Density (porosity)
	2.	Particle size (particle diameter)
	3.	Particle shape
	4.	Particle surface area
	5.	Surface conditions
	6.	Microstructure (crystal grain structure)
	7.	Type and amount of lattice defects
	9.	Gas content within a particle
	10.	Adsorbed gas layer
	11.	Amount of surface oxide
	11.	Reactivity

Table 12.2 — Characteristics of a mass of powder

1.	Particle characteristics (as in Table 12.1)
2.	Average particle size
3.	Particle size distribution
4.	Average particle shape
5.	Particle shape distribution
6.	Specific surface (surface area per 1 gram)
7.	Apparent density
8.	Tap density
9.	Flow of the powder
10.	Friction conditions between the particles
11.	Compressibility (compactability)

Table 12.3 — Characteristics of porosity in a mass of loose powders

1.	Total pore volume
2.	Pore volume between powder particles
3.	Pore volume within powder particles
4.	Number of pores between powder particles
5.	Average pore size
6.	Pore size distribution
7.	Pore shape

ize a three-dimensional object with a single dimension. Although accurate size data may be obtained only for spherically shaped dense particles, particles of more complicated shapes can hardly be characterized by one figure.

The term 'powder particle' refers usually to an object which has a physical boundary regardless of the finite size, and as defined in BS 2955, 'a powder shall consist of discrete particles of any material with a maximum dimension of less than 1000 micrometre–1 mm'.

Instrumentation and techniques for particle characterization depend on the use for which information collected is intended [5]. For the general size spectrum, regular and irregular particles may be classified according to their physical dimensions, as shown in Tables 12.4 and 12.5.

Table 12.4 — Arbitrary classifications of physical methods

Physical criterion	Diameter measured		Size range applicability/μm
1. Image	Projected area	d_a	Optical 0.8–800 Electron 0.002–15
2. Mechanical	Sieve diameter	d_t	Dry 40–1000 Wet 1–40
3. Dynamic	Free-fall diameter	d_f	Gravity 1–100
	Drag diameter	d_d	Centrifugal 0.05–25
	Stokes diameter	d_{st}	
4. Attenuation	Projected area	d_a	X-ray 0.05–100
	Stokes diameter	d_{st}	Centrifugal 0.05–50
5. Scattering	Surface volume diameter	d_{sv}	0.3–50
6. Electrical	Volume diameter	d_v	0.5–300
7. Surface	Surface diameter	d_s	Permeametry 0.1–50 (mean)
	Surface volume diameter	d_{sv}	Adsorption 0.005–50 (mean)

Table 12.5 — Equivalent measurements of diameters

Key	Description	Symbol
i	A circle having the same projected area as the profile of a particle	d_a
ii	The width of a minimum square aperture through which a particle will pass	d_t
iii	A sphere having the same volume as the particle	d_v
iv	A sphere having the same surface area as the particle	d_s
v	A sphere having the same velocity of fall in a fluid if in the Stokes region	d_{st}
vi	A sphere having the same ratio of surface area to volume as the particle	d_{sv}

As can be deduced from these tables, the range normally encountered in compaction technology is 1000–10 micrometres, while 0.001 micrometre is commonly encountered in the sub-micrometre to colloidal dimensional lower limit range.

The aim of Table 12.4 is to aid [5] the scientist or technologist in the selection of the relevant measurement technique, while Table 12.5 gives the more common equivalents, which differ numerically and depend to some extent on the physical principles employed in the measurement process, with the exception of the spherical particles.

Studies on image-forming instruments and techniques [6–14] mechanical diameters and techniques [15–22] attenuation diameter [23,24] scattering diameter [25], electrical resistance diameter [26–31], surface diameter [32–42], as well as, perhaps, the best extensive account of particle size analysis on paint and surface coatings [179], can be found in the literature.

The characterization of the particle shape is even more difficult than that of the size. The best characterization of shape can be only an approximation [2,3,4].

12.2.2 Particle shape

It is known that the shape of particles affects to a great extent the behaviour of the powder. Fig. 12.1 shows a variety of particle shapes [2,3]. Factors at present in use usually refer to the 'average shape' of particles in a mass of powder. Hausner *et al.* [2,3,4] characterized the shape of an individual particle by comparing some of its features with those of a rectangle drawn around the projection of the particle. They established three ratios, resulting in the elongation factor, the bulkiness factor, and the surface factor which characterizes the shape of the particle independently of its size. It was concluded that any discussion of the 'shape' characteristics represents a three-dimensional problem, and for the sake of simplicity and better understanding it should be treated as a two-dimensional problem.

Various types of particle shape have been described in BS 2955 1958, and subsequently defined [43] but it has been pointed out [5] that these verbal descrip-

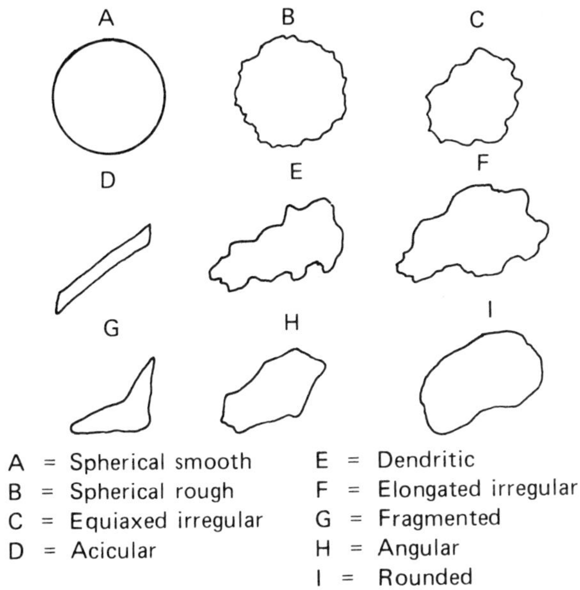

Fig. 12.1 — A variety of particle shapes.

tions are inadequate for incorporation in calculations for ascertaining the effect of particle shape in various particle systems and on particle properties. Particle shape is a fundamental characteristic of a particle, in a manner similar to that of the size of a particle, and can be assessed from individual particles by the measurement of the length or diameter of the particle [44]. British Standard BS 512 of 1966 attempted to designate arbitrary ratios of particle breadth to length, B/L, and particle thickness to breadth, T/B, to various three-dimensional particle shapes such as discs, rods, blades, and equidimensional particles so that numerical values could be assigned to shape rather than literal descriptions. It was, however, Herdan [44] who stated that 'shape factors' have a three-fold function, namely:

(1) They are factors of proportionality between the particle sizes determined by different methods of particle characterization. Thus the particle size determined by microscopical techniques can be compared with that determined by sedimentation.
(2) They are conversion factors for expressing the results of different particle measurement techniques as an 'equivalent sphere' parameter.
(3) They transform the second and third power of the measured particle diameter into the particle surface and particle volume respectively.

The volume or surface of a particle can therefore be written as follows:

Volume of particle ≡ volume coefficient (volume shape factor) multiplied by the cube of some characteristic dimension.

Surface of particle ≡ surface coefficient (surface shape factor) multiplied by the square of some characteristic dimension.

Further, the above definitions can be written algebraically, as

$$V = \alpha_v x^3$$

and

$$S = \alpha_s x^2$$

where x = size characteristic of the method of measurement.

It has been found empirically [45] that the ratio s/x^2, where s is the average particle surface and x the mean particle size, remains sensibly constant for different sizes of the same material. This is the reason why it is at times possible to speak of an average particle shape even though variations of shape with size for some materials have been found [46]. A surface–volume shape coefficient can be obtained by understanding that from a microscope count method the size characteristics in a one-dimensional quantity can either be a Feret's d_F, Martin's, d_M, or project area, d_a, diameter. Because the volume and surface for a given particle are fixed, the numerical values of coefficients of shape α_v and α_s are dependent on the chosen size characteristics and subscripts to α_s and α_v, and must be amplified to denote the method of particle size measurement. Thus, for a projected area diameter, d_a, the volume shape factor becomes α_{va} and the surface shape factor α_{sa}.

$$V = \alpha_{va} d_a^3$$

and

$$S = \alpha_{sa} d_a^2.$$

The two-dimensional quantity, area, can be measured from the microscopical image of the particle, and the three-dimensional quantity, volume, can be measured from either the sedimentation or electrical resistance technique [5]. The surface of an irregularly shaped particle can be represented as $\alpha_{sa} d_a^2$ when the characteristic dimension is the projected area diameter d_a and by definition is equivalent to πd_s^2 when the dimension is that of a sphere of the same surface. Likewise, the volume of an irregular particle is $\alpha_{va} d_a^3$ and equivalent, for a sphere, to $\pi d_v^3/6$.

It follows therefore that the surface and volume shape factors, α_s and α_v, are numerically equal to 3.14 and 0.52 respectively when the particle is spherical because of the relationship of surface and volume of sphere being πd^2 and $\pi d^3/6$ respectively [5].

Use of the above two ratios leads to surface–volume shape coefficient being obtained:

$$\alpha_{sv,x} = \frac{\alpha_{sx}}{\alpha_{vx}}$$

where x is the arbitrary symbol which depends on the method used to characterize the size of particles.

A review of various methods of particle shape analysis by Beddow, *et al.* [47], proposed four classes of evaluation:

I Specific characteristic of individual particles
II Bulk properties of powders
III Mathematical techniques for shape generation
IV Literal definitions of shape.

An alternative proposal by Davies [48] listed, for purposes of clarity, that particle shape can be represented either by

(a) The general particle form, i.e. aspect ratio, proportionality constants, or
(b) The surface topography, i.e. roughness, angularity or roundness or a combination of the above two categories.

The reason for the above clarification is because shape information is needed to describe the behaviour of particles in many industries in which particles are mixed, reacted, stored, or transported, based on the complex interaction of the main parameters of particle size, density, shape, and surface.

As discussed earlier, Hausner [49], in Fig. 12.2, used the concept of surface

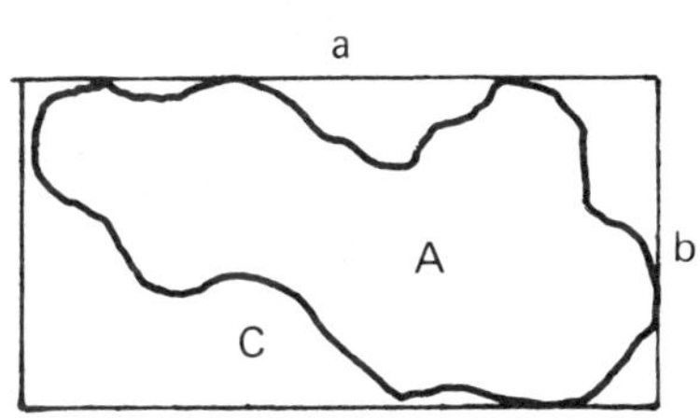

Fig. 12.2 — Characteristics of the particle shape.

shape factor, from the area and perimeter of an irregular particle, to encompass the complicated shape of the particle image with a rectangle. He measured such characteristics as side lengths of the rectangle, a and b, the projected area, A, and the circumference of this area, C.

Given that the cross-sectional area and circumference of a spherical particle is $A = d^2\pi/4$ and $C = d\pi$ respectively, the relationship is $C^2 = 4\pi A$. Therefore, for a spherical particle, which gives the minimum surface for a given volume, a shape factor can be written as [5]:

$$Z = \frac{C^2}{12.6A}$$

which has a value of unity for a spherical particle.

which has a value of unity for a spherical particle.

Fig. 12.3 shows further particle shapes, showing the side lengths a and b, projected particle area A, and the circumference of this area, C. The applicable dimensions for the shapes can be found in Table 12.6. It will be observed for the various shaped particles that Z increases considerably when the surface is corrugated, having values between 1 and 10, with dendritic shaped particles having the highest values.

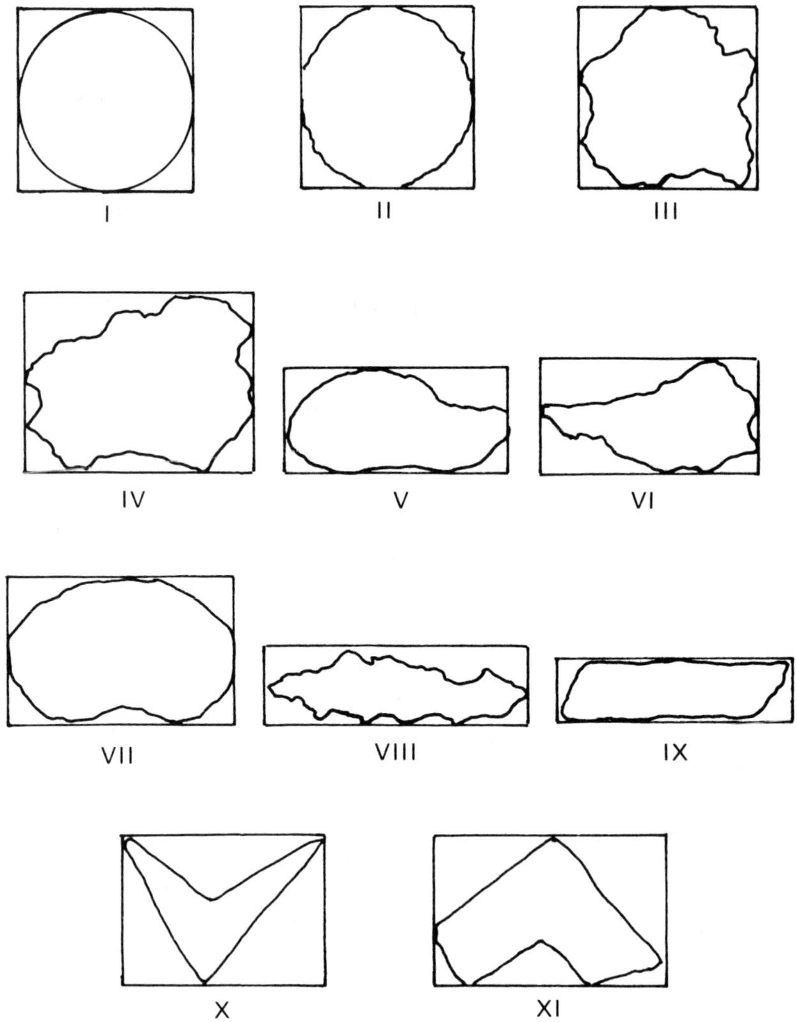

Fig. 12.3 — A variety of particle shapes (see Table 12.6).

Table 12.6 — Relative dimensions and characteristic results of powder particles of various shapes

Shape	Relative dimensions				Particle characteristics		
	a	b	A	C	x	y	z
I	6	6	28.2	18.8	1	0.78	1
II	6	6	28.2	26.5	1	0.78	1.98
III	6	6	25.4	25.5	1	0.71	2.04
IV	8	6	36.2	29.4	1.34	0.76	1.89
V	7.6	3.4	18.7	20.5	2.23	0.72	1.77
VI	7.5	3.8	15.5	21.6	1.97	0.54	2.2
VII	8	5	29.4	22.9	1.6	0.74	1.4
VIII	9	2.5	11.8	33.5	3.6	0.52	7.5
IX	8	2	13.2	17.8	4.0	0.83	1.89
X	7	5	10.7	19.5	1.4	0.31	2.8
XI	8	5	20.2	21.7	1.6	0.5	1.84

The elongation factor,

$$x = a/b$$

defined by Hausner [2, 41], was found not to be informative enough [5] to characterize shape because particles used in powder metallurgy varied between $x = 5$ and 10. It was further found that Hausner's bulkiness factor, expressed as

$$y = \frac{A}{a + b}$$

can be used in relation to the packing density of material, although values above 0.8 are seldom seen, and with values of $y < 0.6$ indicate specimens with large indentations and deep notches.

Other studies to determine shape factors have been carried out [50,51], correlating the relationship between perimeter and the linear diameter of Feret and Martin before and after the advent of the image analyser. Computers using complex mathematical functions are now available [52–59]. Indirect measurement of particle shape from the ratio of surface area can be obtained by photosedimentation techniques [60], and application of shape factors for surface area evaluation are now a matter of routine [5]. Taking, for example, a log-normal distribution number count, this can be converted to a log-normal surface distribution, and the specific surface of the particulate system determined from [5]

$$S_V = \frac{\phi}{x_{sv}}$$

where ϕ is the shape factor which is taken as 6 for spherical particles.

12.2.3 Effect of humidity on the friction in a powder mass

In many cases, the powder behaviour depends on the friction conditions in a powder mass [2]. It is not generally known that the flow of a powder under otherwise closely controlled conditions depends on humidity. Tests made by Lewis *et al.* [4] with mixtures of tungsten powders consisting of particle sizes of 8 and 16 μm with the flow of the powder determined in air of various relative humidities, showed that the flow time for this powder decreases with increasing relative humidity up to 45%, and then increases again, as is evidenced from Fig. 12.4. Up to this optimum humidity the

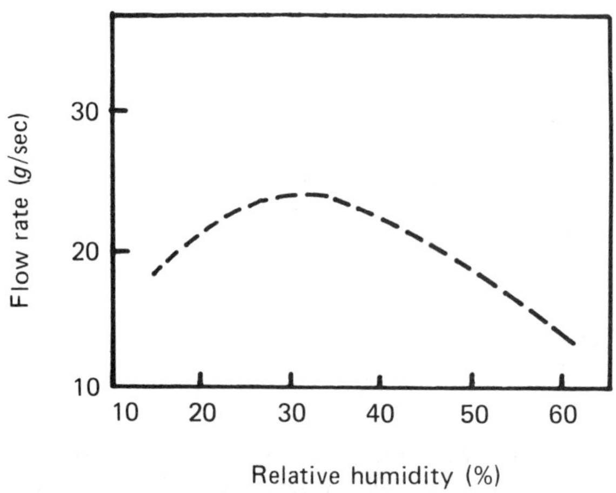

Fig. 12.4 — Effect of humidity on the flow rate of tungsten powder.

moisture in the air acts as a lubricant, lowers friction, and increases the flow rate. Above this optimum humidity the larger amount of moisture acts adversely and decreases the flow rate. The figure of 45% relative humidity is optimum only for the respective powder different with respect to material, particle size and shape. Other characteristics will probably require a different humidity to obtain maximum flow. This is just an example for an environmental effect of which very little is known.

12.2.4 Forces in a mass of powder

Among the new processes for making metal powders, Hausner, *et al.* [2] directed their studies to the fabrication of ultra-fine powders of 0.001 to 0.1 μm sizes. These powders were said to offer interesting aspects, but are still very little used for production purposes because they are not easy to handle and mainly because so very little is known about their characterization. It is clear, however, that many factors which characterize powders with larger particles (sieve size range) cannot be applied in the same way to ultra-fine powders. There are many reasons for the difference in the characterization of these two types of powder. Fig. 12.5 shows the main differences between the ultra-fines, fines, and powders of sieve size range. Larger particles, i.e. heavier particles, follow mainly gravity forces, whereas adhesion, cohesion, and electrostatic forces play a very minor role in the behaviour of most of

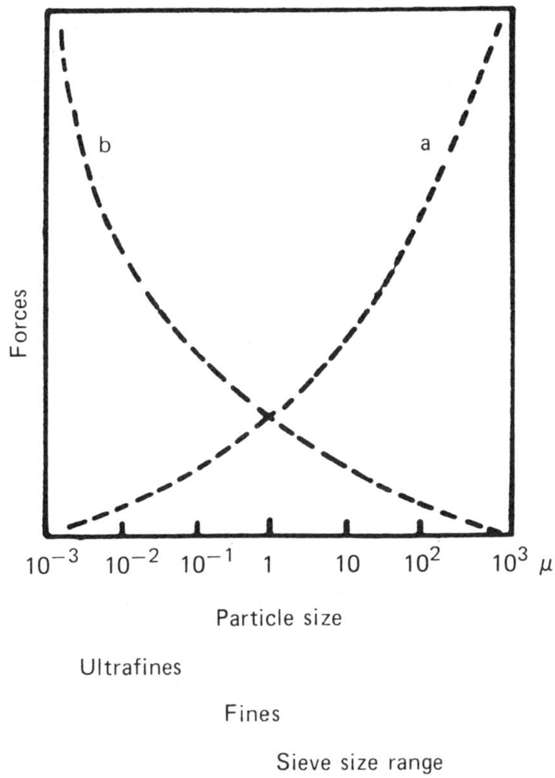

Fig. 12.5 — Forces in a mass of powder, (a) gravity, (b) cohesive, electrostatic.

the coarse powder particles. Gravity forces, however, are so small for ultra-fine powders that they are strongly overshadowed by adhesion, cohesion, and electrostatic forces to which they are exposed. Stokes' law cannot be applied any more. Characteristics such as flow, apparent and top densities for ultra-fine powders have therefore entirely different meanings than for sieve size range powders. Fig. 12.5 indicates clearly that we have to learn to characterize the ultra-fines in a different way from the coarse powders. Friction between ultra-fine particles is basically different from that of coarse particles.

12.2.5 Mixing of powders

The scale of segregation of a mixture measures the size of regions of unmixed or segregated material. The intensity of segregation of a mixture measures the amount of dilution of regions of unmixed material by other mixture components. The mixture quality of a powder mixture will be improved if the scale and intensity segregation are reduced. Unlike gaseous and miscible liquid systems this reduction process will not occur by molecular diffusion, and work must be carried out on the system to generate interparticulate motion. The particles may be tumbled, kneaded, fluidized, sheared, or scooped, but the individual particles must be given the opportunity of relocating themselves within the mixture. The ability of particles to move freely is evidently an important characteristic, and it leads to a general division of powder mixtures into free-flowing mixtures and cohesive or non-free-flowing mixtures.

To a large extent mixing of powders depends on the mobility of the particles, and mobility depends on friction. From Fig. 12.5 we now know that the mobility of ultra-fines depends on entirely different forces from those of the coarser particles, and therefore the mobility of ultra-fines is entirely different from that of coarse particles. In view of this, it may safely be concluded that the mixing problems for ultra fines are different and more complicated than for coarser size powders [2].

An important characteristic of the ultra-fines is their ability to form agglomerates — their 'conglomerance'. This property or characteristic for which the term 'autohesion' was proposed [2], should not be confused with 'adhesion' which refers to the interaction of particles with solid surfaces. In both cases, however, the effect of the environment is much greater on ultra-fines than on larger particles.

12.2.6 Role of surface films

Very little is known about the role of surface films or layers on the behaviour of powders. It has been well established that these films affect strongly the friction conditions in a powder mass and that the type of surface layers or films determines many powder processes, especially the sintering process. The role of these surface layers is, however, entirely different for ultra-fines compared to coarse powders.

12.3 DISCUSSION

The problems mentioned above are not specific just for metallic powders, but are very similar for any powder whether it is metallic or ceramic, whether it is graphite or cement, and the same problems are of concern in the plastic, pharmaceutical, cosmetic, and food industries. The problems are practically identical, and it is only the standpoint from which we deal with these problems which is different.

12.4 CHARACTERIZATION OF THE POWDER COMPACTION PROCESS

It is not sufficient to know solely the particle size, distribution, and shape of the individual particles, as the space within a compact or tablet is not completely occupied with solid material. In-between the solid powders, there are voids filled with liquid or gas, and at times the pores within the individual particles. These void space distributions within compacts or pores within particles play an important role in the characterization of compacts, but it is the cumulative effect of the voids or pores which has a greater effect on the strength, compressibility, and permeability of the product.

How the particles are packed, whether coherent, where sticking of the particles occurs at the particle surface contacts, or non-coherent, where sticking does not occur, and the size of the space within a compact, is a function of particle size, different packing arrangements being achieved by particle orientation and the degree of compaction.

In the compaction of powders in a rigid mould to produce a coherent mass, the fabrication processes are usually considered to occur in three stages [5]:

(a) Particles or powders must flow or be fed to a mould.
(b) Particles are forced into intimate contact on application of a compressive force, forming areas of contact between the particles to produce bonds.
(c) Removal of stress and ejection of the compacted material in the fabrication process depends on the resultant area of contact, fracture, or deformation of the particles, in addition to the shape and dimensions of the discontinuities between particles.

Information on the size, shape, surface, and porosity of the material before and after compaction is required to aid in the characterization of the compaction as well as to quantify the degree and properties of the compaction process and compacted material. These parameters have been the subject of much research work, and may be found in the literature [1–60]. A knowledge of the structural form of the initial packing formation of spherical or irregular shapes is also necessary before detailed analysis of the topography of compressed or uncompressed particles by either nitrogen adsorption or mercury intrusion techniques discussed in the previous chapter, is undertaken.

The closeness of packing is mostly related to some properties of the powder, for example, bulk density, heat conduction, and adhesion, where fusion can occur in a coherent compact by the application of heat to give a sintered product.

When porosity is present in particles, various subdivisions of volume are deduced, giving different densities. In every case, however, force or energy is applied to the particulate system such that porosity is reduced and a densified product is obtained.

The densification of powders is an operation common to a wide variety of operations, and several compaction techniques have been developed to suit particular powder properties or product requirements. For example, vibro-compaction employs gravitational acceleration, slip-casting employs surface tension forces, and for ferrous metals, magnetic compaction forces have been employed. The most widely recognized compaction technique is that of compressing the powder between loaded punches in a die of appropriate cross-section. Fig. 12.6 shows a compression simulator used in pharmaceutical granulation and compaction [61]. The simulator consists of two independently operated, servo-controlled hydraulic actuators which can move standard pharmaceutical punches within a die at velocities up to 400 mm s^{-1} to a positional accuracy of 5 μm. In addition, the simulator can follow a force–time profile with the same relative accuracy up to a maximum compressive force of 55 kN. The two actuators have separate control systems which can be programmed. The required upper and lower punch displacement–time profiles are converted to digital form by use of an ultrasonic graph plotter or a minicomputer. The two profiles are stored separately then fed out simultaneously to the two actuators at any rate between 0.001 and 30 Hz, which gives a range of cycle times of 33.3 ms to 1000 s. A typical recorder trace is reproduced in Fig. 12.7. Hunter [61] states that each actuator is instrumented with a precision load cell and displacement transducer which serve two purposes: first, to feed back force and positional data to

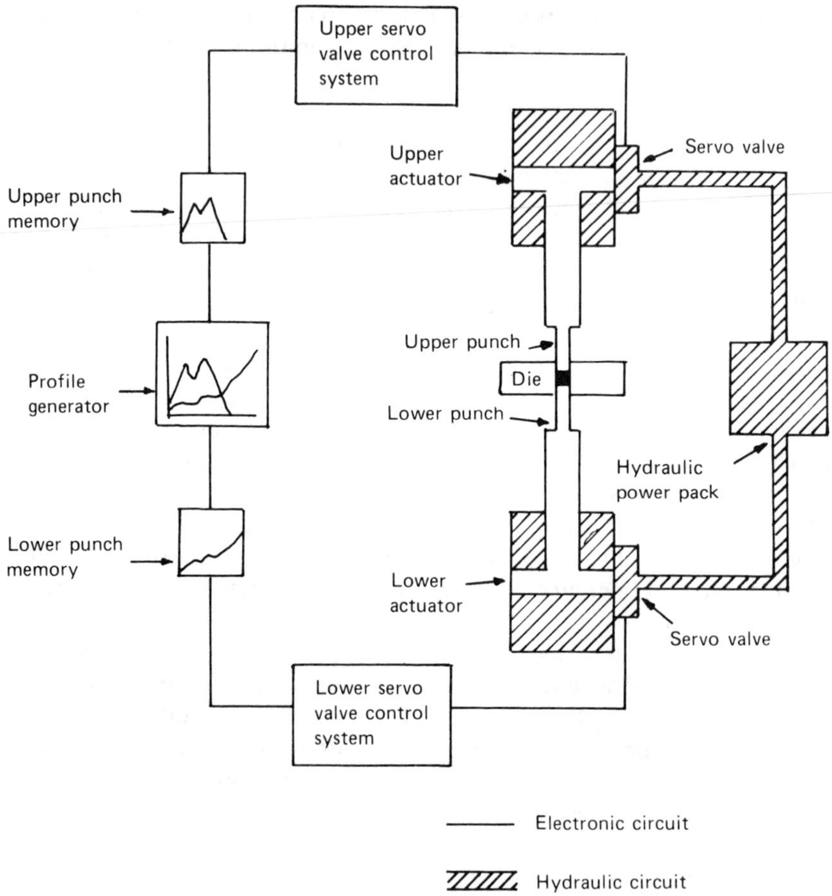

Fig. 12.6 — Compression simulator designed for ICI plc.

the control system; and second, to provide a permanent record of the force and displacement histories of each punch. For the latter purpose the outputs from the transducers are relayed to galvanometers in a multi-channel ultraviolet recorder, and, via A to D converters, to a minicomputer. They may also be monitored on peak-picking digital voltmeters.

Fig. 12.8 shows the forces induced in a cylindrical compact being compressed from one end [62]. MacLeod [62] showed that if the material being compressed were a perfect fluid, force would be transmitted isostatically, and

$$F_A = F_L = F_R .$$

In a particulate system, however, friction is generated at the material–die wall interface as a result of induced radial force, F_R. Radial force F_R is a function of the Poisson ratio, ν, of the material and is related [63] to force F_A by the expression:

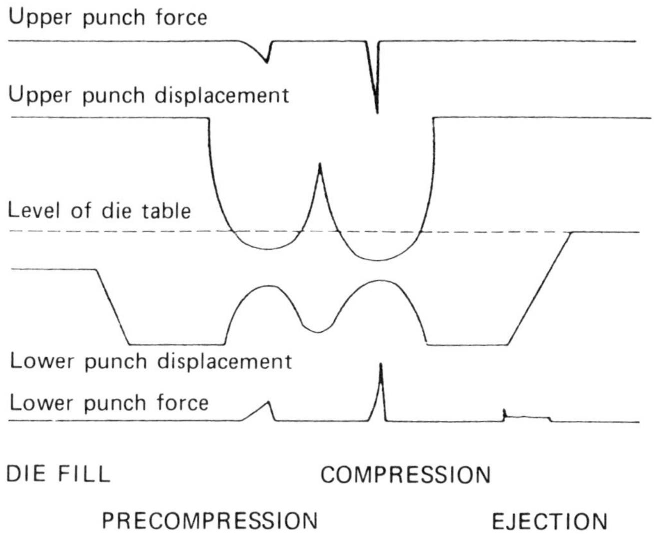

Fig. 12.7 — Typical ultraviolet recorder traces.

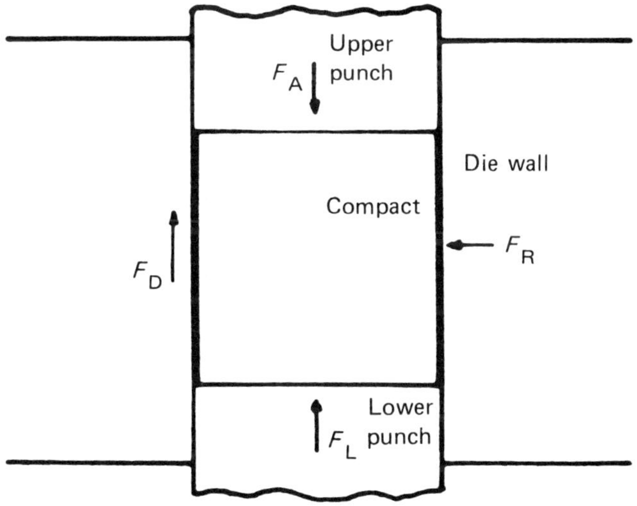

Fig. 12.8 — Forces operating on a powder under compression.

$$F_R = \frac{v}{1-x} \cdot F_A$$

A proportion of the applied force is transmitted to the die body and appears as die wall reaction force, F_D. The remainder of the applied force is transmitted through the powder to the lower punch where it is measured as transmitted force, F_L. The force balance may be written as

$$F_A = F_L + F_D .$$

If F_D is considered as losses due to friction and F_L as force available for densification [64], then F_D incorporates losses due both to die wall fuction, f_w, and interparticle friction, f_i. Interparticle friction would be dominant at an early particle rearrangement stage of densification, and die wall friction for the rest of the process. Each component may be expressed in terms of radial force, F_R. Where μ_w is the coefficient of die wall friction, and μ_i is the coefficient of interparticle friction:

$$\begin{aligned} F_D &= (f_w + f_i) \\ &= F_R\mu_w + F_R\mu_i \\ &= F_R(\mu_w + \mu_i) \end{aligned}$$

and the force balance may be rewitten:

$$F_A = F_L + F_R(\mu_w + \mu_i) .$$

Duwez & Zwell [65] plotted the pressure ratio P_L/P_A against compact H/D ratio and obtained a dimensionless curve of the form:

$$\ln\left(\frac{P_A}{P_L}\right) = K_1\left(\frac{H}{D}\right) .$$

Sheinharz and co-workers [66] confirmed that the ratio F_L/F_A was a constant for a powder, depending on compact mass, die geometry, and lubrication conditions. Later workers [67] used the ratio F_L/F_A, termed the *R*-value, as a measure of lubricant efficiency.

The extent of die wall and interparticle frictional forces can be estimated, by using these simple relationships, from external measurements, but information obtained by a static balance of forces in this way is said [62] to represent values averaged over the compact. It is argued that since F_R and μ_w vary over the compact length, i.e. distance from the applied force, the information gained is of limited value. In particular, since forces are resolved in terms of their axial and radial components, no information is gained about stress distribution, and correlation of force measurements with compact density structure is impossble.

Fig. 12.9 shows the shear apparatus used by Hvorslev. Hvorslev [68] was the first to derive a relationship between the shear stress, τ at failure to the normal stress, δ

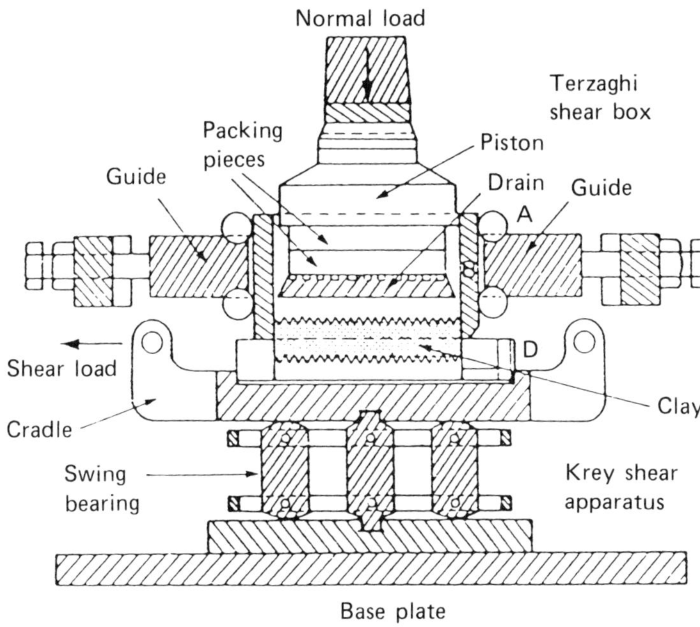

Fig. 12.9 — Shear apparatus used by Hvorslev.

and the void ratio, e, of a particulate solid. The condition of failure is therefore defined [5] by a surface known as the 'Hvorslev surface' in the three-dimensional system whose coordinates are δ, τ, and e.

In the compaction process, it is necessary to know how the following variables can be determined. Remember that the density of a specimen is commonly defined as the mass divided by its volume. When a specimen has no internal pores, then no difficulty arises in density determination but when porosity is present in particles, various subdivisions of volume and therefore different densities are determined.

12.4.1 True density

Put simply, the density of the particle measured by any method that will measure the internal pores of the particulate material gives:

$$\rho_t = \frac{\text{Mass}}{V_t}.$$

An X-ray method [69,70] can be used for the determination of true density, and it is the measurement of the volume of the unit cell of the crystal derived from:

$$\rho_t = \frac{n\,M_w}{NV}$$

where:

n= number of molecules in unit cell
M_w= molecular weight
N= Avogadro's number
V= Unit cell volume.

Also, using Archimedes principles [71,72], single crystal densities can be measured. In the process of this determination, care must be taken to minimize the errors arising from surface tension effects, adsorption of liquid used (like water, xylene), and bubble formation [5].

12.4.2 Particle density

Determination of particle density is carried out by measurement of the volume of a fluid which does not enter the closed internal pores of the material, given by:

$$\rho_p = \frac{\text{Mass}}{V_p}$$

where:

ρ_p= particle density
V_p= particle volume and equivalent to $V_t + V_c$
V_t= volume of solid and true volume
V_c= volume of closed pores
V_o= volume of open pores.

As specified in BS 1902: 1952, it is necessary to remove air from the specimen before admittance of the probe fluid, and if the fluid is liquid, the specimen should be boiled with the probe liquid as stated in the standard. Owing to the compressibility and low density of gases, the use of gases as a probe fluid make volume measurements, at times, more difficult. Using appropriate apparatus, particle densities can be obtained by the evacuation of air or other gases from the specimen pores. By introducing helium and allowing it to penetrate through small pores, the displacement volume can be measured. Because of its small molecular size and assumed lack of adsorption on solid surfaces at room temperature, helium is the most common gas used.

12.4.3 Bulk density or apparent density

When formed under stated, freely poured conditions, bulk density may be defined as the mass of particles per unit pore volume, and may be written as:

$$\rho_p = \frac{\text{Mass}}{V_b}$$

where:

V_b= bulk volume, which is equivalent to $V_t + V_c + V_o$.

By direct volume measurement, open pore filling with liquids or low melting point solids, or a mercury displacement method [73,74], this parameter can be determined. Mercury will penetrate circular capillaries of approximately 15 μm diameter, at one atmosphere pressure.

12.4.4 Tap density

When a stated amount of powder in a container of stated dimensions is vibrated or tapped under definitive conditions, the tap density of a powder is obtained [5].

In considering the space occupied by the packing of particles, it will be observed that the column is not completely filled with solid matter. It follows, therefore, that the degree to which the space is filled with solid matter depends upon the orientation and arrangement of particles and the existence or non-existence of the shape and size of pores within particles. This is the reason why much confusion exists with the terms 'voidage' and 'porosity', and although the interparticular space can be regarded as a voidage, with the intraparticular space being regarded as a porosity, particle porosity, i.e. intraparticle space, when present, has an effect upon the overall space — voidage — in a compact or packing of particles. Voidage can be defined as either [5]:

(a) the ratio of open space volume to the combined open space and solid material volume, or
(b) the ratio of the interparticle space volume to the mechanical volume of a coherent mass.

Similarly, powder or compact porosity can be defined as voidage or as the ratio of the volume of voids plus the volume of open pores to the total volume occupied by the powder or compact mass. Hence:

$$\text{Voidage} = e = \frac{\text{volume of voids}}{\text{volume of bed or compact}} = \frac{V_v}{V_v + V_s}$$

where V_s and V_v are void and solid volume respectively, and

$$\text{Powder porosity} = \frac{\text{total volume} - \text{solid volume}}{\text{total volume}}$$

$$= 1 - \frac{\text{solid volume}}{\text{total volume}}$$

$$= 1 - \frac{V_t}{V_b}$$

where

V_b= mass/ρ_p
V_t= mass/ρ_t
ρ_t, ρ_b= true density of solid and bulk density of bed respectively.

The porosity of a powder or a compact is then equivalent to:

$$e = 1 - \frac{\text{bulk density}}{\text{true density}} = 1 - \frac{\rho_b}{\rho_t}$$

It should be noted that the interparticle volume for a non-porous solid is the difference between the reciprocals of bulk density and the true density, the particle density for a non-porous solid being equivalent to the true density. Stanley–Wood [5] has emphasized that for a porous material there is, in addition to the interparticle volume, an internal pore volume. This internal pore volume is the difference between the reciprocals of true density and the particle density because the particle density does not include internal pores. Further, particle porosity can be defined as the ratio of the space volume within an individual or collection of individual uncompressed particles to the overall displacement volume of the individual or collection of individual uncompressed particles. Hence, for compacts of porous materials the sum of voids and internal volume is the difference between the reciprocals of bulk density and true density.

Finally, when the non-coherent arrangement of monosized non-porous particles is considered, it should be noted that the degree of porosity is independent of particle size, and porosity is solely a function of arrangement. The four commonest regular packings for spheres are shown in Table 12.7 The volume of the unit cell to be

Table 12.7 — Characteristics of some regular packing

Type of packing	Points of contact or co-ordination number n	Volume of unit cell*	Porosity e	Pore volume per cm³ of spheres cm³/cm³	Pore volume per unit cell†	Radius of sphere inscribed in the cavities‡	Radius of circle inscribed in the throats connecting cavities†
Hexagonal close packed (rhombohedral)	12	$0.71d^3$	0.260	0.350	$0.18d^3$	$0.225R$-octahedral $0.414R$-tetrahedral	$0.155R$
Body centred tetragonal	10	$0.75d^3$	0.302	0.430	$0.23d^3$	$0.291R$	$0.265R$ $0.155R$
Primitive hexagonal (orthorhombic)	8	$0.87d^3$	0.395	0.654	$0.34d^3$	$0.527R$	$0.414R$ $0.155R$
Primitive cubic	6	d^3	0.476	0.910	$0.48d^3$	$0.732R$	$0.414R$
Tetrahedral	4	—	0.660	1.94	—	$1.00R$	$0.732R$

†d = diameter of particle.
‡R = radius of particle.

considered and the pore volume per unit cell are, however, dependent on particle size, because a property like a ratio is dimensionless, while a property like an area or volume which includes dimensions will vary with size [5].

12.4.5 Density–pressure relationships in powder compaction

The compaction of powdered materials is carried out primarily to increase the density of the material. If the ultimate goal of the overall process is the attainment of minimum porosity, compaction is responsible for most of the densification. For example, loose powdered metals [75] have porosities of about 65 to 75 per cent. Axial loading in a die or hydrostatic pressure may effectively reduce the porosity of most powders to 20 per cent or lower with pressures of 50 000 psi and greater.

The relationship between the density of a powder compact and the pressure needed to achieve that density is most commonly obtained by pressing several compacts, each at a different pressure. Measurement of the densities of the individual compacts when they are removed from the die then provides the density–pressure relationship.

At present, densification of powders during compaction is variously described by terms such as compactibility, compressibility, and compression ratio. Schwarzkopf [76] proposed that compactibility be defined as the minimum pressure needed to produce a green strength, while compressibility be used to indicate the extent to which the density of a powder is increased by a given pressure. Compression ratio is generally defined as the ratio of the compact density obtained at a given pressure to the apparent density of the loose powder. In other words, if a compressive force of sufficient magnitude is applied to a powder held within the confines of a mould the particles comprising the powder will adhere, and a coherent strong mass having some degree of porosity will be produced.

Hardman & Lilley [77] have reviewed the mechanisms of compaction of powdered materials and its origins. Published work on the subject did not appear until the early nineteenth century when Wollaston [78] described his powder compaction process for the production of billets of pure platinum. Since then, several hundred papers relating to powder compaction have been published. Walker [79] and others [80,81] presented empirical relationship between the compacting pressure and the change in height or volume of the powder mass, and to some extent these have proved to be useful in the selection of optimum compaction conditions, even though the relationships have done little to explain the basic complexities of the process, especially the understanding of the behaviour of powders under applied stress [77].

Seelig and Wulff [82] attempted to rectify the situation by suggesting that compaction proceeded in four main stages, namely,

(1) Particle rearrangement leading to closer packing.
(2) Formation of arches and vaults, protecting voids capable of supporting the applied pressure.
(3) A multifarious process of deformation, cold welding, fragmentation, and percolation resulting in higher density, i.e. elastic and plastic deformation.
(4) Bulk compression of the material itself.

Fig. 12.10 shows a model for the consolidation process. The slope and relative extent of each stage will depend on material properties and process conditions.

Oudemans [83] considered the initial rearrangement state to be of particular importance with very fine powders (< 1 μm particle size). This agrees with the view expressed by Jones [84] that the repacking process will be influenced largely by the

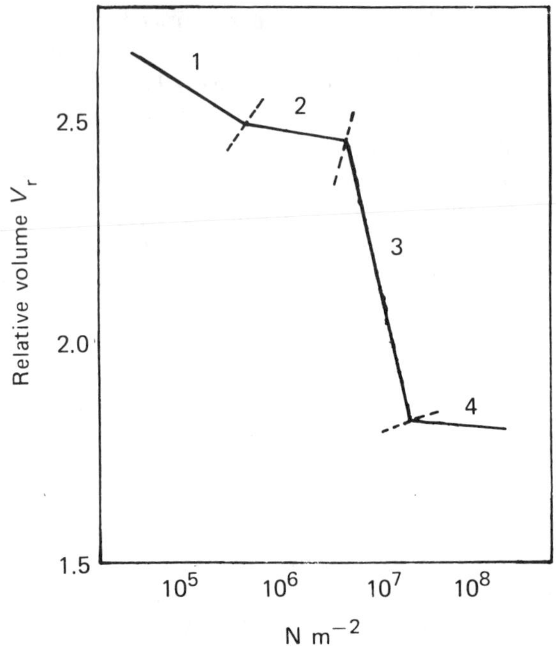

Fig. 12.10 — Stages in the compaction process.

frictional properties of the particles; with smaller particles the number of particle contacts per unit volume will be greater, and the likelihood of fragile arching systems within the loose powder will be correspondingly increased. Cooper & Eaton [85] discussed the relative merits of particle fracture and particle deformation as pore-filling mechanisms. They concluded that higher ultimate porosities will always be associated with powders undergoing compaction by fracture, since as pores become smaller the probability of fragments being produced which will just fill these pores is progressively reduced.

These compaction mechanisms, are, however, purely speculative [77], and in one instance only has visual observation of particles before and after compaction been attempted seriously. Newman [86] attempted to verify the theory that thc stress conditions encountered in the briquetting process can give rise to plastic deformation of coal particles of approximately 500 μm linear dimension. Examination of the briquetted samples by an optical microscope led him to conclude that plastic flow had taken place, but the subsequent suggestion of Gregory [87] that deformation may be due to the formation of micro-cracks in the coal particles cast some doubt on this conclusion. However, since optical microscopy does not reveal details much smaller than 1 μm, it was impossible to draw firm conclusions regarding the plastic flow of coal. The use of the scanning electron microscope with its high resolving power by Hardman & Lilley [77] showed that considerable particle fracture occurs during compaction of coal, and that some particle fragments remain wedged between larger particles to form chain bridges which maintain a more open structure.

Despite the antiquity of the compaction process the fundamental question of the mechanism whereby the compressed powder becomes a coherent compact remains unanswered. Newitt & Conway-Jones [88] examined closely the capillary adhesion due to the presence of moisture between the particles, but detailed study of the dry-bonding process has not been undertaken. As long ago [77] as 1721 Sir Isaac Newton proposed that 'the parts of all homogeneal hard bodies which fully touch one another stick together very strongly'. This statement is in striking agreement with theories of friction developed by workers in the present century [89–91], who consider that where true contact occurs there is no interface between the two faces, and the one solid effectively becomes a part of the other. Seelig [92] suggested that this true contact between the surfaces of particles is responsible for the cohesion of compressed dry powders. Turba & Rumpf [93] challenged this theory, and considered that such regions of intimate contact were unlikely to be produced, and presented evidence to show that bonding was due principally to the short range van der Waals forces. The argument is due, at least in part, to the fact that the mechanism of adhesion is dependent on the type of powder being considered. While it is unlikely that the brittle, chaffy barium sulphate particles used by Turba & Rumpf underwent extensive plastic deformation, this kind of behaviour clearly occurs with soft metal powders and results in the formation of large regions of intimate contact [77]. Wulff [94] pointed out that the greater the plasticity of the metal powder the greater would be the possibility of forming atomic bonds during the pressing operation.

Seelig [95] suggested that electrical conductivity measurements might shed light on the nature of the adhesion in compresssed metal powders. Klar & Shafer [96] made such measurements and found a strong correlation between strength and electrical conductivity of unsintered metal powder compacts. They considered that the bonding between particles arises from the areas of true contact, which also determines the electrical conductivity of their specimens. Bowden & Tabor [97] have pointed out the difficulties of using measurements of electrical conductivity as a means of measuring contact area, and suggested that this remarkable correlation by Klar & Shafer must, therefore, be treated with caution. In addition, the strength of a powder compact will be determined not only by the area of contact between particles, but also by the shapes and dimensions of the discontinuities between the particles. On the basis of compaction mechanisms, Hardman & Lilley [77] found that the compaction of both sucrose and coal powders is dominated by particle fracture. For sodium chloride the area of contact, A, is related to the compacting pressure, P, by the equation

$$P = kA + K$$

where k and K are constants.

Very extensive literature reviews on the fundamental aspects of the consolidation of powders [98] and pressing of metal powders [99] have been carried out by both James [98] and Seelig [99].

12.4.6 Some applications of the compaction process

12.4.6.1 Pharmaceutical tablets

Compressed powders or granules in the form of pharmaceutical tablets are a common method of delivering drug substances as medicines. The compressed tablet was evolved from the moulded pill as a solid dosage form which is both elegant and convenient for the patient to carry and swallow. It maintains the drug in a dry, stable presentation, yet disintegrates rapidly into a finely divided form when ingested, thus ensuring rapid dissolution of the active agent. It is also easy and cheap to pack and transport, and can be produced on high-speed tabletting machines, with suitable emboss marks to effect product identification.

However, the physical properties of the tablets may influence the manner and means by which the drug is released *in vivo*, so that factors such as the rate in which a liquid, e.g. water or gastro-intestinal fluid, penetrates the compacted mass may be critical to the biological behaviour of the delivery system.

Pharmaceutical tablets are known to be porous to a varying degree, depending on the materials used, the method of granulation, and the compressional force used to prepare the compact. In general, it may be concluded that an increase in compressional force will result in a decrease in the pore size of the resultant compact, but the relationship between pressure and pore size is evidently complex, as has been demonstrated in the literature review.

Fig. 12.11 shows a typical pharmaceutical granulation process. The granulation

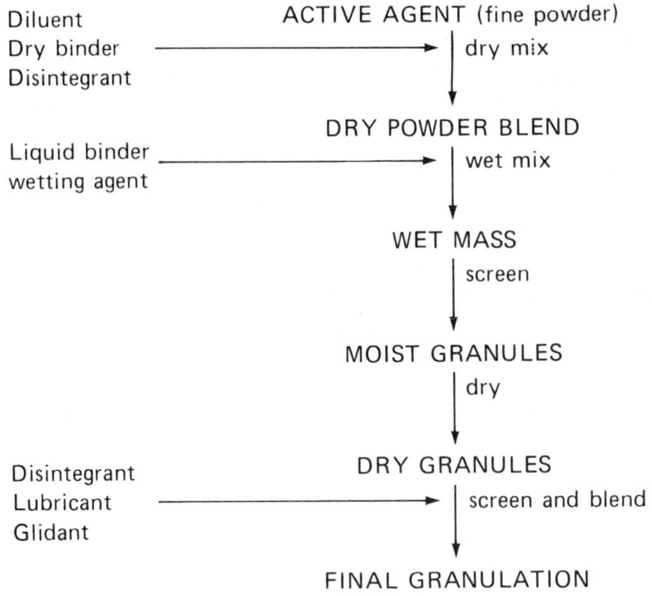

Fig. 12.11 — Typical pharmaceutical granulation process.

process comprises massing the powders with a liquid binder to form moist granules which are then dried and finally force-screened to produce a granulation with the required size distribution. Further observation of Fig. 12.11 shows that a pharmaceutical tablet contains many ingredients or excipients, in addition to the active agents.

The binders are materials selected for their ability to form strong bonds even at high compression–decompression rates. The solid disintegrants conversely enable the tablet to disrupt quickly when wetted, i.e. when ingested. Other materials with specialized functions are similarly incorporated such as wetting agents, solid lubricants, and glidants. The mean granule size and size distribution are fixed by the process conditions, assuring optimal flow and packing characteristics with minimal dust generation. The granular intermediate also has improved compression properties because the process conditions and excipients are selected to achieve a friable granulation which deforms plastically to produce strong tablets with uniform surface appearance, using relatively low compression pressure, approximately 200–300 MPa. Uniform distribution of the wetting agent and liquid binder also determines that the formed tablets have a hydrophilic pore structure, which allows rapid penetration of liquid and subsequent disintegration when ingested [61].

12.4.6.2 Granulation mechanisms

Rumpf [100] has reviewed the basic mechanisms. There are three types of force available:

(1) Bonds due to free liquids.
(2) Bonds due to solid bridges.
(3) Surface forces.

12.4.6.2.1 Free liquid bonds

In considering bonds due to free liquids, Newitt & Conway-Jones [88] identified four types of liquid bonding, depending on the proportional amount of liquid present. Knowing that when a liquid is in contact with a solid the predominant force becomes that due to surface tension, the four states of liquid bonding are as illustrated in Fig. 12.12, i.e. pendular, funicular, capillary, and droplet.

In the pendular state, the liquid bridges between the particles are all separate and independent, being centred on points of contact between the particles. Because these forces are strong, the surface tension draws the particles together. If the proportion of liquid is increased the liquid can now freely move, and the bonding force is less strong. At higher liquid contents all the interstices are filled with liquid, and the strength of the granule is due to curvature of the liquid between particles on the outside. There is little internal strength in this case, but if the liquid is removed by evaporation the particles are drawn together and can form a successfully strong granule. In the final state the particles are not in contact, and the droplet has very little strength [101].

Rumpf [102] calculated the strength of the bond due to the pendular bridge, and showed that the strength of the bond was not very dependent on the amount of moisture present until the funicular state was reached, but the resistance of the bond to rupture increased with increased liquid content because the particles could be pulled further apart without individual bridges breaking. This means in effect that spraying a liquid on to a dry powder will create pendular bridges which will form strong granules. The quantity of liquid used is not critical, but too much should be avoided [101].

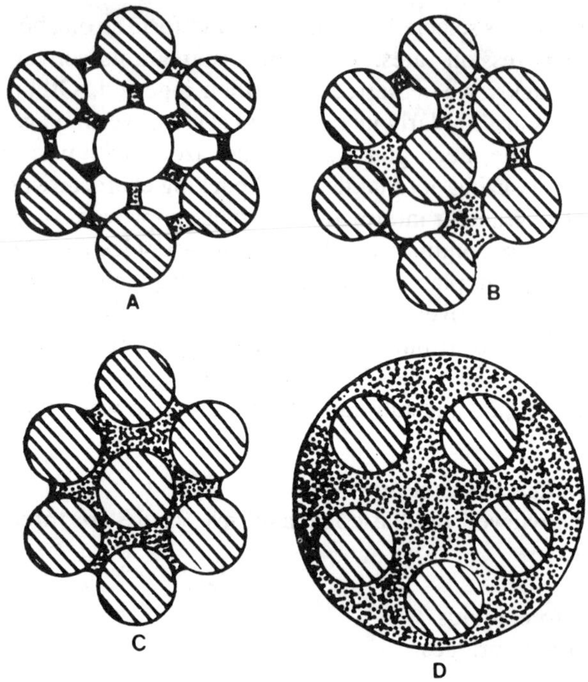

Fig. 12.12 — Types of liquid bonding in granules. A: pendular, B: funicular, C: capillary, D: droplet.

12.4.6.2.2 Bonds due to solid bridges

Crystalline bridges, use of liquid binders, and use of solid binders are more permanant structures. Although liquids form a very important part in the initial stages of granulation, it is not possible to rely on these bonds to form a strong granule. It is preferable to convert the bridge into a more permanent structure, and there are various possibilities.

12.4.6.2.3 Crystalline bridges

If the powder is slightly soluble in the liquid state, consequent evaporation of the pendular ring will cause any dissolved solid to be thrown out of solution, and since in the process the pendular ring will be shrinking, a strong bridge will be formed.

12.4.6.2.4 Use of liquid binders

If the powder is not sufficiently soluble for crystalline bridges to be formed, another method is to add to the liquid a binder such that when the liquid is evaporated the binder effectively glues the particles together. The binder may set solid or may remain as a high-viscosity liquid.

12.4.6.2.5 Use of solid binders

An alternative to the liquid binder is the use of a finely ground solid such as cement which reacts with water. The solid binder is intimately mixed with the powder before granulation. On the addition of liquid during the process the chemical reaction takes place and cements the particles together. This has been found to be particularly useful in the granulation of iron ore [101].

12.4.6.2.6 Surface forces
It is generally known that there exist between all solids molecular-derived forces known collectively as van der Waals forces. The magnitude of these is known for some materials, and they are on the whole very small compared with those discussed above. These forces are further weakened by surface contamination and adsorbed gases, and so are unlikely to be significant in granulation.

12.4.6.3 Compacted catalysts
In the formation of catalysts and catalyst supports for use in the manufacture of gases, for example in the production of ammonia, a range of compacted catalysts each carrying out a particular reaction and made in several different shapes from a variety of materials, is required. Some examples of these pellets are shown in Fig. 12.13. The pelleting or compacting requirements for catalysts are critical for three specific reasons [103,104–110].

(a) The metal oxides used are very hard.
(b) Catalysts must be simultaneously strong and porous.
(c) Catalysts must also have a high internal surface area.

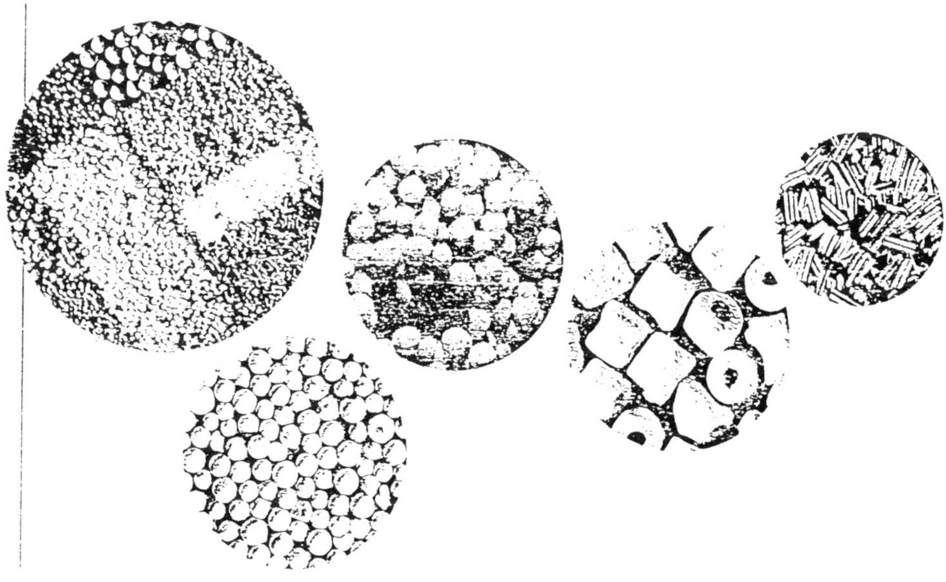

Fig. 12.13 — Catalysts in a variety of shapes and sizes.

Fig. 12.14 shows a typical catalyst manufacturing plant, and the asterisk (*) in Fig. 12.15 illustrates typical points in a refinery complex utilizing various catalysts. Table 12.8 shows typical chemical markets for catalysts.

12.4.6.4 Compacted ceramics

Fig. 12.16 shows a typical ceramic fabrication route. Precipitation conditions define the primary particle size and the particle shape, all of which affect later stages in the fabrication process [62]. Precipitation variables are:

(1) Solution concentrations
(2) pH
(3) Nature of anion
(4) Method of mixing
(5) Temperature

The size of precipitate particle is usually chosen to permit ease of filtration (see Table 4.1), and precipitation conditions are fixed accordingly. In direct strike precipitation, the acid is added to the bulk metal-salt solution, and metal-salt solution to bulk acid in reverse strike precipitation. Particle and crystallite size additionally depend on solution temperature, aging, and agitation after precipitation. The most common precipitate anions are hydroxide, peroxide, or oxalate, while other organic acid radicals may be chosen for special circumstances. The criterion applied in choosing the anion is said to be for ease of decomposition to form a pure oxide [62].

Calcination entails decomposition of the precipitate under an oxidizing atmosphere to form the pure oxide. Air is the natural choice of atmosphere for most applications, but pure oxygen, carbon dioxide, and gas mixtures are also employed. During calcination the precipitate is heated to 500°C or above, and powder properties can be adjusted at this stage by varying the temperature and atmosphere. The basic particle shape is conferred during precipitation, and the relict structure is retained through calcination. During calcination the primary particles sinter together to form strong aggregates, resistant to breakdown during sieving operations or ultrasonic dispersion for particle size analysis, to an extent determined by the calcination temperature. If the specific surface area is too high the quantity of binder required is high, the green compact density consequently low, and the subsequent shrinkage on sintering high and often non-uniform. An optimum specific surface area exists for each powder, and in the case of nuclear ceramics [62] this is typically in the range 3–7 $m^2\,g^{-1}$.

Spray-drying of the wet slurry under fluidized bed or batch reactor conditions is also widely used for commercial production of ceramic powders, with or without an intermediate wet milling stage for adjustment of particle size. Thermal conversion of the precipitate to the oxide may be effected during spraying or as a subsequent stage. The granular product is then suitable for press feeding without further size adjustment [62].

Alternatives to precipitation reactions for oxide formation are freeze drying, thermal decomposition of salts, and oxidation of the metal. In freeze drying the salt solutions are sprayed into a freeze chamber, and the resultant frozen droplets are converted into the anhydrous salts by sublimation of the ice under controlled

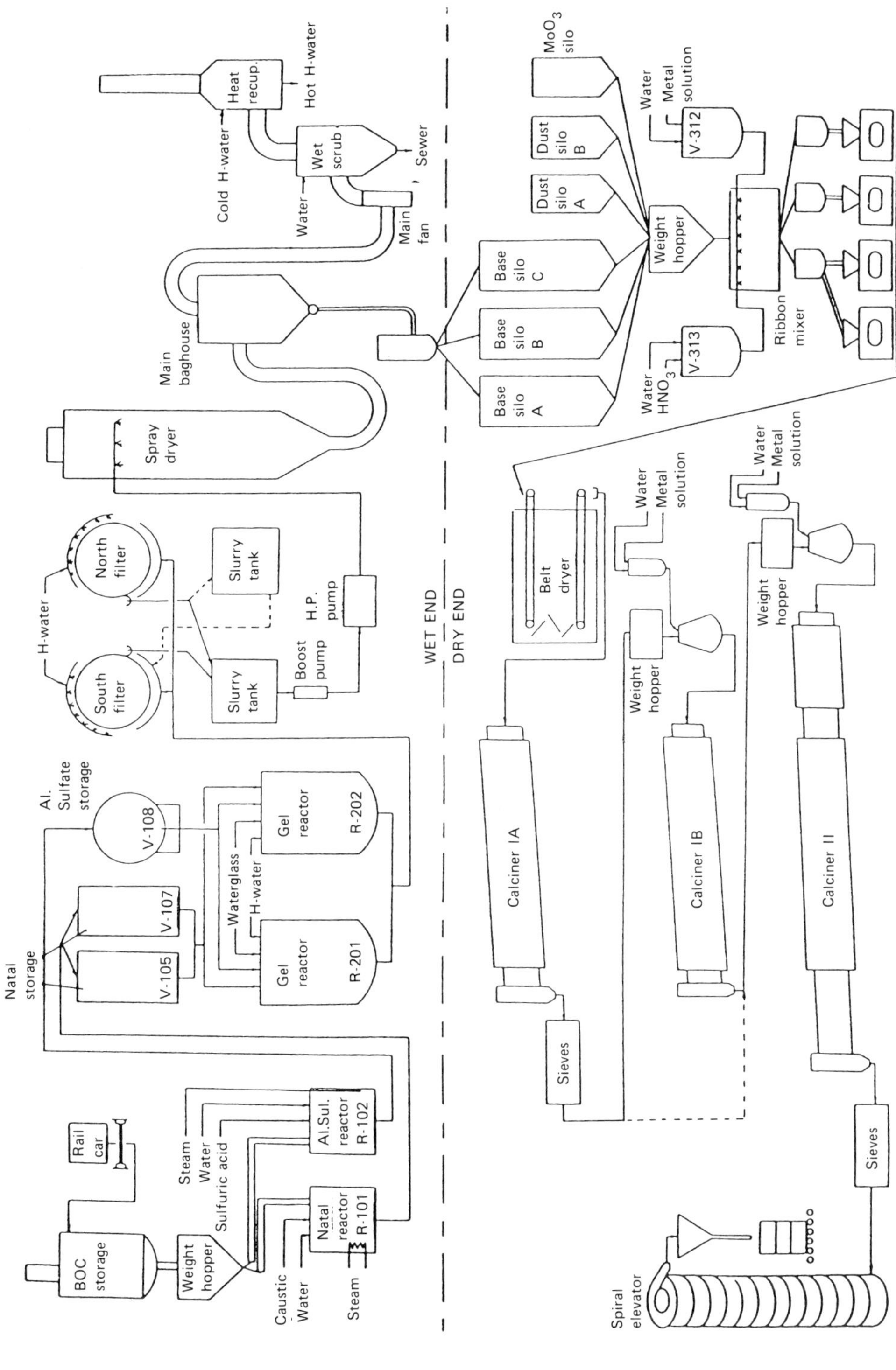

Fig. 12.14 — Hydrotreating catalyst plant.

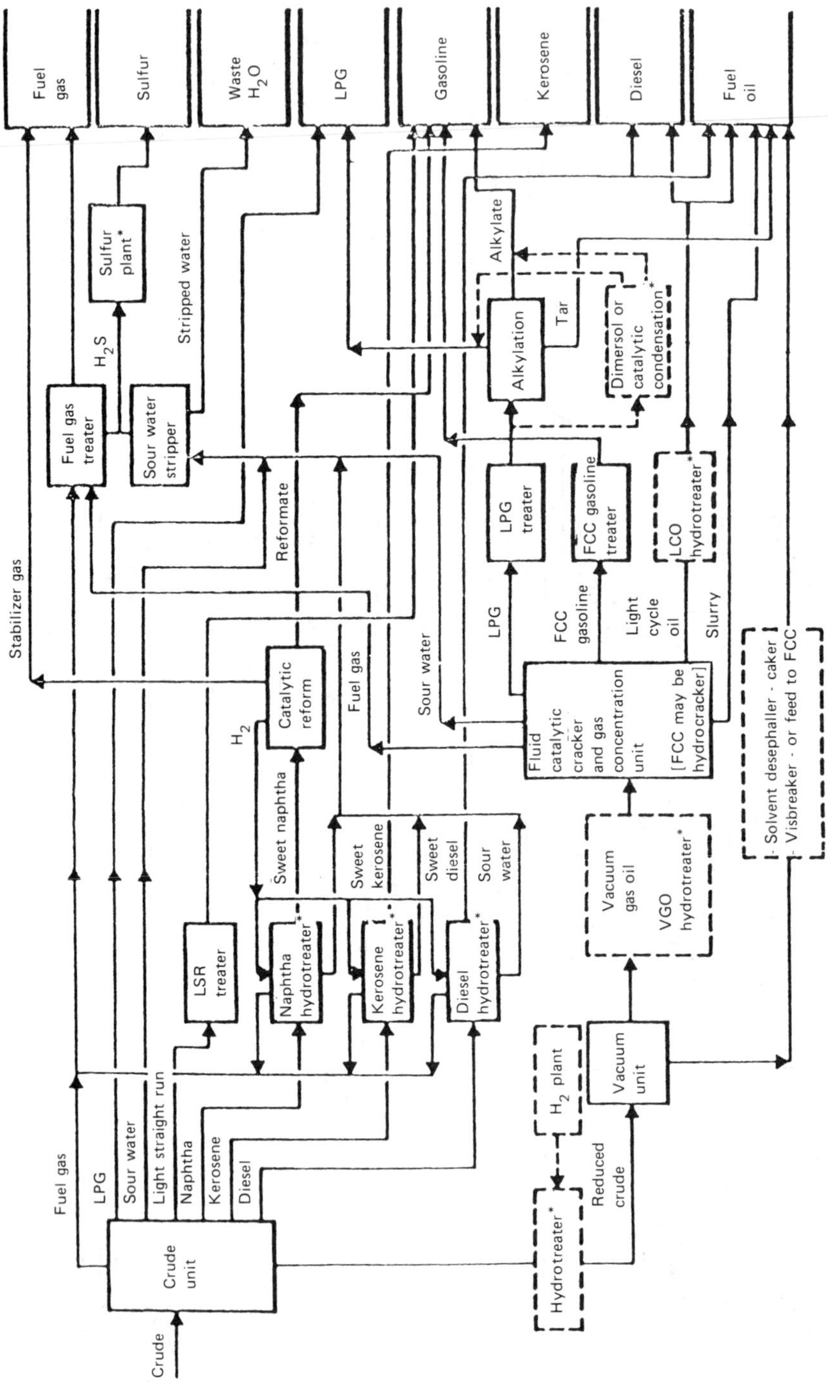

Fig. 12.15 — Typical complex refinery block flow diagram and product state.

Table 12.8 — The chemical market for catalysts

Category	Thousand lb 1987s	1995s	Type
Oxidation catalysts			
Maleic and phthalic anhydrides	1 050	1 250	Vanadia
Formaldehyde	83	95	Iron-molybdenum
Ethylene oxide	1 900	2 400	Silverson support
Acrylonitrile	2 050	2 400	Molybdenum, bismuth, and other metals
Vinyl chloride monomer	3 300	4 200	Copperschloride-alumina
Sulphuric acid	3 400	3 500	Vanadia on silica
Total	**11 783**	**13 845**	
Steam reforming, ammonia and methanol catalysts			
Reforming, pretreatment	690	860	Zinc oxide-alkali halide
Primary reforming	4 100	5 150	Supported nickel
Secondary reforming	630	800	Supported nickel
High-temperature shift	3 400	4 250	Iron-chromium oxides
Low-temperature shift	3 700	4 600	Copper and zinc oxides on alumina
Methanation	630	850	Supported nickel
Ammonia	2 600	3 300	Iron and other oxides
Methanol	800	800	Copper-zinc with alumina or chromia
Total	**16 550**	**20 610**	
Hydrogenation catalysts			
Cyclohexane	300	360	Nickel or platinum
Edible fats and oils	4 700	5 100	Nickel slurry
Inedible fats and oils	2 050	2 400	Nickel
Alcohols	900	1 300	Copper or rhodium salts for oxo alcohols; copper oxides and chromites for fatty alcohols
Amines	1 200	1 400	Silica-nickel aluminas, cobalt, activated nickel
Total	**9 150**	**10 560**	
Dehydrogenation catalysts			
Butadiene (oxydehydrogenation)	225	350	Chrome alumina
Styrene monomer	3 500	4 200	Iron oxide and promoters
Total	**3 725**	**4 500**	
Organic synthesis catalysts (listed by active substance)			
Precious metals[1]	5 400	5 750	
Cobalt (as contained Co)[2]	290	350	
Phosphoric acid[3]	13 500	16 000	
Anhydrous alumium chloride[4]	67 000	75 000	
Total	**86 230**	**97 100**	
Polymerization catalysts			
Ziegler-Natta	29 000	35 000	Transitional metal compound plus metal alkyl (e.g. $TiCl_4 + (C_2H_5)_3Al$)
Reaction initiators	15 000	17 000	Organic peroxides and percarbonates
Urethane catalysts	6 000	7 000	Usually organotins
Total	**50 000**	**59 000**	

[1] Includes catalysts for acetic and teraphthalic acids, caprolactam, hydrogen peroxide and vinyl acetate monomer. [2] Includes adipic and taraphthalic acids. [3] Including polymer gasoline, cumene, propylene trimer, and tetramer. [4] Including detergent alklate, ethyl benzene, dyes. Source: Business Trend Analysts (Commack, N.Y.).

conditions of vacuum and temperature. The porous salt particles are then calcined to form the oxide. Thermal decomposition of the salts may also be applied directly to the liquid droplets or by pyrolysis of salts absorbed on a combustible medium such as

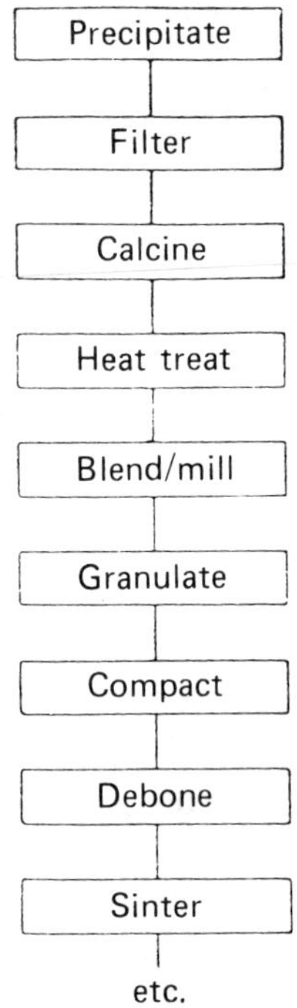

Fig. 12.16 — Typical ceramic fabrication process.

cellulose or activated carbon. For high-purity oxides or for powders with ultra-fine primary particle size, direct oxidation of the metal is employed, often in the vapour phase [62].

Milling is necessary to break down coarse aggregates which may have high levels of included porosity. This is because in most cases ceramic powders are non-segregating, and for extremely hard ceramics such as nitrides it may be necessary to employ high-energy vibro-milling in mills lined with tungsten carbide.

Finally, to provide a free-flowing, dust-free feed to the compaction process to ensure uniform die filling and uniform compaction behaviour, the milled powder must be granulated.

12.4.6.5 Summary

We may summarize in general that the principal steps in uniaxial die compaction are as follows:

(1) Flow of the correct amount of powder into the die.
(2) Application of pressure.
(3) Compaction, i.e. reduction of voids and increase in bulk density.
(4) Bond formation.
(5) Removal of pressure.
(6) Ejection of compact.

Typical strain and strain rate curves as functions of stress are shown in Fig. 12.17.

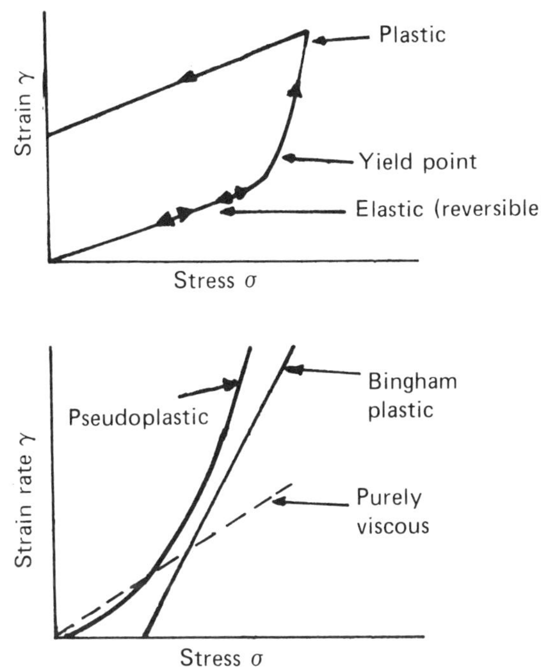

Fig. 12.17 — Typical strain and strain rate curves as functions of stress.

The main effects which govern tabletting can be divided into material properties and machine conditions as follows:

(1) Intrinsic material properties:
   (a) Mechanical nature of the material to be tabletted, i.e. viscous, plastic, or elastic.
   (b) Material properties, i.e. values of its viscosity, plasticity, hardness, etc.
   (c) Properties and amounts of additives, lubricants, and binders.

(2) Particulate properties
   (a) Mean size and size distribution
   (b) Shape
   (c) Agglomerate porosity
   (d) Moisture content.

(3) Nature of applied load
  (a) Axial compression, isostatic compression, or shear
  (b) Rate of load application and removal. Punch pressure versus time, or position versus time.
(4) Die geometry
  (a) Length
  (b) Diameter
  (c) Shape complexity, regions of different depth or thickness, etc.

The interactions between the material characteristics and the applied mechanical conditions determine the properties of the product.

12.5 CHARACTERIZATION OF CLAYS, AND CARBON BLACK POWDERS

12.5.1 Introduction

The objective of this brief characterization is to highlight the source, at times the processing or uses of some of these powders. For further in depth reading, the references will provide a domino-effect for other more exhaustive references.

12.5.2 Clays

It is known that natural clays are the products of weathering of rocks. They are found either at the location of their formation (residual clays) or at areas of deposition (sedimentary clays) after being carried either by wind (aeolian clays) or by glaciers (glacial clays) or by water [111].

In geology and in soil science, clays are mostly identified with a given small particle size fraction, below 2 μm, of mixtures of minerals or soils, irrespective of composition or structure [112]. When a ceramist is defining clays, the emphasis is usually on the development of plastic properties in the wet condition, and of a hard, solid material when dried or fired [113]. Clay, technologically, is identified as clay mineral or a layer silicate (or phyllosilicate — 'phyllo' — leaf like) of colloidal dimensions, defined as having in at least one direction a dimension roughly in the range of 1 nm to 1 μm [114].

Granquist & Pollack [115] discussed synthetic clays. Synthesis from an appropriate mixture of oxides and hydroxides in water at normal pressure under reflux conditions is relatively easy. It has very desirable properties for application in speciality products, and sources of the natural mineral are said to be limited. Special features of synthetic clays are homogeneity, batch reproducibility, and absence of undesirable impurities.

In the processing of mined natural clays which normally contain coarse impurities which reduces the whiteness of the clay, the coarse material is removed by either dry or wet processing. In the dry process, the ground clays are separated from coarse material by air flotation, while in the wet process, the raw clay is suspended in water, using peptizing agents, and coarse material is removed by sedimentation. The clay is then flocculated, and the sedimented flocs are collected and dried [116].

Particulate iron compounds are removed by chemical solution processes using hydrosulphites, although techniques using electrophoresis or dielectric separation have been developed [117].

12.5.2.1 Uses

Because clays provide a relatively cheap source of particulate matter in the colloidal size range, the plate-like shape of the particles and large surface area make them often uniquely suitable for a number of applications. The dry powder is used as a carrier for insecticides, in agricultural dusting operations, as a filler for rubber, plastics and adhesives, and as a catalyst support. Ceramic clays are those in the kaolinite group, but the largest tonnage use of kaolinites is in paper, both as fillers and as coating pigments. In the paint industry, clays are used as pigments in both water and and oil base paints. In the petroleum industry drilling fluids are paired with high concentrations of clays. Acid-tested clays have been used as cracking catalysts, and highly adsorptive clays are applied as adsorbents in refining. In lubricating greases, clays are used as gelling agents.

It must be emphasized that the performance of clays depend largely on their surface properties as they affect interaction with chemical compounds and suspension media as well as particle to particle interaction. Hence surface characterization is an essential part of applied research on clay systems. [118]. Crystallographic and morphological characterizations are beyond the scope of this book, and may be found elsewhere [119–128].

12.5.3 Carbon black

Because of their chemical nature and colloidal structure, carbon blacks occupy a special position in the world of pigments and fillers. In the family of pigments, they are the strongest absorbers of light over the entire visible spectrum, making them the preferred pigments for jet black colour or for darkening other pigment compositions [129]. As fillers they impart the highest degree of reinforcement in rubber, and are vital for the performance of pneumatic tyres [130].

It is known that carbon blacks are composed of nearly pure carbon in units of colloidal dimensions [131]. The carbon atoms are arranged in graphitic layers, or, put differently, in poly-nuclear arrays containing dozens of 6-carbon aromatic rings [132–135]. All other forms of carbon, except diamond, are also composed of graphitic layers similar to those of carbon black, but of different size, degree of perfection, and degree of ordering [136]. Many massive forms of carbon contain pores, and by suitable oxidative treatment the surface area of adsorbent charcoals, for example, can be raised to a level exceeding that of the finest carbon blacks, but because such carbons cannot be dispersed into colloidal entities, they cannot function as pigments or fillers [137].

The colloidal units of carbon black are called aggregates because they appear to be composed of many spheroidal particles. Formerly, the particles were thought to be the ultimate units of carbon black; however, it is now known that an aggregate is a single carbon entity, made up of graphitic layers which are continuous from one particle to the next [138].

Differences in average particle size are the primary structural basis for classifying the 50 or more currently produced grades of carbon black. At one extreme, the finest carbon blacks, which are used for their jet-black colour, are of the order of 5–10 nm in particle size, while at the other extreme, the coarsest blacks are of the order of 500 nm. Other distinctions between grades are based on the so-called 'structure' or bulkiness of the aggregates, the surface chemistry, and the porosity, present in a few pigmentary grades only.

In the dry state, groups of aggregates are held together by secondary forces forming agglomerates. These may be very loose, as in the smoke streamers formed by fresh soot, or somewhat more compact as in the flocs of which fluffy grades of carbon black are composed, or still denser as in the pelletized grades. In applications of carbon black the agglomerates must be separated into the smallest entities, i.e. the aggregates, and these must be dispersed in the medium or vehicle [139].

Table 12.9 shows the typical manufacturing processes which are used at present for the production of carbon black [140]. Of these, the furnace process is by far the most important, in terms of annual production, future prospects, and versatility [141]. In the oil furnace process, an aromatic residual oil or feedstock is sprayed into a hot gas–air flame in a specially designed furnace. After formation of the carbon black, the hot smoke is cooled rapidly (quenched) with water, and the carbon black, now in the form of agglomerates, is collected by means of bag filters, cyclone separators, or other devices [142].

Table 12.9 — Manufacturing processes for carbon black

Process or type	Process feature	Raw material
Furnace	Partial combustion and cracking in refractory furnace	Petroleum residue Natural gas
Impingement (channel or roller)	Diffusion flame impingement on cold surface of channel iron or of steel roller	Natural gas Town gas Coal tar
Thermal	Thermal cracking in super-heated refractory chamber	Natural gas
Lamp black	Partial combustion in open pans	Coal tar Petroleum residue
Acetylene	Thermal cracking	Acetylene

The impingement (channel and roller) and thermal processes are of relatively minor importance today, and their use will undoubtedly diminish still further because of changing economics and because of pollution problems [143]. In the past, the impingement process was used to produce grades of the finest particle size, while the thermal process produced grades of the largest particle size, but the furnace process has recently been extended to make blacks of comparable performance and properties at both ends of the particle size spectrum [144].

Acetylene black is manufactured in a continuous reactor similar to the furnace process, but because of the exothermic nature of the formation of carbon from acetylene, it is not necessary to burn gas to supply heat. Lamp blacks are made by burning liquid hydrocarbons in open shallow pans [145].

The mechanism of formation of carbon black has not been completely established [146]. A scheme for which considerable evidence exists is shown in Fig. 12.18

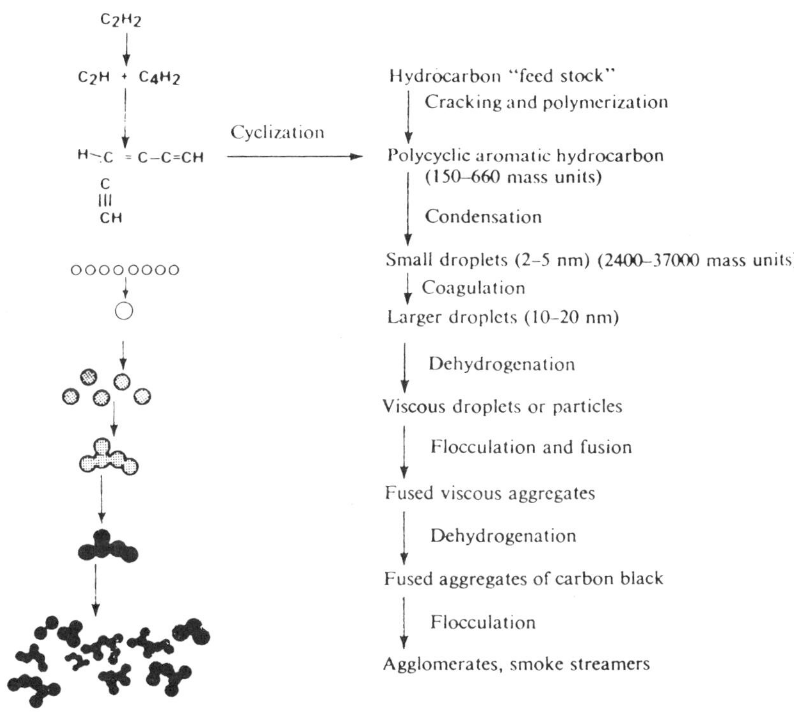

Fig. 12.18 — Mechanism of carbon black formation.

Polycyclic aromatic molecules and radicals, which owe their stability to a high resonance energy, are formed from hydrocarbon raw materials by cracking, polymerization, and cyclization. The stable, planar polycyclics condense into very small droplets which collide and coalesce to form larger droplets or semi-liquid particles. The polycyclic species tend to orient predominantly concentrically parallel to the surface. Progressive dehydrogenation and linking together of the polycyclics leads to an increase in viscosity, so that on further collision the droplets stick together in the form of aggregates, but do not coalesce into spherical form. Further dehydrogenation converts the polycyclic species to carbon. Some authors [147–149] believe that further deposition of carbon takes place on the surface of the carbonized aggregate. Each aggregate is formed within a gas pocket or 'turb' which is undergoing a particular thermal history over a period of the order of milliseconds at temperatures of 1100–1800°C, depending on the process [150–153].

By suitable furnace design and adjustment of operating conditions, the manufacturer can control the average size of the particles and, to a lesser extent, the average number of particles per aggregate, associated with structure. The latter property can also be controlled or lowered by addition of traces of potassium salts to the flame. Dannenberg [154–156] has reported progress being made toward controlling the uniformity of particle and aggregate size by the development of furnaces with more uniform thermal and compositional history of the 'turbs'.

After the carbon black is formed and collected, it is generally conveyed to a pelletizer, where the low-density fluffy black is converted to free-flowing pellets of around 0.3–1 mm in diameter. A typical pelletization process uses stirring of the fluffy black with enough water to wet it; the wet pellets are then dried in a rotary drier. A dry pelletization process may also be used, in which the fluffy black is simply tumbled for a long period in a large rotary drum. The pelletized carbon black, of 20–30 lbs per cu. ft. (0.3 to 0.5 g cm^{-3}) in density, can be handled by bulk methods and shipped in hopper cars for tonnage applications, and is also sold in 50 lb bags for smaller volume uses. For special applications where dispersion of the black may present exceptional problems, such as with very fine blacks used for colouring enamels or plastics, the fluffy black is sold in 25 lb or 50 lb bags at 6–15 lbs per cu. ft. density. The fluffy black is difficult to handle, and requries a long time to incorporate in a vehicle, although it may ultimately give a better dispersion than a pelletized black [157].

12.5.3.1 Uses

Table 12.10 shows properties and principal applications of some representative grades of carbon black [158]. The grades are identified by ASTM† number, by description, and by their letter designation. The description is based on their principal application and behavior; for example, a long flow channel (LFC) black is one which is used principally in lithographic inks, to which it imparts excellent viscosity characteristics (long flow).

The terminology used for rubber-grade blacks reached its nadir in the early 1950s with the introduction of the 'intermediate super abrasion furnace' or ISAF grade. In the mid-1960s ASTM introduced a numbering system for rubber-grade blacks (D1765-67) in which the first number refers to the particle size classification, and the other numbers are set arbitrarily. It was pointed out that each manufacturer uses furnaces of different design, different raw materials, etc. so that the carbon black of a particular grade made by one manufacturer is generally not identical with that made by another manufacturer, although they may be essentially interchangeable in their principal applications [159]. In the non-rubber fields, there is little attempt at standardization of grades; instead, each manufacturer has one or more proprietary trade-named blacks within the broad classifications indicated in the table.

† The American Society for Testing and Materials.

Table 12.10 — Typical properties and principal applications of representative grades of carbon black

GRADE		PROPERTIES				APPLICATIONS
ASTM number or description	Letter designation	Particle size (nm)[a]	Surface area (BET) $m^2 g^{-1}$	DBP absorption cc per 100 g[b]	Volatile content %	
Oil furnace blacks						
N220	ISAF	22	115	115	1.5	Tyre treads, conveyor belts, inks, plastics, synthetic fibres.
N330	HAF	27	80	102	1.0	Tyre treads and carcasses, news inks.
N326	HAF-LS	27	80	72	1.0	Tyre carcasses, inks, paints, plastics.
N347	HAF-HS	26	90	125	1.0	Tyre treads, extruded rubber articles.
N550	FEF	42	42	120	1.0	Tyre carcasses, inner tubes, cable jackets, paints.
N774	SRF-NS	80	27	70	1.0	Similar to N550; tinting paints, plastics, cements.
N472	XCF	30	254	180	2.5	Electrical conductivity, cable connectors, coated fabrics, dry cells.
High colour furnace	HCF	14	240	60	2.0	Replaces HCC grade.
Medium colour furnace	MCF	18	200	120	1.0	Replaces MCC grade.
Long Flow Furnace	LFF	24	115	69	5.0	Replaces LFC grade.
Medium flow furnace	MFF	25	95	72	2.5	Replaces MFC grade.
Thermal blacks						
N880	FT	200	12	33	0.5	Cable insulation, molded rubber articles, tinting.
N990	MT	450	8	36	0.5	Similar to N880; high-temperature insulation.
Channel blacks						
S301	MPC	24	120	100	5.0	Off-the-road tyres, plastics, paints.
High colour channel	HCC	10	850	175	13.0	Automotive enamels, lacquers, plastics.
Medium colour channel	MCC	16	320	110	5.0	Plastics (UV protection).
Long flow channel	LFC	24	330	80	12.0	Lithographic and letterpress inks.
Medium flow channel	MFC	24	200	100	7.0	Inks, paints, fibres.
Miscellaneous blacks						
Lampblack	LB	200	n.a.	120	0.7	Paints, inks, textile finishes.
Acetylene	ACET	40	67	260	0.3	Similar to N472.

[a] Approximate number-average particle size (d_n) in nm, based on electron microscopy.
[b] Dibutylphthalate absorption of pelletized grades.

The total production of carbon black in the non-communist countries in 1973 exceeded six thousand million pounds, of which 90–95% was used in rubber [160]. About half of this is of the so-called reinforcing grades of furnace blacks, of about 70–140 $m^2 g^{-1}$ surface area. These blacks confer on rubber vulcanizates a high modulus of elasticity, high tensile strength, and greatly improve abrasion resistance, a combination especially suitable for tyre treads, which have been reinforced with carbon black since the early 1900s. Tensile strength and abrasion resistance depend primarily on particle size (or surface area), while the modulus of elasticity depends primarily on 'structure'. Typical formulations include about 50 parts of reinforcing black per 100 parts of rubber (by weight). Carbon blacks of lower surface area are called 'semi-reinforcing' because they do not raise the tensile strength and abrasion resistance of rubber to as great an extent as the reinforcing grades. They are used in applications where these properties are not important, such as tyre carcasses, extruded rubber mouldings and gaskets, etc. [161–163].

Non-rubber applications of carbon black are generally based on their absorption of light and ultraviolet radiation, the electrical conductivity, and to a minor extent, their high surface area and adsorptive capacity. In black paints and enamels, the highest degree of jetness is associated with the finest particle size. In these applications the concentration of carbon black is of the order of 6% or less. In lithographic inks, in addition to jetness, the flow properties are very important. Blacks which have been oxidized after manufacture (aftertreated) are used in this application because they permit a relatively fluid ink to be formulated at concentrations of 20–25% black, as required for maximum covering power in the relatively thin layer of ink which is transferred. In many other types of ink, such as news ink, where cost is a major factor, less expensive grades of black are used, including the same reinforcing and semi-reinforcing furnace blacks as those used in the rubber industry. The price of the darkest blacks used in coating and plastics may be as much as five times the cost of the less jet (medium colour) grades, which in turn are about three times the cost of the semi-reinforcing grades [164].

In vinyls and other plastics where carbon black is added solely as a black colourant, the concentration may be of the order of 1–2%, and the same grades may be used as for black enamels; however, for ease of dispersion in the plastic, the carbon black may be used in the fluffy form. In polyethylene used for covering electrical cables and other outdoor use, carbon black adsorbs ultraviolet and other radiation and thus prevents degradation and embrittlement. Here medium-colour channel blacks or furnace blacks in the reinforcing or finer particle size range may be used at about 2% concentration. In some plastics applications, a high degree of electrical conductivity is required; this requires a fairly high loading of black, preferably a black of high structure and low volatile content such as acetylene black or a special conductive grade of furnace black. Conductive blacks are also used in dry cell batteries, as well as in some rubber applications [165].

The adsorptive capacity of carbon black is applied in the decoloring of sugar and in sewage treatment; however, massive forms of carbon such as adsorptive charcoals are generally preferred because their physical form facilitates separation of the carbon from the liquid being clarified. Smith [166], Stokes [167], and Garrett [168] have further described a host of miscellaneous applications for carbon blacks. As stated earlier, the surface chemistry and adsorption techniques are beyond the scope of this book, but further information may be found elsewhere [169–178].

REFERENCES

[1] Allen, T., *Particle size measurement*, 3rd ed., Chapman and Hall, New York, London (1981).

[2] Hausner, H. H., Friction conditions in a mass of metal powder. *International Journ. of Powder Met.*, **3**, 4, 7–13 (1967).

[3] Grey, R. O. & Beddow, J. K., On the Hausner ratio and its relationship to some properties of metal powders. *Powder Technology.*, **2**, 6, 323–326 (1969).

[4] Lewis, H. D., Ready, T. L., & Hausner, H. H., Some observations of the effect of atmospheric humidity on the flow of metal powders. *Los Alamos Report LAMS-2773* (Nov. 14, 1962).

[5] Stanley-Wood, N. G. (ed.) *Enlargement and compaction of particulate solids.* Butterworths Monographs in Chemical Engineering, Butterworth & Co. (Publishers) Ltd, London (1983).

[6] British Standards Institution, British Standard House, 2 Park Street, London.

[7] Metal Research Ltd., Melbourne, Royston, Cambridge, England.

[8] Millipore (UK) Ltd, Millipore House, Abbey Road, London, NW10 75P.

[9] Walton, W. H., *Bur. J. Appl. Phys. Suppl.,* **3**, s121 (1954).

[10] Imanco (Image Analysing Computer Ltd.) Metals Research Ltd, Melbourne, Royston, Cambridge, England.

[11] Vickers Instruments Limited, Haxby Road, York, Yorkshire, England.

[12] Fairs, G. L., *J. R. Microsc. Soc.,* **71**, 209 (1951).

[13] May, K. R., *J. Scient. Instrum.,* **22**, 187 (1965).

[14] Hamilton, R. J., Holdsworth, J. F., & Walton, A., *Br. J. Appl. Phys. Suppl.* **3**, s101 (1954).

[15] Heywood, H., *Trans. Instn. Min. Metall.,* **55**, 373 (1945/46).

[16] Mullin, J. W., *Chem. and Ind.*, 1435 (1971).

[17] Suhm, H. D., *Powder Technol.,* **2**, 6, 356 (1969).

[18] Endecotts Ltd, Lombard Road, London SW 19, England.

[19] Allen, T., *Particle size measurement*, 2nd ed., Chapman and Hall, New York, London (1975).

[20] Kamack, H. H., *Analyt. Chem.,* **23**, 844 (1951).

[21] V.D.I. Richtlinie Nr. 2031, Ausgabe (1962).

[22] DIN 6611, Sedimentationsanalyse in Schwerfeld, Grundlage (April 1971).

[23] Allen, T., *Powder Technol.,* **2**, 134, 141 (1968).

[24] Ladal (Scientific Equipment) Ltd, Warley Edge Farm, Warley, Halifax, Yorkshire, England.

[25] Olaf, J. & Robcock, R., *Staub. Reinhalt, Luft.,* **21**, 495 (1901).

[26] Berg, R. H., *Am. Soc. Test. Mater. Sp. Pub.* 234 and 245 (1965).

[27] Kubitschek, H. E., *Nature,* **182**, 234 (1958).

[28] Allen, T. & Marshall, K., *The electrical sensing zone method*, University of Bradford Press, Bradford (1972).

[29] Princen, L. H. & Kwolek, W. F., *Rev. Scient. Instrum.,* **36**, 646 (1965).

[30] Spielman, L. & Goren, S. I., *J. Colloid Interface Sci.*, **26**, 175 (1968).
[31] Wales, M. & Wilson, J. N., *Rev. Scient. Instrum.*, **32**, 1132 (1961).
[32] Herdan, G., *Small particle statistics*, 2nd ed., Butterworths Scientific Publications, London (1960).
[33] Allen, T., *Proceedings of Second International Conference on Particle Size Analysis, Society for Analytical Chemistry, London,* (1970).
[34] Rose, H. E., *Powders in industry*, p. 130, Society for Chemical Industry, London (1961).
[35] Davies, R., *Powder Technol.*, **12**, 111 (1975).
[36] Davies, C. N., *Nature,* **201**, 905 (1964).
[37] Pipel, N., *Paint Manuf.*, July 23 (1969).
[38] Flook, A. G., *Powder Technol.*, **21**, 295 (1978).
[39] Erlich, B. & Weinberg, B., *J. Sediment. Petrol.*, **40**, 205 (1970).
[40] Meloy, T. P., *Powder Technol.*, **17**, 27 (1977).
[41] Hausner, H. H., *Proceedings of Conference on Particle Size Analysis, Society for Analytical Chemistry, London,* (1966).
[42] Heywood, H., *J. Pharm. Pharmac., Suppl.*, **15**, 56T (1963).
[43] British Standards Institution BS 2955 (1958).
[44] Herdan, G., as Ref. [32].
[45] British Standards 4359, Part 3, 1970.
[46] Allen, T., as Ref. [33].
[47] Beddow. J. K., Vetter, A. F., & Sissan, K., *Powder Metall. Int.*, **8**, 2, 69 (1976).
[48] Davies, R., as Ref. [35].
[49] Hausner, H. H., as Ref. [41].
[50] Church, T., *Powder Technol.*, **2**, 1, 27 (1968).
[51] Dwyer, J. L., Manalon, D. A., & Merbon, R. R. A., *Proceedings of the 2nd Intern. Conf. on Particle Size Analysis, Society for Analytical Chemistry, London* (1970).
[52] Schwartz, H. P. & Shane, K. C., *Sedimentology*, **13**, 273 (1969).
[53] Naylor, A. G. & Wright, C. D., *Proceedings of the 3rd Intern. Conf. on Particle Size Analysis*, Ed. Groves, Heyden, London (1977).
[54] Meloy, T. P., as Ref. [40] .
[55] Beddow. J. K., Vetter, A. F., & Sissan, K., *Powder Metall. Int.*, **8**, 69, 107 (1978).
[56] Ehrlich, B . & Weinberg, B., as Ref. [39].
[57] Beddow, J. K., Philip, G. C., & Vetter, A. F., *Powder Technol.*, **18**, 19 (1977).
[58] Kaye, B. H., *Powder Technol.*, **21**, 1 (1978).
[59] Mandelbrot, B. P., *Fractals, form, chance and dimensions*, Freeman, San Francisco (1977).
[60] Allen, T., see Ref. [33].
[61] Hunter, B., Pharmaceutical granulation and compaction. In: *Enlargement and compaction of particulate solids*. (ed.) Stanleywood, N. G., Butterworth & Co. (Publishers) Ltd, London (1983).
[62] MacLeod, H. M., Compaction of ceramics. In: *Enlargement and compaction of particulate solids*. (ed.) Stanleywood, N. G., Butterworth & Co. (Publishers) Ltd, London (1983).

[63] Long, W. M., *Powder Metall.*, **6**, 73 (1960).
[64] James, P. J., *Powder Metall. Int.*, **4**, 2 (1972).
[65] Dowez, P. & Zwell, L., *J. Metals*, **1**, 2, 137 (1949).
[66] Sheinharz, I., McCullogh, H. M., & Zambrow, J. L., *J. Metals*, **6**, 515 (1954).
[67] Nelson, E., Naqvi, S. M., Busse, L. W., & Higuchi, T., *J. Am. Pharm. ASS. Sci. Ed.*, **43**, 596 (1954).
[68] Hvorslev, M. J., *Ingeniorvidenskabelige Skrifter A*, No. 45 (1937).
[69] Smakula, A. & Kalnajas, J., *Phys, Rev.*, **99**, 1737 (1955).
[70] Smakula, A., Kalnajas, J., & Sils, V., *Phys, Rev.*, **99**, 1747 (1955).
[71] Adam, N. K. A., *Physical chemistry*, Clarendon Press, Oxford (1956).
[72] Graton, L. C. & Fraser, H. J., *J. Geol.*, **43**, 785 (1935).
[73] Maggs, F. A. P., Schwake, P. H., & Williams, J. H., *Nature*, **186**, 956 (1960).
[74] Budworth, D. W., *J. Br. Ceram. Soc.*, **1**, 448 (1964).
[75] Heckel, R. W., Density pressure relationships in powder compaction. *Trans. of the Metallurgical Society of AIME*, **221**, 671–675 (1961).
[76] Schwarzkopf, P., *Powder metallurgy*, MacMillan (1947).
[77] Hardman, J. S. & Lilley, B. A., Mechanisms of compaction of powdered materials. *Proc. of Royal Society, London*, A. **333**, 183–199 (1973).
[78] Wollaston, W. H., *Phil. Trans. Royal Soc. London*, **119**, 1 (1828).
[79] Walker, E. E., *Trans. Faraday Soc.*, **19**, 73 (1923).
[80] Bal'shin, M. Yu, *Vestnik Metalloprom.*, **18**, 124 (1938).
[81] Habberjam, G. M., *Br. J. Appl. Phys.*, **15**, 1238 (1964).
[82] Seelig, R. P. & Wulff, J., *Trans. Am. Inst. Min. Metall. Engrs.*, **171**, 492 (1947).
[83] Oudemans, G. J., *Sci. Ceram.*, **2**, 133 (1965).
[84] Jones, W. D., *Fundamental principles of powder metallurgy.*, Edward Arnold, London (1960).
[85] Cooper, A. R. & Eaton, L. E., *J. Am. Ceram. Soc.*, **45**, 97 (1962).
[86] Newman, P. C.., *Br. J. Appl. Phys.*, **6**, 348 (1955).
[87] Gregory, H. R., *Trans. Instn. Chem. Engrs.*, **40**, 241 (1962).
[88] Newitt, D. M. & Conway-Jones, J. M. *Trans. Instn. Chem. Engrs.*, **36**, 422 (1958).
[89] Ernst, H. & Merchant, M. E., *Conf. on Friction and Surface Finish*, M.I.T. (1940).
[90] Holm, R., *Electrical contacts*. Uppsala: Almquist and Wiksells (1946).
[91] Bowden, F. P. & Tabor, D., *Friction and lubrication of solids.*, Vol. 2, Oxford, Clarendon Press (1964).
[92] Seelig, R. P. *Trans. Am. Inst. Min. Metall. Engrs.*, **171**, 506 (1947).
[93] Turba, E. & Rumpf, H., *Silic. Ind.*, **30**, 409 (1965).
[94] Wulff, J., In: *Powder metallurgy*. American Society for Metals, Cleveland, Ohio (1942).
[95] Seelig, R. P., as Ref. [92].
[96] Klar, E. & Shafer, W. M., *Int. J. Powd. Met.*, **5**, 2, 5, and **5**, 4, 5 (1969).
[97] Bowden, F. P. & Tabor, D., *Proc. R. Soc. London*, A **169**, 391 (1939); also, *Friction and lubrication of solids*, Vol. 1, Oxford Press (1954).
[98] James, P. J., Fundamental aspects of the consolidation of powders, *Powder Metallurgy International*, **4**, Nos. 3 and 4, 145–149, 193–197 (1972).

[99] Seelig, R. P., Introduction to seminar on review of literature on pressing of metal powders, *Metals Technology* TP.2236, and *Powder Metallurgy*, 506–518 (1947); See also, *Metals and Alloys*, **12**, 6, 744–748 (1940); *Trans. AIME*, **166** (1946); *Metals Tech*. TP.2044 (1946).
[100] Rumpf, H., In *Agglomeration*, ed. Knepper, W. A., Interscience, New York (1952).
[101] Lloyd, P. J., Mechanisms of size enlargement, In: *Enlargement and compaction of particulate solids*. (ed.) Stanley-Wood, N. G., Butterworth & Co. (Publishers) Ltd, London (1983).
[102] Rumpf, H., *Chemie-Ingr.-Techn.*, **30**, 144 (1958).
[103] Benbow, J. J., Mechanisms of compaction, In: *Enlargement and compaction of particulate solids*. (ed.) Stanley-Wood, N. G., Butterworth & Co. (Publishers) Ltd, London (1983).
[104] Igwe, G. J. I., Catalysts in oil and petrochemicals. *Proceedings of the 17th Annual Conf. of the Nigerian Society of Chemical Engineers*, 141–163 (1987).
[105] Igwe, G. J. I., Allen, T., & Stanley-Wood, N. G., Surface area measurement and gas permeametry at sub-atmospheric pressures. *Particle size analysis* (ed. Lloyd, P. J.), John Wiley & Sons, Chichester, New York (1988).
[106] Igwe, G. J. I., Gas and particle surface reaction at sub-atmospheric pressures. *Proceedings of the 1st National Fluid Dynamics Congress of the American Institute of Aeronautics and Astronautics* (88–3630), USA (1988).
[107] Igwe, G. J. I., Diversification from crude oil to petrochemicals. *Proceedings of the Conference: Petrochemicals in Nigeria in Year 2000*: 39–45, published by Al Minbar, 32 Rue Washington, 75008, Paris (1987).
[108] Igwe, G. J. I., Ewili, J. M., & Wami, E. N., Problems and prospects of natural gas transmission and distribution in Nigeria. *Proceedings of the Society of Petroleum Engineers, Gas Technology Symposium, Dallas–Texas, USA*, 279–293 (1988).
[109] Igwe, G. J. I., Improved apparatus for measurement of surface area of powders. *Journal of Industrial and Engineering Chemistry Research*, **26**, 6, 1254–1259 (1987).
[110] Igwe, G. J. I. & Allen, T., Advances in gas permeability measurements. *Journal of Chemical Engineering & Technology*, **10**, 2, 86–92 (1987).
[111] Andrews, D. E., Schmidt, P. W., & van Olphen, H., *Clays and Clay Minerals*, **15**, 321 (1967).
[112] Busenberg, E. & Clemency, C. V., *Clays and Clay Minerals*, **21**, 213 (1973); Cruz, M., Jacobs, H., & Fripiat, J. J., *Intern. Clay Conf. Madrid*, Pre-prints, **59** (1972).
[113] van Olphen, H., *An introduction to clay colloid chemistry*, Interscience, New York (1963).
[114] Marshall, C. E., *The physical chemistry and mineralogy of soils*, Vol. 1, *Soil materials*, Vol. 1, John Wiley and Sons, New York (1964).
[115] Grandquist, W. T. & Pollack, S. S., *Clays and Clay Minerals*, **8**, 150 (1960).
[116] Diamond, S. & Kinter, E. B., *Clays and Clay Minerals*, **5**, 334 (1958).
[117] Grim, R. E., *Clay mineralogy*, McGraw-Hill, New York (1953).
[118] Giese, R. E., *Clays and Clay Minerals*, **21**, 145 (1973).

[119] Theng, B. K. G., *The chemistry of clay-organic reactions*, John Wiley and Sons, New York and Adam Hilger, London (1974).
[120] Hang, Pham Ti. & Brindley, G. W., *Clays and Clay Minerals*, **18**, 203 (1970).
[121] Kahn, A., *Clays and Clay Minerals*, **6**, 220 (1959).
[122] Keenan, A. G., Mooney, R. W., & Wood, L. A., *J. Phys. Colloid Chem.*, **55**, 1462 (1951).
[123] Martin, R. T., *Clays and Clay Minerals*, **6**, 259 (1959).
[124] Mooney, R. W., Keenan, A. G., & Wood, L. A., *J. Amer. Chem. Soc.*, **74**, 1367, 1371 (1952).
[125] Theng, B. K. G., *Clays and Clay Minerals*, **19**, 383 (1971).
[126] van Olphen, H., *Clays and Clay Minerals*, **2**, 418 (1954).
[127] van Olphen, H., *J. Colloid Sci.*, **17**, 660 (1962).
[128] van Olphen, H., In: *Surface area determination* (D. H. Everett & R. H. Ottewill, eds), p. 255, Butterworths, London (1970).
[129] Medalia, A. I. & Rivin, D., (In: *Characterization of powder surfaces*) ed. G. D. Parfitt & K. S. W. Sing, Academic Press, 279–345 (Carbon Black), London, New York (1976).
[130] Medalia, A. I., *Rubber Age*, **93**, 580 (1963).
[131] Medalia, A. I., *J. Colloid Interface Sci.*, **32**, 115 (1970).
[132] Medalia, A. I., In: *Surface and colloid science* (E. Matijevic, ed.) **4**, 1–92, New York (1971).
[133] Medalia, A. I. & Heckman, F. A., *J. Colloid Interface Sci.*, **36**, 173 (1971).
[134] Medalia, A. I. & Hornik, G. J., *Pattern Recognition*, **4**, 155 (1972).
[135] Medalia, A. I. & Richards, L. W., *J. Colloid Interface Sci.*, **40**, 233 (1972).
[136] Medalia, A. I., Dannenberg, E. M., Heckman, F. A., & Cotten, G. R., *Rubber Chem. Technol.*, **46**, 1239 (1973).
[137] Rivin, D., *Proceedings of the 5th Conf. on Carbon*, **2**, 199, Pergamon Press, New York (1963).
[138] Rivin, D., *Rubber Chem. Technol.*, **44**, 307 (1971).
[139] Rivin, D. & Aron, J., *Paper presented at 6th Conf. on Carbon, Cleveland; Abstract in Carbon*, **3**, 358 (1965).
[140] Rivin, D. & Aron, J., U.S. Patent 3479299 (18 November 1969, to Cabot Corp.) (1969a).
[141] Rivin, D. & Aron, J., U.S. Patent 3479300 (18 November 1969, to Cabot Corp.) (1969b).
[142] Rivin, D. & Illinger, J. L., *J. Colloid Interface Sci.*, **31**, 85 (1969).
[143] Rivin, D., Troiano, P. F., & Perkins, J. S., *Paper presented at 8th Conf. on Carbon, Buffalo, New York* (1967).
[144] Rivin, D., Aron, J., & Medalia, A. I., *Rubber Chem. Technol.*, **41**, 330 (1968).
[145] Rivin, D., Aron, J., & Richards, L. B., U.S. Patent 3442679 (6 May 1969 to Cabot Corp.).
[146] Rivin, D., Aron, J., & Perkins, J. S., *Paper presented at 3rd North American Meeting of the Catalysis Society, San Francisco, Calif.* (Feb. 1974).
[147] Adamson, A. W., *Physical chemistry of surfaces*, Chap. XI, Interscience, New York (1967).

[148] Avgul, N. N. & Kiselev, A. V., In: *Chemistry and physics of carbon*. (P. L. Walker, Jr. ed.), **6**, 1–124, Marcel Dekker, New York (1970).
[149] Barton, S. S., Gillespie, D., & Harrison, B. H., *Carbon*, **11**, 649 (1973).
[150] Baugher, J. F. & Ellis, B., *J. Colloid Interface Sci.*, **38**, 658 (1972).
[151] van Beek, L. R. H. & van Pul, B. I. C. F., *J. Appl. Polymer Sci.*, **6**, 651 (1962).
[152] Boehm, H. P., *Adv. Catalysis*, **16**, 198 (1966).
[153] Broekhoff, J. C. P. & Linsen, B. G., In: B. G. Linsen, *Physical and chemical aspects of adsorbents and catalysts*, Academic Press, London and New York (1970).
[154] Dannenberg, E. M., *Ind. Eng. Chem.*, **44**, 813 (1952).
[155] Dannenberg, E. M., *Trans. Inst. Rubber Ind.*, **42**, T26 (1966b).
[156] Dannenberg, E. M., *J. Inst. Rubber Ind.*, **5**, 190 (1971).
[157] Burgess, K. A., Scott, C. E., & Hess, W. M., *Rubber Chem. Technol.*, **44**, 230 (1971).
[158] Cotten, G. R. & Dannenberg, E. M., *Tyre Sci. Technol.* **2**, 211 (1974).
[159] Ehrburger, P. & Donnet, J. B., *Carbon*, **11**, 309 (1973).
[160] Ergun, S., In: *Chemistry and physics of carbon* (P. L. Walker, Jr., ed.), **3**, 45–120, Marcel Dekker, New York (1968b).
[161] Ergun, S. & Mentser, M., In: *Chemistry and physics of carbon* (P. L. Walker, Jr., ed.), **1**, 203–265, Marcel Dekker, New York (1965).
[162] Given, P. H. & Hill, L. W. *Carbon*, **7**, 649 (1969).
[163] Given, P. H. & Hill, L. W. *Carbon*, **6**, 525 (1968).
[164] Gregg, S. J. & Sing, K. S. W., *Adsorption, surface area and porosity*, Academic Press, London and New York (1967).
[165] Honak, E. R., *Farbe und Lack*, **70**, 791 (1964).
[166] Smith, W. R. & Kasten, G. A., *Rubber Chem. Technol.*, **43**, 960 (1970).
[167] Stokes, C. A., In: *Kirk-Othmer encyclopedia of chemical technology* (A. Standen, ed.), 2nd ed., Suppl. Vol., 91, Interscience, New York (1971).
[168] Garrett, M. D., In: *Pigment handbook* (T. C. Patton, ed.), 709–743, Wiley, New York (1973).
[169] Holmes, J. M., In: E. A. Flood, *The solid–gas interface*, **1**, Chap. 5, Marcel Dekker, New York (1967).
[170] Findenegg, G. H., *J. Chem. Soc. Faraday I*, **69**, 1069 (1973).
[171] Voll, M. & Boehm, H. P., *Carbon*, **9**, 473, 481 (1971).
[172] Voet, A., *J. Colloid Interface Sci.*, **30**, 264 (1969).
[173] Tomita, A. & Tamai, Y., *J. Colloid Interface Sci.*, **36**, 153 (1971a).
[174] Sing, K. S. W., *Proc. Int. Sym. Surface Area Determination*, 25, Butterworths, London (1970).
[175] Schaeffer, W. D., Smith, W. R., & Polley, M. H., *Ind. Eng. Chem.*, **45**, 1721 (1953).
[176] Schaeffer, W. D. & Smith, W. R., *Ind. Eng. Chem.*, **47**, 1286 (1955).
[177] Robson, D., Harker, H., & Whynne-Jones, W. F. K., *Surface Sci.*, **9**, 246 (1968).
[178] Rose, H. E., *J. Appl. Chem.*, **7**, 244 (1957).
[179] Lambourne, R. (ed.), *Paint and surface coatings: theory and practice*. Ellis Horwood Ltd, Publishers, Chichester, England (1987).

13

Fine particles in the plastic industrial process: instrumentation, and maintenance

13.1 INTRODUCTION

A thorough and extensive review of the plastic industrial process has been carried out by many workers [1–22]. There are many different types of plastics just as there are different types of metals and ceramics. But where the latter are composed of individual atoms or small molecules in crystalline or amorphous lattices, the essential ingredients of every plastic is a high molecular weight polymer, a long chain molecule containing thousands of repeating small molecular units.

Most commercial polymers are synthesized from such simpler molecules, or monomers, as ethylene and styrene; and, while some polymers exhibit a degree of crystalline structure, most are wholly or mainly amorphous. The cellulosic plastics alone are derived from a natural high polymer — cotton cellulose.

13.1.1 The difference between polymers and plastics

Synthetic large molecules are made by joining together thousands of small molecular units known as monomers. The process of joining the monomers together is called polymerization, and the number of these units in the long molecule is known as the degree of polymerization. The names of many polymers consist of the name of the monomer with the prefix 'poly'. For example, the polymers polypropylene and polystyrene are produced from propylene and styrene respectively.

The words 'polymers' and 'plastics' are often taken as synonymous, but in fact there is a distinction. The polymer is the pure material which results from the process of polymerization and is usually taken as the family name for materials which have long chain-like molecules; this includes rubber. Pure polymers are seldom used on their own, and it is when additives are present that the term plastic is applied. Polymers contain additives for a number of reasons. In some cases impurities are present as a result of the polymerization process, and it may be uneconomic to remove these to get the pure polymer. In other cases additives such as stabilizers, lubricants, fillers, pigments, etc. are added to enhance the properties of the materials. By tradition the term plastics excludes rubber.

The main objective of this chapter is simply to highlight the fact that powder particles, whether as resins or pigments, are used in the plastic industries. Many engineering polymers or resins which contain fillers and extenders, catalysts, pelletized additive systems or concentrates, organic or inorganic pigments, etc. are powder particulate composites. A classic example is vulcanized rubber. Carbon black consists of tiny carbon spheroids only 50 Å to 5000 Å in diameter. The carbon black improves the strength, stiffness, hardness, wear resistance, and heat resistance of the rubber.

Other special properties may be obtained from particulate materials. For example, polyethylene may contain metallic powders, such as lead, to improve adsorption of fission products in nuclear application, or bronze, to improve the electrical conductivity of the polymer and permit the polymer to be electroplated.

Pigments are black, white, coloured, metallic, or fluorescent organic or inorganic solids which are insoluble and remain essentially unaffected by the medium into which they are dispersed or incorporated. They are small in particle size, generally in the range of 0.01–1.0 μm diameter. Pigments produce colour by selective absorption of light, but because they are solids, they also scatter light. Light scattering is undesirable in a transparent material, but desirable if opacity is the goal. Some organic pigments that are extremeley small in particle size scatter very little light and therefore act like dyes; for example, Benzimidazolone Carmine HF3C, with a particle size of 0.5–0.07 μm. Some colourless pigments, relatively large in particle size (up to 100 μm) are used as fillers or extenders.

The processing of resins or pigments or plastics generally, into useful products cannot take place without an underlying appreciation of the following fundamentals and processing techniques.

There are two important classes of plastics: thermoplastics and thermosets.

13.1.2 Thermoplastics

In a thermoplastic material the long chain-like molecules are held together by relatively weak Van der Waals forces. A useful image of the structure is a mass of randomly distributed long strand of sticky wool. When the material is heated the intermolecular forces are weakened so that it becomes soft and flexible, and eventually, at high temperatures, it is a viscous melt. When the material is allowed to cool it solidifies again. This cycle of softening by heat and solidifying when cooled can be repeated more or less indefinitely, and is a definite advantage in that it is the basis of most processing methods for these materials. It does have its drawbacks, however, because it means that the properties of thermoplastics are heat sensitive. A useful analogy which is often used to describe these materials is that like candle wax they can be repeatedly softened by heat and will solidify when cooled.

Examples of thermoplastics are polyethylene, polyvinyl chloride, polystyrene, nylon, cellulose acetate, acetal polycarbonate, polymethyl methacrylate, and polypropylene.

13.1.3 Thermosets

A thermosetting material is produced by a chemical reaction which has two stages. The first stage results in the formation of long chain-like molecules similar to those present in thermoplastics, but still capable of further reaction. The second stage of the reaction takes place during moulding, usually under the application of heat and pressure. The resultant moulding will be rigid when cooled, but a close network structure has been set up within the material. During the second stage the long molecular chains have been interlinked by strong bonds so that the material cannot be softened again by the application of heat. If excess heat is applied to these materials they will char and degrade. This type of behaviour is analogous to boiling an egg. Once the egg has cooled and is hard, it cannot be softened again by the application of heat.

Since the cross-linking of the molecules is by strong chemical bonds, thermosetting materials are characteristically quite rigid materials, and their mechanical properties are not heat sensitive. Examples of thermosets are phenol formaldehyde, melamine formaldehyde, urea formaldehyde, epoxies, and some polyesters.

13.1.4 Viscoelastic behaviour of plastics

In a perfectly elastic (Hookean) material the stress, δ, is directly proportional to the strain, ε, and the relationship may be written, for uniaxial stress and strain, as

$$\delta = \text{constant} \times \varepsilon \tag{13.1}$$

where the constant is referred to as the modulus of the material.

In a perfectly viscous (Newtonian) fluid the shear stress, τ, is directly proportional to the rate of strain, $\dot{\gamma}$, and the relationship may be written as

$$\tau = \text{constant} \times \dot{\gamma} \tag{13.2}$$

where the constant in this case is referred to as the viscosity of the fluid.

Polymeric materials exhibit mechanical properties which come somewhere between these two ideal cases, hence they are termed viscoelastic. In a viscoelastic material the stress is a function of strain and time, and so may be described by an equation of the form

$$\delta = f(\varepsilon, t) \tag{13.3}$$

This type of response is referred to as non-linear viscoelastic, but as it is not amenable to simple analysis it is often reduced to the form

$$\delta = \varepsilon \cdot f(t)\ . \tag{13.4}$$

This equation is the basis of linear viscoelasticity and simply indicates that, in a tensile test, for example, for a fixed value of elapsed time, the stress will be directly proportional to the strain.

The most characteristic features of viscoelastic materials are that they exhibit a time-dependent strain response to a constant stress (creep) and a time-dependent stress response to a constant strain (relaxation). In addition, when the applied stress is removed the materials have the ability to recover slowly. These effects can also be observed in metals, but the difference is that in plastics they occur at room temperature whereas in metals they occur only at very high temperatures.

13.2 FEEDSTOCK AND REACTIONS

The chains of the long plastics molecules usually consist of carbon atoms linked together, though some, like acetal resin (polyoxymethylene) and the silicones, have oxygen or silicon atoms in the chains. The simple chemicals from which monomers and polymers ultimately derive are usually obtained from crude oil or natural gas, but they sometimes come from coal, sand, salt, and air. For example, the feedstocks for making polystyrene, the central polymer of a widely used plastics family, are benzene and ethylene, both of which are today derived from petroleum. The two are reacted to form ethyl benzene, which is then cracked to remove one molecule of hydrogen, restoring the ethylenic double bond, and yielding one molecule of the monomer styrene. With the aid of a catalyst, one bond of each styrene double bond is used to link, in sequence one to another and so on, until a stray radical terminates the chain. Polystyrene is one of a great class called addition polymers. They constitute the largest plastics families — the polyolefins, the polystyrenes, acrylics, vinyls, fluoroplastics, and others.

Other polymers, such as the nylons, are made by the reactions of two different molecules, both of which have two reactive end groups. As the chain grows, the formation of each link is accompanied by the production of a small molecule such as water, which must be removed to achieve a high degree of polymerization. These are called condensation polymers, and they make up most of the other commercial plastics.

13.3 PROCESSING TECHNIQUES

One of the most outstanding features of plastics is the ease with which they can be processed. More often than not finished articles which may be quite complex in shape are produced in a single operation. The processing operations of heating, shaping, and cooling may be continuous (e.g. production of pipe by extrusion) or a repeated cycle of events (e.g. production of a telephone housing by injection moulding), but in most cases the processes may be automated and so are particularly suitable for mass production. There is a wide range of techniques by which plastics articles may be produced. In most cases the choice of method is based on the shape of the component and whether it is thermoplastic or thermosetting, and whether complicated shape and intricate details can be made in one operation with little or no further post-fabrication requirement. We can classify these principal processing methods by first defining who the processors are.

13.4 PROCESSORS

Polymers are produced by chemical companies and shipped to processors — fabricating companies that melt thermoplastics or formulate and mix the components of thermosets to form plastic products (see Table 13.1) in diverse shapes. The

Table 13.1 — Petrochemical products

Ink	Life jackets	Plywood adhesive
Heart valves	TV cabinets	Cameras
Crayons	Car battery cases	Anesthetics
Parachutes	Insect repellants	Artificial turf
Telephones	Ice buckets	Artificial limbs
Enamel	Fertilizers	Bandages
Transparent tape	Hair coloring	Dentures
Antiseptics	Toilet seats	Mops
Vacuum bottles	Denture adhesive	Beach umbrellas
Purses	Loudspeakers	Ballpoint pens
Deodorant	Movie film	Boats
Panty hose	Fishing boots	Golf bags
Rubbing alcohol	Candles	Nail polish
Shag rugs	Water pipes	Tool boxes
Epoxy paint	Car enamel	Tape recorders
Oil filters	Shower curtains	Curtains
Pajamas	Credit cards	Vitamin capsules
Upholstery	Aspirin	Dashboards
Hearing aids	Permanent-press clothes	Putty
Car sound insulation	Golf balls	Percolators
Dresses	Detergents	Skis
Cassettes	Sun glasses	Tool racks
Motocycle helmets	Glue	Slacks
Pillows	Fishing rods	Yarn
Clothesline	Linoleum	Insecticides
Soap dishes	Plastic wood	Fishing lures
Shoes	Soft contact lenses	Perfumes
Tobacco pouches	Dice	Shoe polish
Refrigerator linings	Trash bags	Petroleum jelly
Electrician's tape	Hand lotion	Faucet washers
Model cars	Shampoo	Caulking
Folding doors	Shaving cream	Food preservatives
Floor wax	Safety glass	Antihistamines
Sweaters	Ear-rings	Cortisone
Sports car bodies	Salad bowls	Dyes
Tyres	Tennis rackets	LP records
Hair curlers	Drinking cups	Vaporizers
Dishwashing liquids	Housepaint	Solvents
Unbreakable dishes	Roller skate wheels	Cigarette filters
Toothbrushes	Guitar strings	Roofing
Combs	Ammonia	Cold cream
Toothpaste	Eyeglasses	Synthetic rubber
Tents	Ice chests	Glycerin
Lipstick	Umbrellas	Rubber cement
Ice cube trays	Paint rollers	Nylon rope
Electric blankets	Luggage	Fan belts
Antifreeze	Refrigerants	Typewriter ribbons
Footballs	Paint brushes	Balloons
Syringes		

polymer is shipped in pelletized, granulated, powdered, or liquid forms and is at this stage called resin. With thermoplastics, the predominant resin form is that of pellets, which are usually unpigmented (natural) and may incorporate colorants or basic additives such as stabilizers. With a few exceptions, most resins are not used in the natural state, but are compounded — combined with other materials, typically by means of mechanical mixing or melt-state blending, to yield a compound which is ready to be used by the processor. The compounding may be done by the resin company, by an independent compounding company, or by the processor. Like resin, compound may be pelletized, granulated, powdered, or liquid.

13.5 VARIATION OF POLYMER PROPERTIES

Polymer properties can be varied during polymerization (the basic chemical process carried out at the resin company) in which polymer is formed under the influence of heat, pressure, catalysis, or a combination, inside vessels called reactors. One special form of property variation uses two or three different monomers as comonomers, i.e. copolymerizing them to produce copolymers (two comonomers) or terpolymers (three). Properties of such multipolymers are sometimes intermediate between those of the homopolymers that may be made from the monomers used alone, and sometimes superior to them.

13.6 COMPOUNDING

Even more versatile than these in-reactor methods of varying properties, is compounding. In this post-reactor technique, resin is combined with any one or a combination of the following.

(a) Additives such as colorants, flame retardants, heat or light stabilizers, lubricants, and others.
(b) Fillers or reinforcements such as glass fibres, particulate minerals, or hollow glass spheres.
(c) Other polymers, to form a polymer blend or alloy.

13.7 COMPOSITES

Filled or reinforced plastics are widely referred to as composites. Some composites, however, are not produced by means of simple mixing (compounding) of polymer and reinforcement, but by means of more complex techniques involving weaving, winding, or otherwise aligning reinforcement in special patterns. In advanced composites, moreover, both the polymer matrix and the reinforcement material have unusually high performance properties.

Still other composites have neither filler nor reinforcement. In packaging laminates, building panels, and other materials systems, semi-finished plastic products like sheet or film are in some way made to adhere to non-plastic products such as aluminium foil, plywood, or structural components like wood or metal assemblies.

13.8 FOAMS

Another special type of material system is that of plastic foam, which is produced either by compounding with resin an additive called a foaming agent, or by injecting air or a volatile fluid into the still-liquid polymer while it is being processed into a finished product. The object is to introduce into the internal structure of the finished product a network of gas spaces, or cells, that make it less dense than the solid polymer. Foams may be rigid or flexible, open-celled or closed-celled. A special type of foam is structural foam, which undergoes less density reduction than other types and has a solid skin integral with its cellular core.

13.9 ELASTOMERS

A third special material is that of elastomers, which are copolymers or blends that, unlike flexible plastics, are truly resilient, easily and quickly recovering their original shape after being bent, stretched, or otherwise deformed. For reasons having as much to do with established commercial practice as with underlying technology, thermoset elastomers, including both natural and synthetic rubber, are considered part of the rubber industry, whereas thermoplastic elastomers, a relatively new sector, are considered part of plastics.

13.10 PLASTICS AND TEXTILES

A similar combination of tradition and technology is the cause of yet another division between industries: that of plastics and textiles. Although many textile fabrics are woven of fibres produced from essentially the same polymers as those used in plastics, most fibre applications of polymers are not regarded as part of the plastics business.

What distinguishes plastics from other organic materials, however, is not so much the boundaries between industrial sectors as a capability more or less unique to plastics: that of being modified or diversified seemingly endlessly, as has been suggested, to meet vastly diverse end-use needs. Thus, each of the nearly 500 types of compounds presently being manufactured has its own set of properties and prices.

13.11 HISTORY OF PLASTICS, AND HIERARCHY

The history of plastics goes back over 100 years to the introduction of cellulose nitrate as a substitute for ivory in billiard balls. In 1909, there appeared the 'Bakelite' phenolic compounds, from which black telephones and auto distributor caps were made for several decades. Since then, many new polymers have appeared, with development accelerating after World War II. The vast bulk of annual plastics industry tonnage since the war has been accounted for by commodity resins, each of which, at least so far as its basic or workhorse grades are concerned, is available in relatively standardized formulations from many resin companies. The proliferation of polymer types, however, has also included engineering resins, which even in basic grades, typically exhibit performance of a distinctly higher order than that of commodity materials. Indeed, among commercial thermoplastics we now perceive a hierarchy of four tiers, namely, advanced, engineering, intermediate, and commodity. As we move down the hierarchy of Table 13.2, we generally observe greater ability of materials to carry loads, to withstand impact and high temperature, and to resist attack by solvents and chemicals.

Table 13.2 — Hierarchy of plastics

Plastic group	Typical member	Price, \$/lb
Commodity (volume)	Polystyrene (PS)	0.30
Intermediate	Polyethylene terephthalate (PET)	0.60
Engineering	Nylon	1.75
Advanced	Polyetheretherketone (PEEK)	25.00

13.12 EFFECT OF PLASTICS ON OTHER MATERIALS

Before 1930, most household goods and industrial components were made of metals, woods, glass, papers, leathers, and vulcanized natural rubber. Plastics have since invaded all of the markets for these materials, and, in addition, because they are readily moulded into complex shapes that are very costly to produce from the above materials, they have created vast new markets of their own.

Traditional engineering design for load-bearing elements was based on allowable deflection and/or failure by yielding or breaking. The fundamental material property governing deflection under load is the elastic modulus, E (Young's modulus), defined as the tensile stress required to produce 100% strain. Since most materials start yielding long before the strain reaches 100%, E is usually found by measuring the stress at some small strains and calculating the ratio of stress to strain. Steels have moduli of about 29 million lb/sq. in. (29 M.p.s.i,), or, in SI units, 200 gigapascals (GPa), the highest of all common materials save tungsten, and roughly 60 times the typical moduli of rigid plastics. Zinc, brass, and aluminium have moduli about one-third that of steel. The yield strengths of steel and brass are also high, about four times those of the 'advanced' polymers (unfilled), while aluminium and zinc are comparable to them. How, then, can plastic compete for load-bearing service?

In most applications where deflection is important, it is deflection in bending that is under consideration. The stiffness of a structure in bending depends directly on the material modulus and also on the volume of the structure. For many applications where the weight of the structure is a design consideration, the ratio of modulus to density or strength to density is more relevant than modulus or strength *per se*. Let us look at typical values for these properties below, in Table 13.3.

A more subtle factor in bending is how easy it is to place most of the material relatively far out from the neutral axis of the bending member. This is the principle typified by the steel I-beam used so widely in large structures. Since plastics are easily moulded into any desired shape, variations on this principle are widely employed in plastics products to enhance their stiffness while being economical of material.

Table 13.3 — Typical properties of materials

Material	Modulus, E, (M.p.s.i.)	Strength, S, (K.p.s.i.)	Specific gravity	E/d	S/d
Mild steel	29	60	7.8	3.7	7.7
Aluminium	10	18	2.7	3.7	6.7
Brass	13	50	8.4	1.6	6.0
Zinc	13	13	6.8	1.9	1.9
Douglas fir*	1.4	10	0.5	2.8	20.0
PS	0.51	8	1.05	0.49	7.6
PET	0.29	9	1.35	0.21	6.7
Nylon**	0.42	8	1.14	0.37	7.0
PEEK	0.56	12	1.31	0.43	9.2

*=air-dried, with the grain.
**=as-moulded, 0.2% moisture.

An important way in which plastics differ, near room temperature and above, from metals and ceramics is their tendency to creep under load. Put simply, this means that plastics subjected to stresses of the order of one-third or more of their full strength will gradually stretch or flow in the direction of the imposed stress. The behaviour exists in metals and ceramics, too, but at much higher temperatures. Plastics reinforced with refractory fibres of glass or carbon (RP) have much lower creep rates than the neat resins. In contrast, highly plasticized or flexible resins will creep rapidly even at low stresses.

13.13 RUST-PROOF

We are all familiar with the rusting of ferrous metals and the need for costly protective procedures such as pickling, galvanizing, priming, and multi-coat painting to prevent rusting. Plastics do not have that problem; by incorporating some colour into the moulding compound, the surface is finished permanently as the part is made. Most plastics have excellent resistance to inorganic acids, bases, and water solutions of most salts and concentrations. Many are also highly resistant to organic solvents, though, for particular solvents, some are much better than others.

13.14 OTHER PLASTIC PROPERTIES

Perhaps the most felicitous property of plastics as a class is their ability to be processed, by one method or another, into almost any conceivable shape with any desired surface finish from mirror gloss to complex patterns and textures, requiring little or no finishing after primary manufacture. To make such a simple object as a smooth, true sphere 2 in. in diameter from even the most close-grained wood requires a great many operations and much time and meticulous care in finishing.

Once one has invested an equivalent effort in making a tool-steel mould cavity, thousands of nearly perfect spheres are easily produced by injection moulding of the thermoplastic of choice.

Whereas woods all have very pronounced grain and exhibit a great range of properties along and across the grain, even from a single log, moulded plastics are more nearly isotropic in properties. Most of them absorb and desorb very little moisture as the ambient humidity ranges through the four seasons. Who has not experienced a wooden door that let dust blow in when the weather was cold and dry, but swelled and became hard to open and close in rainy times? Finally, to make any but standard rectangular shapes of wood requires cutting away and discarding much, somethimes most of the wood with which one starts. With plastics, only the material in the final product is required. With thermoplastics, even trimmings and rejected parts can be ground up and recycled into saleable products. While some lower-melting metals can be die-cast, a process similar to injection moulding, steel parts often must be machined from sheet and bar stock, with the similar wastage factors of woodworking.

The application properties of plastics, including fibre-reinforced composites, encompass the whole range of properties of all materials at temperatures up to about 500°F. As pointed out earlier, recently developed thermoplastic elastomers even have properties much like those of the old standby rubber varieties. The thermoset polyurthane family alone can be formulated for high strength and rigidity or for high extensibility, resilience, and abrasion resistance better than that of the toughest rubbers. Or it can be regenerated as a liquid that, poured between inner and outer walls, becomes a rigid foam of high insulation quality. Other PUR formulations are injection moulded to make structural-foam furniture parts, parts not possible with traditional materials.

13.15 MULTI-MATERIAL STRUCTURES

Plastics are used in combination with many other materials to produce structures not possible with any one of them. Sandwich panels composed of two thin sheets of glass fibre-reinforced polyester, joined by a thick layer of PUR foam, cemented balsa wood, or aluminium honeycomb, provide great resistance to bending with low weight and good economy. Tremendous tonnages of papers, foils, and clothings, laminated with plastics films, are used in packaging and furniture making. In many plastics industrial components, threaded inserts are moulded in, making it possible to utilize the higher shear strength and greater compactness of metal screws in joining plastics components to others. Large amounts of plastics are consumed in adhesive bonding of all materials. For example, the layers of most plywoods are bonded with phenolic-based adhesives, while the heat-, chemical-, and mar-resistant counter and table-top surfaces of the ubiquitous decorative laminates are melamine resin. Unlike old-style water-based glues and solvent cements, two-component, high-strength epoxy adhesives do not dry but harden by chemical crosslinking with little shrinkage, so they can completely fill and bond even imperfect joints.

13.16 PLASTICS FROM RESINS

The primary resins of which plastics are compounded are mde in solution, as fine particles in suspensions, or as melts. Whatever their initial form, most thermoplastics are converted by the resin-makers into BB-sized pellets known as moulding powder. During this finishing operation, small amounts of stabilizers and other conditioners are added. Some of this moulding powder is shipped, in bags, drums, large cartons, truckloads, or hopper cars, directly to processors who convert it to such finished products as sheet, pipe, film, monofilaments, coatings on paper, foil, and cloth; bottles and other items are produced in large quantities. Much of the resin, however, goes to compounders who blend it with more additives to produce compounds with special properties needed by their customers, or for their own products. Compounding usually entails blending the components, then extruding them to again make pellet compound, which may be shipped to manufacturers of finished products. In captive compounding operations, processors buy raw materials and do the blending themselves.

Some thermoset resins are delivered in liquid form for use in moulding operations in which reinforcing mats, cloths, or chopped fibrous-glass roving are combined with the catalyzed resin on forms or in moulds, then cured in place. Again, large quantities of these liquid resins (polyesters and epoxies are typical), also go to compounders who combine them with fillers, pigments, fibrous reinforcements, and heat-activated catalysts to produce thermosetting moulding compounds and prepregs, which are pre-impregnated, formable, heat-curable sheets.

13.17 CONVERTING COMPOUNDS TO PRODUCTS

There are two broad classes of shapes that include nearly all finished products made of plastics. In the first class, the cross-section perpendicular to the length is the same throughout. It includes such products as garden hose, pipe, sheet, film, wire coatings, rods, and a myriad of special profiles such as refrigerator gaskets and house siding, products that are cylindrical in the geometer's general sense, with any desired cross-section, and which can be cut to any desired length. All others, those that are not cylindrical, make up the second group: moulded, cast, and fabricated products.

13.17.1 The extrusion processes

The cylindrical products, especially the thermoplastic ones, are mostly made by extrusion, using mainly single-screw extruders. However, large quantities of floor coverings and plasticized film are made by calendering, in which the heat-softened compound is dropped onto a stack of four massive rolls. As it passes between the successive roll nips it is squeezed into its final flat shape and thickness.

Since an extrusion machine is the most common means of converting reactor product (polymer) into resin pellets, it is safe to say that about 90% of all thermoplastics have been extruded at least once. Inside the extruder barrel is a flighted screw that is turned by a hefty motor and speed reducer. Rotation of the screw, deep at the intake end and shallow at the delivery end, causes friction and compression which melt and mix the raw material, then force it through an opening of the appropriate shape, called a die, at the head end. Downstream are shape-holding and cooling devices to support and chill the hot extrudate until it has hardened. In making resin pellets, the spaghetti-like strands are hot-cut at the die face by rotating knives or pulled away, chilled, and cut cold in a strand chopper. Pipe usually is pulled through a ring that imparts the final dimension to the outside diameter, then through a long trough in which cooling water circulates.

13.17.2 Sheet-extrusion line

Sheet is made on a sheet-extrusion line. As the molten sheet leaves the die it quickly passes onto a chilled, chrome-plated roll, one of three that impart a high gloss to the sheet, cool it to the point where it is firmly set, then discharge it onto a roller conveyor where it cools further before being edge-trimmed and cut to length.

13.17.3 Extrusion-cast film

Extrusion-cast film is made in a similar manner except that, being thinner, it is more completely cooled on the casting roll and instead of being cut into fixed lengths, it is wound into rolls. In the blown-film process, the die is a cylindrical block rather than a long straight one, and the film emerges as a tube from the annular die lip, passing through one or more rings that cool it. Internal cooling also is used to increase production rates. As the molten tube emerges, usually in an upward direction, it is quickly inflated by internal pressure and simultaneously rapidly stretched in the lengthwise direction. It is frozen while this two-directional tension is maintained, causing the polymer chains to become oriented and tensed in all directions parallel to the film surface. The orientation makes the film about equally strong in both directions, whereas cast film, which usually is oriented in only the machine direction, is weaker in the transverse direction. The familiar dry-cleaning and super-market produce bags are made from blown film. Between the zone of inflation and final wind-up into a roll, the new film is first flattened out, corona-treated to enable it to accept printing inks, then printed with whatever labels and instructions the dry cleaner or produce manager wants.

13.17.4 Co-extrusion

In the last two decades great advances have been made in co-extrusion which, as it sounds, means to make an extrudate of two or more layers, often of different resins, by feeding a single die from two or more layers, often of different resins, by feeding a single die from two or several extruders. The technique has been applied mostly to films and sheet, to produce a laminate that combines the properties of the several resins, properties not available in any one resin.

13.17.5 Sheet thermoformers

A large percentage of the sheet made on sheet lines is shipped to secondary processors, called sheet thermoformers. They re-heat the sheet to just below the melting range in swingaway infrared sandwich heaters, then apply vacuum to draw it down into open, chilled moulds that convert it into trays, refrigerator door liners and inner boxes, automobile parts, packages of great variety, bath tubs, toys, and so forth and so forth. For small items, many identical pieces are formed from a single large sheet. The desired pieces are cut from the sheet while the trim is ground by rotating cutters with heavy blades, into small pellets that are recycled to the sheet extruders. In some large thermoforming shops dedicated to producing a single product, the sheet extrusion and thermoforming lines may be coupled. By feeding the fresh-made sheet while still warm to the thermoforming machine, one saves much energy and gets a substantial reduction in overall thermoforming time.

13.17.6 The injection moulding machine

More kinds of plastics parts and products are made by injection moulding machines than by any other process. For many years resin was melted in injection machines by being rammed through a heated cylinder with a metal core called a torpedo. Eventually it was realized that this conduction heating not only was slow, but also sometimes burned the plastic, and the screw-plasticating principle, already widely used in extruders, was adapted to injection moulding.

All modern injection moulding machines use this method, but not all use the screw itself as the injection ram. In this machine, the screw rotates rapidly for part of the moulding cycle while the melt just previously injected into the chilled mould cools and freezes, and the part is ejected. As the mould closes, screw rotation ceases, and a hydraulic ram thrusts the screw forward to inject the collected melt into the waiting mould. As this occurs, the nozzle of the machine makes high-pressure contact with the sprue bushing of the mould, sealing the connection.

Moulds may be opened hydraulically or mechanically. By the use of side-draws, multi-plate moulds, and automatic unscrewing devices in the mould, it is possible to produce extremely complex parts with deep recesses and screw threads. Some moulds have only one or two cavities; for large-volume production of small parts, moulds containing from several to hundreds of cavities have been successfully operated.

Moulded items with large projected areas at the parting surface of the mould exert large mould-opening forces when the melt is injected at high pressure, typically 5 to 15 k.p.s.i. If the mould-clamping force is not safely greater than the mould-opening force, the mould may open slightly and melt will leak and form an unsightly fringe called flash along the edges of the part. Flash removal is costly and disfiguring.

Two-colour moulding, e.g. numerals on business-machine keys, can be made by injecting one colour from one injection machine, allowing the first shot to harden, then shifting the mould plate and injecting the second material into the cavities moulded into the first. Co-moulding, analogous to co-extrusion, is also done.

13.17.7 The blow moulding machine

Blow moulding is used to make hollow items — bottles are the dominant product — that can be completely closed at the ends. Most used is extrusion-blow moulding. In this process, a tube of hot melt, called a parison, is extruded downward between the open halves of a chilled aluminium mould. As the tip of the parison passes the bottom of the mould, the mould halves are clamped together, pinching off the tube at the ends. The mould is moved aside, making room for another that will catch the next parison. Air or air and water mist is injected at the top, inflating the soft tube against the mould surfaces. By circulating air or cold CO_2 within the formed tube, one cools the melt from both sides and greatly reduces the cooling time and the whole moulding cycle.

13.17.8 Rotational casting moulding

A rather different way to make hollow parts is rotational casting, also called rotomoulding. A measured charge of finely powdered resin or liquid solution or suspension of resin, which may be pre-warmed, is loaded into each of one or more moulds that are then completely closed. The moulds are mounted on a rotator, within an oven, which revolves the moulds on two or three axes at once, distributing the contents evenly over the inside mould surfaces. Heating is maintained long enough to fuse the compound and form an even, non-porous coating. Rotation is continued while the moulds are cooled until the coating freezes, after which they are opened. Some very large hollow objects have been made by the rotomoulding process.

13.17.9 Plastic foams processing

Plastic foams constitute a sizeable fraction of the total volume of plastics product. Foams are made from a wide variety of plastics, in densities from almost solid down to about 0.05 g./cm^3, by several methods. Polystyrene in the form of tiny beads has the property of absorbing pentane under pressure and retaining the solvent for weeks, even months. Pentane, a very volatile liquid soluble in PS, vapourizes and acts as a blowing agent when the beads are heated. In a two-stage process, the beads first are pre-expanded by heating to make fluffy little balls, which are allowed to sit for a day to even out the distribution of residual pentane. The balls are then poured into a low-pressure mould, flushed for a few minutes with live steam, then cooled with air blasts to make anything from small moulded packaging parts to huge logs up to $2\times4\times16$ ft in size. The live steam partly resoftens the PS, completes expansion of the foam by vaporizing the pentane, and fuses the expanded beads into a cohesive, strong structure. The logs are then sliced with hot wires to create standard rectangular shapes. PS foam is also made by extrusion, using an inert gas such as nitrogen as the blowing agent.

Polyethylene foams also are produced by extrusion into sheet and into wire coatings, usually as the interlayer in coaxial cable. These foams, too, are used widely as cushioning in industrial packaging.

As discussed earlier, polyurethane foams can be generated in place by thoroughly mixing two reactive liquid components, typically a linear polyester and an isocyanate. Carbon dioxide gas generated in the condensation reaction inflates the foam, and crosslinking locks up the bubbly structure, which may have open or closed cells. Foam moulding machines provide extremely intense and rapid mixing of liquid components, the mix then being injected into a closed mould where it expands and cures to make the desired product.

13.17.10 Compression moulding

Compression moulding is the simplest and oldest plastics moulding process, and has long been used to make products from thermoset plastics. With phenolic compounds, for example, moulding compounds or preformed slugs are preheated close to their curing temperature and loaded into a heated mould cavity. Because the phenolic-resin molecules contain charged atoms, they preheat quickly and uniformly in microwave ovens. The force, or male member of the mould, descends, squeezing the softened compound into all parts of the gap between force and mould. After the moulding has cured under pressure, the mould is opened and the part removed.

13.17.11 Transfer moulding

In transfer moulding, also used to mould thermosets, the preheated compound is loaded into a chamber connected by runners to the mould itself. When the ram moves into the pot, the plastic mass is forced through the runners into the mould and pressure is held during cure. This process is better suited than compression moulding to making small parts in multi-cavity moulds. Recently, thermosetting formulators have developed compounds that can be held for long times at temperatures slightly below the curing range without setting up. These are injection moulded just as thermoplastics are, except, of course, that the mould is heated rather than cooled. By insulating the runners from the mould proper and maintaining them at a lower temperature, the soft resin trapped in them is kept from cross-linking and moves out with the next shot.

13.17.12 Fibre-reinforced plastics

Most thermoplastic resins are available in compounds containing reinforcing fibres, particularly glass fibres. These are made by mixing in glass or other fibres that have been chopped to very short lengths. They are injection moulded in the usual ways, with some special considerations because of the glass fibres. In the same way, chopped glass fibres can be compounded into thermosetting moulding materials for compression, transfer, or injection moulding.

The best effects of fibre reinforcement are realized when the fibres are not chopped to $\frac{1}{8}$-in. lengths but are incorporated in the moulding in long continuous lengths, or in the form of felted mats or woven cloths. This is frequently possible in compression moulding, often called matched metal moulding, by precutting the mat or cloth to the desired shape and laying it in the mould before adding resin, the latter usually free-flowing at room temperature. One of the simplest methods, hand lay-up, is to lay out the mat or cloth on a convex mould, impregnate the fibrous structure with catalyzed resin mix using squeegees and rollers, and then allow the mass to cure at room temperature or in an oven or autoclave.

One of the variants on the hand lay-up method is called spray-up. In this technique the operator uses a gun that has three heads, two for resin and one for chopping glass roving into 2-in. lengths. Pumps deliver catalyzed resin from one tank to one spray nozzle, and resin-containing accelerator from a second tank to the second spray nozzle. The chopped glass is flung forward into the focus of the two sprays, and the intimate mixture of resin, catalyst, accelerator, and glass fibre adheres to the mould surface wherever it is directed. It is easy to control the local thickness of the deposit with this method, so stiffness and strength can be imparted to the degrees needed in different areas of the part. Many small and large boats have been made by spray-up. Standard 4×8-ft panels are made automatedly on long conveyors and cut to length after curing.

13.17.13 Filament winding

Filament winding, as its name implies, is a way of making reinforced plastics by winding a resin-soaked filament containing about 200 individual glass fibres onto a rotating male form or mandrel. The early, simplest shapes made this way were cylindrical. Today, computer-controlled filament winders rotate the mandrel in complex movements that deposit the filament in precalculated patterns and thicknesses to make pressure vessels and more complex structures.

13.17.14 Pultrusion

A process somewhat similar to extrusion is pultrusion, useful for making continuous lengths of RP products having constant cross-sections. The fibre may be oriented entirely in the length direction, or some of it can be brought in from the side to provide multi-directional strength. Glass or other roving, a bundle of filaments, is pulled through a resin bath, then through a long, heated steel die that both shapes the bundle and cures it. As in extrusion, the final product is cut to the desired length.

13.17.15 Fabricating and decorating

Most semi-finished plastic shapes, as well as finished parts, can be machined, sawed, and processed with abrasives in the way that metals are, some more easily than others. Cutting tools generally are ground with larger rake angles to permit free escape of chips. Tools must be kept quite sharp, and feed rates and speeds must be adjusted to prevent heat-softening of the cut surface, which gums up the operation. Thick sheets and bar stock are available in most thermoplastics and some thermosets, permitting fabrication of large pieces.

Most plastics are readily decorated, as extruded or moulded, by painting (seldom necessary), printing, hot stamping, vacuum metallizing, adhesive labelling, and so on. Polyolefins and fluoropolymers have non-polar surfaces of low energy to which few substances adhere well. The former usually are pre-treated by some method, such as flaming or exposure to corona discharge, that oxidizes the surface and increases its level of polar energy. The fluoropolymers, particularly polytetrafluoroethylene (Du Pont's Teflon), require even more drastic pretreatment.

Embossing may be done with many thermoplastics; it is probably used most with flexible PVC film and laminates to impart attractive textures.

13.18 SUMMARY

This extensive study has been carried out to highlight the broad and main aspects of plastics, their place in the spectrum of materials of manufacture, and the ways in which they are converted to useful products of commerce.

Most importantly, the study has deliberately been exhaustive to demonstrate that for every inch of the process, small-scale operations form the backbone of every developed and industrial society, which is of course, transferable to developing Third World countries, especially those having oil and gas based natural resources.

13.19 INSTRUMENTATION

The broad field of instrumentation covers that branch of science which concerns itself with design, fabrication, measurement, calibration, and data reduction of instruments used in process control, communication, remote sensing, data processing, instrumental analysis, and a host of other technical applications. The majority of these are highly technical in nature, and instrument science is expanding rapidly [23–30].

Put simply, instrumentation is the group of devices used directly or indirectly to measure or control a variable. The term includes primary elements (sensors), final control elements (control valves and pumps), and switches, push buttons, controllers, annunciators, and related devices used to manipulate the variable.

Instrumentation, required for proper operation of petrochemical and plastic plants, is often criticized for failure to meet reliability and usefulness expectation. However, the same instrumentation can perform satisfactorily in other locations.

Instrumentation is actually what really works in the field. Answers to what makes instrumentation reliable and maintainable lie outside manufacturer's manuals. Too often this knowledge is not shared outside the process plant because the persons responsible do not consider their solutions unique or important. They are hard-working and creative individuals who have overcome many obstacles to develop fixes and procedures for improved instrumentation performances.

13.19.1 Instrumentation for extrusion and injection moulding

The use of feedback controls, properly applied, can greatly affect the economics of extrusion and moulding operations. It is generally accepted that the use of material handling systems and other automated devices reduces direct labour costs. Instrumentation can also reduce reject and improve quality through accurate control of processing variables such as temperature and pressure.

To speak of a specific industry like blow moulding, it is considered that only the most highly automated plants can compete in high-volume competitive markets of detergent bottles, bleach, and aerosols. Large film converters use the latest instrumentation to obtain the last $\frac{1}{10}$ per cent in efficiency.

Small moulders, likewise, should periodically review existing machines, especially older models, to determine if the instrumentation is responsible for moulding limitations.

It must be pointed out that owing to advanced control dynamics employed in thermoplastic processes and to save time, machine manufacturers and reliable control specialists should be contacted in cases of serious instrument failure.

Extruders have evolved into complex, integrated machines, and the control instrumentation entails heating, cooling, pressure, speed, horsepower, take-off controls, and thickness monitoring which must be operated as a system because of interplay between these transient phenomena. The relationship between temperature and pressure of the melt is complex and has considerable effect on the viscosity and output of a given extruder or injector.

13.19.2 Extruder instrumentation

The primary control instruments for a medium-sized extruder would normally consist of a device to monitor both heating and cooling requirements in the cylinder. The cooling equipment and controls are expected to be quite elaborate with flexibility to handle various materials and screw conditions. There will be a timer for spray cooling regulation and selector switches to preset the degree of maximum heat or cooling potential available to each zone. Temperature indicators should be available for monitoring purposes as well. A melt temperature recorder is supplied for a permanent operating record. A drive motor power recorder is used to log input power. These records are invaluable for duplicating performance and observing temperature and power trends, indicating degree of screen pack plugging, viscosity changes, die adjustments, and many other machine variables. The percentage of heater power per zone, as well as relationship with time, is deemed important enough by the user to permanently instal a cycle time print-out device on each cylinder zone to analyze incremental power input.

Instrument control panels at times have six zones of cylinder control. The cylinder temperature instruments are of time-proportioning control, but for either heating or cooling, by means of a manually set selector switch for each zone. Of interest is the use of simple ammeters showing actual current to the heaters. These are important to determine if the heaters are drawing rated power, are deteriorating, or burned out, in which case they would indicate reduced amperes when the instrument pilot lamp shows a demand for heat. A rev/min tachometer would indicate screw speed, and is usually controlled by a manual run knob through a dynamatic drive. The percentage timer is present for maximum available spray cooling of the cylinder.

Extruders may be designed to have a melt pressure controller. This approach in automation of this important variable receives input from an electrical pressure transducer and has an output compatible with many variable-speed drives to control melt pressure via screw rev/mins. The device would compensate for screen pack plugging and other influential variables. The intention of this system is to provide a melt of constant pressure at the die which determines final output, with temperature and viscosity held constant.

13.19.3 Injection moulding machines instrumentation

Generally, the advent of the reciprocating screw has improved the injection machine. The screw, having greater plasticizing capacity, has nearly replaced the simple ram for the melting function. The forward push of the screw with an end valve approaches the ram method by forcing the plastic into the mould. The barrel temperature is controlled by time-proportioning instruments from thermocouple sensors. The nozzle is controlled manually by a variable autotransformer at appropriate times. The hydraulic relief valves, limit switches, and sequence cycle timers will actually programme the machine throughout the complete cycle, or automatically, if the moulded part rejection is foolproof. The cylinder temperature control consists of three zones, the nozzle being manually controlled by means of a variable autotransformer. Depending upon feed rate, temperature, and material, a selector switch is available, one per zone, to obtain high or low power, the low power level being one-third of the high power maximum. The high power is used for starting up, and possibly for the highest production rates. The low power level is used for small shots, and when longer plasticizing times are available. These descriptions are for machines produced more than twenty years ago. New control instrumentations tell a different story, and it is thought that this need not be discussed further.

13.19.4 Temperature sensing elements and instruments

Thermocouples with quick disconnect locks and jacks are by far the most common means of sensing temperature. Each zone of control on the barrel, cylinder, and die will contain a temperature sensing thermocouple, control instrument, and heater. Other types of sensing elements are thermistors, resistance bulbs, and filled systems, but they are in a minority because of either cost or size. Thermocouples give a relatively linear output signal and will easily drive a galvanometer-type instrument which gives an accurate temperature indication as well as a control function. The three most common control modes are two-position time proportioning, three-position with time proportioning, and full proportioning. The most popular use of these controls are:

Two-position T/P
  Injection machines-barrel
  Extruder machines-die

Three position T/P
  Injection — not used
  Extruder — cylinder (heating and cooling)

Full proportioning
  Injection — barrel
  Extruder — die

On-off, two, and three position time proportioning controls most commonly drive contactors. A proportioning effect is obtained by anticipating the control point and operating the contactor in advance of the control temperature. This frequently entails rapid cycling of the contactor, and, hence shorter life. Full proportioning is defined as a modulated power flow of a magnitude proportional to the difference between the desired and actual temperature. The power modulating device is a saturable reactor or silicon controlled rectifier. The silicon controlled rectifier is capable of practically zero turndown. Saturable reactors normally turndown 3 per cent; however, 1 per cent minimum power reactors are available at premium prices. Heater turndown is important in case of extruders that may run under autothermal conditions.

13.19.5 Pressure techniques

Pressure is a very important variable influencing the performance of an extruder. The die output is directly proportional to the pressure. While temperature and apparent viscosity are also extremely important, the pressure must be known and held constant. The use of a bourdon tube pressure gauge is common, but it must be sealed, usually with a high-temperature grease, to prevent solidification of the melt in the gauge. With the advent of small size, high-temperature, sealed diaphragm pressure transducers, pressure measurement has been simplified.

The electrical output is conveniently transmitted to indicators, controllers, and recorders.

Melt pressure of an injection machine is necessarily of a rapid, cyclic nature. While numerous attempts have been made to measure this variable, it still remains a laboratory process.

13.19.6 Dimensional measurements

Considerable progress in sheet and film thickness has been accomplished during the last ten years. The monitoring of radiation absorption by the material is an accurate measurement of the thickness. The data are presented in a convenient manner, both as to the cross die variations, and length variations due to machine operation. This information is necessary for good quality control.

13.19.7 Horsepower

This variable is important in that it is a function of the extruder energy producing plasticization and pressure, and is generally recorded in large installations. Valve and screen pack conditions are immediately reflected in horsepower. Large extruders operating autothermally require power recorders for the most efficient operation.

13.19.8 Shaft speed

The revolution per minute record of extruder screw is an important record for past performance, either for shift operation or technical analysis. When analyzed with drive horsepower, it reveals concrete information regarding melt and screw conditions. Film take-off speed recorders are easily recorded. Lineal feet per minute is also recordable by using similar techniques.

13.19.9 Feedback control

All automatic control systems use some form of feedback control. Thermocouples measure and feed back to the controller a signal proportional to the temperature. Pressure transducers measure pressure and signal the controller which increases or decreases screw rpm. The dynamic control depends upon many things. For improved performance, it is best to design a thermal process with minimum heat lags and single capacity storage. This demands intimate heater to barrel contact and favours induction heating. Thermocouple locations are a compromise between proximity to the heater and melt.

Pressure control is far different from temperature control. The non-yielding container of an extruder or injection press offers near zero capacity change per input. Therefore, lags and storage are at a minimum and are favourable for good control with properly designed control equipment. Pressure changes can be of the order of hundreds of pounds per second, and dictate use of all electronic control devices.

13.19.10 Present controls and the future

One of the latest control systems for the injection moulding machine is the Van Dorn's microprocessor-based Pathfinder.

Some other new analytical instruments at present in use are:

(1) Moisture analyzer — Computrac TMX, for moisture analysis.
(2) Tinius Olsen Plastomer — for measuring melting index.
(3) Schott automated Intrinsic Viscosity System:
The Schott AVS (Automated Viscosity System) makes it possible to measure the intrinsic viscosities (I.V.) of multiple samples of polyesters, using the solution viscosity method. (I.V. represents a measure of the molecular chain length of a

polyester; thus, it is a measure of the average molecular weight.) Once samples are prepared and dissolved, they can be run automatically with no operator attention.

(4) Rheometrics Solids Analyzer RSA II. This is a microprocessor-based control computer which analyzes and calculates elastic (storage) modulus, viscous (loss) modulus, and mechanical damping.

13.19.11 Analytical instruments

A rapidly expanding field of plastics applications coupled with steady changes in the chemistry and technology of plastics manufacture have made necessary the establishment of analytical techniques consistent with the development of the industry.

13.19.11.1 Infrared spectrophotometry

To date, the technique which has been most widely applied to plastics analysis has been infrared spectrophotometry.

The infrared spectrum of any molecule is a fingerprint characteristic of that molecule only, and may be used both for characterization (qualitative) and for analysis (quantitative).

The information given by any spectrophotometer is dependent on the manner in which radiant energy interacts with matter. In the case of absorption spectrophotometry, a record is produced of the intensity of radiant energy which is transmitted by the sample as a function of wavelength.

Principal attributes of infrared which have contributed to its wide acceptance as an analytical tool in the plastics industry are:

Uniqueness:

No two substances exhibit identical infrared spectra. This makes possible positive identification of unknowns and facilitates decisions on similarity or dissimilarity among different but very similar compounds, for example, differentiation between atactic and isotactic polypropylene.

Sensitivity to structural changes:

Slight structural changes will often produce significant alterations in the infrared spectrum of a molecule. A very subtle example of this is the detection of crystalline/amorphous ratio in solid polyethylene.

Specificity for functional groups:

Since each functional group exhibits absorption in a specific spectral region, detection of functionality may be quite specific, for example, detection of carbonyl absorption in polyethylene as a result of oxidation.

These applications represent just a few typical situations to which infrared spectroscopy may be applied in a plastics analytical laboratory.

13.19.11.2 Ultraviolet spectrophotometry

Instrumentation of similar principle is used to obtain the ultraviolet and visible absorption spectrum of a substance. Although instrument components will differ because of the different spectral range, the fundamentals remain essentially the same.

Ultraviolet spectrophotometric methods, because of their lesser specificity and sensitivity to structural changes, have not found as wide application as has infrared to plastics analysis. However, many materials associated with plastics and polymers may be determined quantitatively, either in the original matrix or by extraction of the matrix with a suitable solvent. Specific among materials of this type are the phthalate ester plasticizers and butylated phenols used as anti-oxidants. These materials are relatively strong UV absorbers and may be detected at very low levels (0.1 per cent) in polymers such as polyethylene and polypropylene.

13.19.11.3 Visible and emission spectrophotometry

Applications dealing with maintenance of colour uniformity of, say, pigmented plastics may be approached by using either visible absorption or reflectance methods. Little has been published concerning the application of visible absorption and emission spectroscopic methods.

Other emission techniques, such as X-ray fluorescence, have been restricted to the same class of applications as has optical emission. X-ray diffraction, however, may be used for determination of crystallinity in polymorphic materials and for elucidation of crystal structure in crystalline materials. Applications of these methods to plastics have also been somewhat limited.

13.19.11.4 Gas chromatography

The other instrumental technique of appreciable significance in plastics analysis has been gas chromatography. The ability of the gas chromatograph to separate complex mixtures into pure components with speed and high sensitivity has placed at the analyst's disposal a tool of significant proportions.

13.19.11.4.1 Principles and instrumentation

Fig. 13.1 illustrates the basic chromatograph flow system. Capillaries separate the

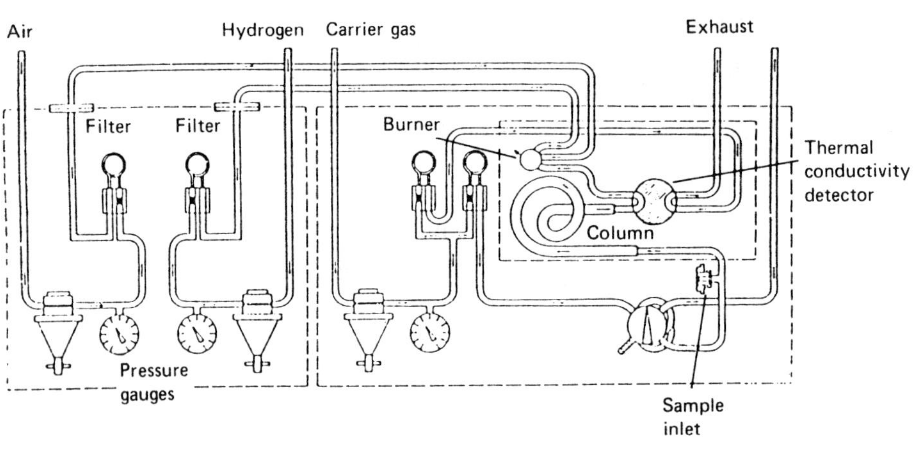

Fig. 13.1 — Gas chromatograph flow diagram.

carrier gas (usually helium) flow into two streams. One stream flows through the gas sample valve, the sample inlet, and the column, to the sample-measuring side of the detector cell. The other stream goes to the reference side of the detector cell.

A gas sample, or a liquid sample of suitable vapour pressure, is introduced into the carrier gas stream by means of the gas sample valve or by syringe injection. The heated sample inlet vaporizes the sample. The carrier gas pushes the sample through the chromatographic column which separates it into its component parts. Fractions are adsorbed or absorbed by the column's packing and pass through the column at characteristic individual rates. They emerge from the column in a time sequence directly proportional to their affinities for the packing.

The fractions of least affinity emerge first. Those of greater affinity follow. With proper resolution in the column, the stream emerging from the column is composed of carrier gas and dilute plugs or bands of components.

In most instances the presence of sample components is detected by a thermal conductivity cell. The elements of the thermal conductivity detector are arranged in a Wheatstone bridge configuration and measure the difference between the sample-carrying stream of carrier gas and the reference stream of pure carrier gas. Sample gas passes into the sample-measuring side of the detector cell, altering the transfer of heat from a heated filament to the cell wall. This change in the temperature of the filament alters its resistance and creates a voltage imbalance between the reference side and the sample side of the detector. The imbalance is recorded on a standard strip-chart recorder. The area under the resulting peak serves as a quantitative index of the sample component. The retention time, i.e. the interval between the sample component and the elution of a sample component, can be interpreted to identify the component.

13.19.11.4.2 Applications

Gas chromatography has found application to many facets of plastics analysis. Principal reasons for this wide acceptance are simplicity of construction and operation of gas chromatographic equipment, the versatility in terms of types of sample to which it may be applied.

Samples ranging in boiling point from −250°C to +400°C may be analyzed, providing that they can be vapourized or converted to a volatile form by reaction or pyrolysis. The principal purpose for which gas chromatography has been used to date has been quantitative analysis.

In plastics analysis, two areas of prime importance have been investigation of monomer purity before polymerization, and determination of monomer ratio and degradation profiles of homo- and co-polymers. A good example of monomer analysis is the detection of alkyl benzenes in high-purity finished styrene. Detectable limits for these materials in some cases are as low as 5 ppm.

13.20 MAINTENANCE

Care is another word for preventive maintenance, and good care is essential to extend the useful life of anything at all, including equipments. Good care delays the day when the object, be it a human being or equipment, after long service, must be refurbished.

We can also say that maintenance is that series of activities done to restore an equipment, plant, or machinery which has broken down back to its normal original state. This may be a simplified statement of what maintenance is, but it attempts to give the total picture.

13.20.1 Third World countries and maintenance

One of the most difficult problems facing developing Third World countries is their inability to service and maintain the few equipments bought from developed countries with their scarce foreign exchange reserves [31].

Whether small or large scale, industries would need:

(a) Individuals with technical knowledge.
(b) Mechanical experience or training.
(c) Electrical/electronic experience or training.
Such individuals must know about motor failures, gear failures, how to fix bearings, valve clearance, agitators, pumps, compressors, how to fix o-rings and gaskets when they are out of place, tube failures as a result of high pressure, fatigue due to wear and tear, and some vibrational analysis.
(d) Workers in maintenance must train on a particular equipment as technology and systems are continuously changing. Many learn on the job.
(e) Preventive and predictive maintenance daily, weekly, monthly, quarterly, semi-annually, and annually.

For those concerned with plastics, the following maintenance and trouble-shooting parts should be monitored: hydraulic systems — hydraulic oil, temperature, reservoir capacity, oil level indicator, pump inlet strainer, hydraulic pump, heat exchanger, hydraulic cylinder, hydraulic motors, manifolds and valves, system pressure relief valve, clamp reducing pressure relief valve, cleaning, and inspection. *Electrical systems:* Wiring systems, electrical control elements, timers, lubrication time on delay, no back pressure time on delay, screw pull back timer, contactor relays, fuse replacements, heater bands, adjustments.

The Pathfinder control system should also be monitored. It is a microprocessor based programmable logic controller optimized for decision oriented tasks like controlling the machine sequence of operation. In addition to controlling the machine sequence, the Pathfinder has an automatic fault locator system to help troubleshoot the machine sequence and pinpoint the source of trouble.

(f) A good maintenance workshop must have:
Machine-tool shop
Machinists
Pipe fitters
Welders
Instrumentation technicians
Constant supply of spare parts
A good work order system
Statistical quality control system
A good safety measure system.

When a sensible maintenance system has been established, the maintenance department could then do what it should actually be doing, i.e. fabrication and construction of materials, equipment manufacturing, and development engineering.

13.20.2 Causes of wear and tear in machines

Neglect causes damage and wear resulting in a greatly reduced moulding life. In transfer and injection moulding of thermosetting materials, wear is often detected in the high flow areas such as in the sprues, runners, gates, and portions of the cavities and cores that are directly opposite the gates. Some troubleshooting guides are given in Table 13.4.

Most often, damage results from continuing to run the mould after flashing occurs. Major damage results from closing the mould on the material itself, bits of flash, chips from sheared undercuts, cracked parts of runner or torn gates, stringers from nozzles or from sprues that are too hot or improperly controlled, and the inexcusable neglect of allowing the mould to clamp on granules of raw materials.

Other sources of damage include:

(a) The result of closing the mould on foreign objects such as inserts, tools, screws, nuts, broken ejector pins, etc., caught between the parting line surfaces. In many cases the smallest bits of foreign objects cause the greatest damage, especially if they are caught at the cavity edge. These small bits are not detected by properly set low-pressure closing protective devices. Clamping on a tiny object, even if it is a plastics granule, concentrates the entire press clamp tonnage on this very small area, exceeding the elastic limit of any mould material regardless of quality and hardness.

(b) The result of the use of screwdrivers, knives, or cutters, etc. used to assist in the removal of sticking parts, flash, short shots, etc. from non-automated moulds.

(c) The result of contact with water on unplated surfaces. Water forms in the moulds from condensation, seepage through porous metals, leaky pipe fittings, and 'O'-rings. Careless handling of hoses and feed lines during hook-up leaves water on the mould surface. This is not harmful if detected immediately and carefully removed. Corrosion is progressive, and even if the moulds are stored after being sprayed with an anti-oxidant, a few drops of water or condensation can cause tremendous and costly damage.

(d) Attack from acids after exposure to corrosive materials which may form when some thermoplastics are decomposed by overheating. Overheating can occur in the plasticizing cylinder, the hot runner system, or in the mould cavities, as the result of too small gates, and inadequate venting or cooling systems.

(e) The result of accidents caused by a number of things. Mistakes in mould installation, continuing to produce in a malfunctioning machine, continuing to operate a mould which has started to squeak or squeal will result in damage. Damage will be done to moulds that refuse to close or open normally or will not eject properly if their operation is allowed to continue without ending the problem. An alert and well-trained foreman can hear trouble starting as he passes through the moulding room. Tool-room technicians contribute their

Table 13.4 — Troubleshooting guide

Problem	Probable cause
Weld lines	1. Stock too cold
	2. Injection speed too slow
	3. Mould too cold
	4. Gate too small
	5. Main runner too small
Burning	1. Stock too hot
	2. Frictional burning — enlarge gates
	3. Inadequate venting (trapped gas)
	4. Overheated heater band
	5. Incorrect screw rev/min
Sinks	1. Injection pressure too low
	2. Incorrect mould temperature
	3. Poor part design
	4. Entrapped gas — vent mould
	5. Gate too small
	6. Nozzle too restrictive
	7. Land length too long
Flash	1. Clamping pressure too low
	2. Injection pressure too high
	3. Inadequate mould supports
	4. Excessive mould temperature
	5. Excessive injection time
	6. Excessive vent depth
Warpage	1. Improper gate location
	2. Gate too restrictive
	3. Poorly designed part
	4. Inadequate packing
	5. Too low injection pressure
	6. Extreme variation in wall thickness
	7. Fibre orientation
Brittleness	1. Melt temperature too high
	2. Residue time too long
	3. Undried material
	4. Contamination
	5. Gates or nozzles too small
	6. Poor part design

share by the improper assembly of moulds, failure to tighten screws, leaving metal chips, grinding dust, or polishing abrasive on or near sliding surfaces.

(f) Fatigue is a major cause of mould damage which leads to a breakdown of those mould components that are subjected to the maximum stresses while cycling correctly over long periods. This usually occurs, if it is going to occur at all, after 100 000 and before 300 000 moulding cycles. In a multiple-cavity mould, components identical to those that break first may last very much longer. Fatigue can manifest itself in components subject to compressive loads as well as those in torque, tension, or bending.

(g) Thermal shock makes a questionable contribution to the deterioration of moulds, unless rapid heating and cooling of the mould is part of the cycle. There is ordinarily only a small effect in comparison with hydraulically and mechanically induced stresses. Moulds run in the higher temperature range — 500 to 800°F deteriorate more rapidly; this is an historical problem with the die-casting dies.

REFERENCES

[1] Crawford, R. J., *Plastics engineering. Progress in polymer science.* Pergamon Press, Oxford, New York (1981).

[2] Cauwenberghe, V., Instrumentation and automation in the paper, rubber, plastics, and polymerization industries. *Proceedings of the 4th IFAC Conference* (1986).

[3] Salomon, D. P. & Biegel, J. E., Economics of small batch flexible manufacturing systems. *Journal of Industrial Engineering,* **16**, 6, 88–96, (June, 1984).

[4] Ogorkiewicz, R. M., (ed.) *Thermoplastics. Properties and design.* John Wiley & Sons, London, New York (1974).

[5] MacDermott, C. P., *Selecting thermoplastics for engineering applications.* Marcel Dekker Inc., 270 Madison Ave., New York, NY 10016, USA (1984).

[6] Morton-Jones, D. H. & Ellis, J. W., *Polymer products, Design materials and processing.* Chapman and Hall, London, New York (1986).

[7] Rosanto, D. V. & Schwartz, R. T., (ed.) *Environmental effects on polymeric materials.* Interscience Publishers and John Wiley & Sons, New York, London (1968).

[8] Mroczkowski, R., Connector contacts: critical surfaces: *Advanced materials and processes,* **134**, 6, 49–54 (1988).

[9] Törnell, B., *Recent developments in PVC polymerization: polymer — plastics technology and engineering* **27**, 1–36 (1988).

[10] Carfagna, C., D'Amore, A., Nicodemo, L., & Nicolais, L., Dimensional stability and orientation in amorphous polymers. *Polymer News,* **13**, 11, 334–336 (1988).

[11] Cavaterra, E., Catalysts and chemicals to make dichloroethane. *Hydrocarbon processing,* 63–67 (Dec. 1988).

[12] Addleman, R. & Barrie, I. T., Measurement and testing: how to assess the flow of molten plymers in real situation. *European Plastic News,* 16–28 (Feb. 1988).

[13] Super thermoplastics. *Modern plastics,* Vol. **65**, No. 13, p. 14–30 (1988).

[14] Scrap recycling. *European Plastics News,* 25–32 (Dec. 1988).

[15] Huang, T. A., Instrumentation development for sensing modes of bubble motion of a blown-film process. *Advances in Polymer Technology,* **8**, 1, 69–75, John Wiley & Sons Inc. (1988).

[16] Liau, J. Y., Jang, B. Z., Hwang, L. R., & Wilcox, R. C., Toughening composites by matrix modification. *Plastics Engineering,* 33–38 (Nov. 1988).

[17] Latest development in machinery and equipment. *Plastics and Rubber,* **13**, 5, 14–24 (1988).

[18] Reher, E. O., Bothmer, D., & Schnabel, R., Forshung und Entwicklung, Energetische Untersuchungen zum Wandgleitverhalten von Polymerschmelzen. *Kunstoffe,* **78**, Heft 11, 1109–1112 (1988).

[19] Akporiaye, D. K. & Price, G. D., Systematic enumeration of zeolite frameworks. *Zeolites,* **9**, 23–32 (Jan, 1989).

[20] Krey, J. & Friedrich, K., Crack layer analysis of fatigue fracture of polyethylene terephthalate. *Polymer,* **29**, 1433–1440, Butterworth & Co. (Publishers) Ltd (1988).

[21] Ruckstein, E., & Kim, K. J., Polymerization in gel-like emulsions. *J. Appl. Polym. Sci.,* **36**, 907–923, John Wiley & Sons Inc. (1988).

[22] Abdel-Azim, A. A. A. & Huglin, M. B., Derivation of solvent-solvent interaction parameters from data on ternary polymer solutions. *Polymer Communications,* **29**, 12, 360–362 (Dec. 1988).

[23] Skrentner, R. G., *Handbook for water and wastewater treatment plants,* Lewis Publishers Inc., 121 South Main Street, Chelsea, Michigan 48118 USA (1988).

[24] *Practical process instrumentation and control.* (ed.) Staff of Chemical Engineering. McGraw-Hill Publications Co., New York, NY, USA (1980).

[25] *Principles of unit operations.* 2nd ed (Ed. Foust, A. S., Wenzel, L. A., Clump, C. W., Maus, L., & Anderson, L. B.), John Wiley & Sons, New York (1980).

[26] Gailey, J. A., Determination of molecular orientation in polymer films by infrared spectrophotometry. *Anal. Chem.,* **33**, 1831 (1961).

[27] Luongo, J. P. & Wood, D. L., Infrared spectrum of polyethylene, *Modern Plastics,* (March, 1961).

[28] Eddy, R. D. & Spell, H. L., Determination of additives in polyethylene by absorption spectroscopy. *Anal. Chem.,* **32**, 1811 (Dec. 1960).

[29] Hollis, O. L., Gas-liquid chromatographic analysis of trace impurities in styrene using capillary columns. *Anal. Chem.,* **33**, 352 (1961).

[30] Ettre, K. & Varadi, P. F., Pyrolysis-gas chromatographic technique for direct analysis of thermal degradation products of polymers. *Anal. Chem.,* **34**, 752 (1962).

[31] Igwe, G. J. I., *Instrumentation and maintenance of equipment in small and medium-scale process industries in Third World Countries.* World Bank Economic Development Institute, Washington, D.C. (August 1989).

14

Particle hazards

14.1 INTRODUCTION

Particle hazards can be classified in a different number of ways, for example, when they are finely divided they can cause dust exposions and fires, health hazards, and desertification. Particle hazards are also insidious in nature. Consider a plant containing large quantities of dust on exposed horizontal surfaces for many years. In such a condition most people tend to discount the danger of a dust explosion, but if a primary explosion occurs from whatever cause, the previously quiescent dust layers may be disturbed, and newly formed dust clouds may then act as energy sources for secondary explosions. The same thing happens with our becoming accustomed to breathing in quantities of finely divided materials in our workplaces and also in our homes and streets. It takes many years before the effects of such inhalations can become evident as indicated by health problems, mortality expectancy differences, or both, as compared with the bulk of the unaffected population. In a carbon black factory, the mixture of carbon powder, sulphuric acid, formic acid, oxylic acid, lead, and zinc pumped into the atmosphere may react with water to form a corrosive and carcinogenic acid medium. For the population, this raises the levels of cancer, skin disease, respiratory problems, hypertension, and premature births, more than the average. Similarly, the problem of desertification may be caused by overgrazing, drought, and many other concomitant factors, but the actual advance of the desert is carried out quietly, spasmodically, and insidiously, except for sand storms, by sand particle saltation under the action of the prevailing winds [1].

14.2 THE CONCEPT OF DUST EXPLOSION

The concept of dust explosion of matter in finely divided form can be seen as being a rapid or uncontrolled thermal expansion of the gas initially present and of additional gaseous matter generated by the chemical reactions occurring therein. There occurs a pressure wave ahead of the flame which is assumed not to reach more than about

100 m/s, which is approximately the velocity of sound in air. This phenomenon is termed 'deflagration', and it is the usual way in which dust explosions occur. It should be differentiated from detonation in which the flame and the shock wave move together. There is no well-documented case of a dust explosion occurring by detonation [2,28].

The flame speed varies with the conditions of the system and is a complicated function of the material, its characteristics of size and shape, the plant or apparatus, the duration of burning, and many other factors. In gases, the maximum flame speed occurs at the stoichiometric ratio, and in dust clouds one can have the maximum flame speed at a large excess of the fuel, i.e. the dust.

To provoke the particles to move so that they produce an explosion, two well-defined explosion types must exist: the primary explosion and the secondary explosion. The primary explosion may be small, simply because there is insufficient material floating around in the vicinity to whip up a really serious explosion. However, the primary explosion, although relatively tame, does agitate the dust into a cloud and this in turn can set the stage for the big secondary explosion. This series of events is common in dust explosions, and it is the main reason why various authors [1,2,3] showed that the danger of dust explosions is insidious.

So long as the dust layers remain undisturbed in out-of-the-way places like pipes, ducts, and the like, or in solid form when it poses no explosive hazard, there may be little or no problem. The danger of these same materials (thin dust layers amount to thousands of millions and even millions of millions of particles) in dust or fine powdered form should not be discounted because ignition is much easier as the particle size decreases. In major industries like agricultural, chemical (including dyestuffs), mining, foodstuffs, metals, pharmaceutical, plastics, and woodworking, the sum total of these factors extends the dangers of dust explosions. Table 14.1 shows cumulative totals of elevator and feed mill explosions.

14.3 MEASURABLE QUANTITIES

There are some dust properties that can be measured or evaluated [4,5]. Ignition properties that can be measured include the minimum ignition temperature (MIT), the minimum ignition energy (MIE), and the minimum explosion concentration (MEC). Because these are not, 'fundamental' properties, the apparatus specification within which the tests are made combined with the test procedures are important and should be adhered to strictly.

The MIT test is one in which a known amount of dust is introduced into an elevated temperature chamber. The lowest temperature to cause ignition is taken as the MIT. The Godbert–Greenwald furnace [2] (modified) is a standard apparatus for this test, and is shown in Fig. 14.1.

The MIE is measured by noting the least electric spark energy required to ignite. This test is carried out in a vertical tube apparatus.

The MEC test is conducted in a Hartman (vertical tube) apparatus on a known sample. A diagram of this equipment [2] is shown in Fig. 14.2.

Palmer [3] has shown the procedure that can be followed in the United States to evaluate dust explosibility, and this is shown in Fig. 14.3.

It is shown that when appropriate tests are selected from the procedures outlined

Table 14.1 — Cumulative totals of elevator and feed mill explosions in the United States during the period 1958–1978 (data abstracted from a report entitled 'Prevention of dust explosions in grain elevators — an achievable goal, USDA, April 1979)

Year	Cumulative explosions in		Cumulative deaths	Cumulative injuries
	Elevators	Feed mills		
1958	8	2	2	27
1959	16	4	5	45
1960	24	8	9	63
1961	32	10	9	80
1962	40	11	12	131
1963	48	17	15	161
1964	50	33	18	183
1965	56	36	20	188
1966	65	41	22	210
1967	77	46	23	224
1968	86	53	35	262
1969	87	58	39	275
1970	97	58	40	289
1971	106	59	44	303
1972	108	65	51	326
1973	114	67	53	336
1974	124	72	66	373
1975	130	75	70	392
1976	148	79	92	474
1977	151	87	157	558
1978	158	92	164	605
Av per year	7.9	4.6	30.25	

in Fig. 14.3, the rating of the particulate matter as a hazard can be determined in accordance with the following scheme.

First, five properties of the dust are determined:

Minimum ignition temperature	MIT
Minimum ignition energy	MIE
Minimum explosion concentration	MEC
Maximum explosion pressure	MEP, and
Maximum rate of pressure rise	MRPR.

Then, the above set of five properties is also determined for a sample of Pittsburgh coal dust to serve as a standard for comparison. Two parameters are then calculated from the above test data, the ignition sensitivity and the explosion severity. Both of these parameters are obtained by comparison with the Pittsburgh coal as standard.

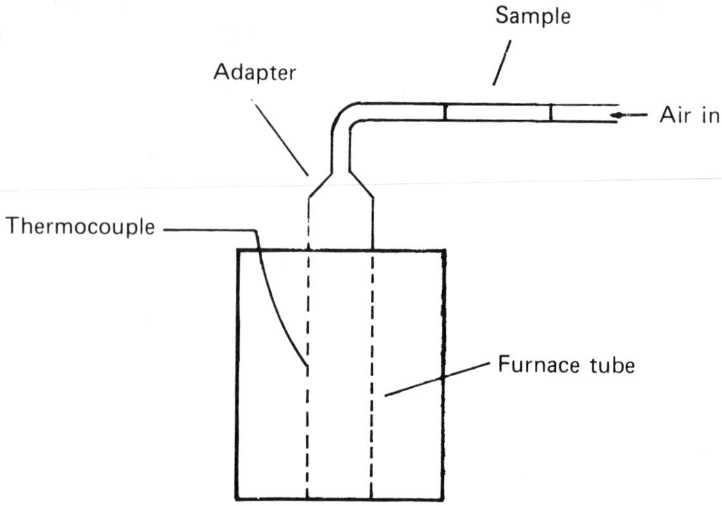

Fig. 14.1 — Apparatus for minimum ignition temperature determination.

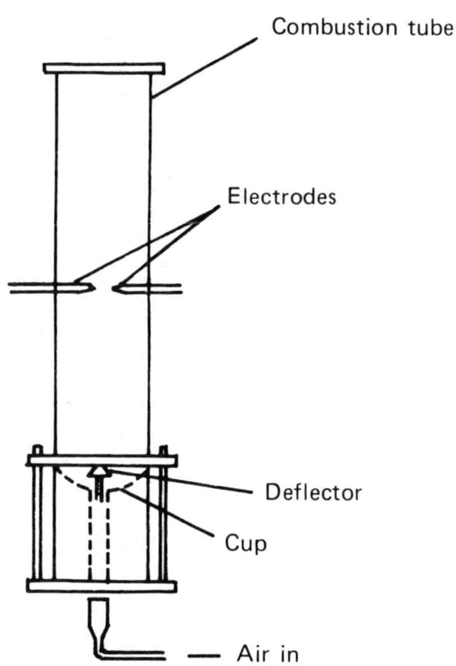

Fig. 14.2 — Vertical tube apparatus for minimum explosion concentration determination.

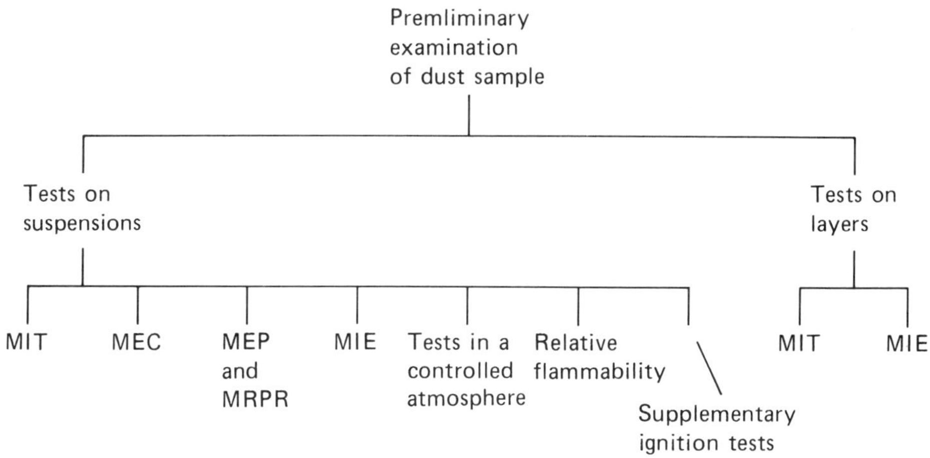

Fig. 14.3 — Basic scheme of tests in USA.

$$\text{Ignition sensitivity} = \frac{\text{MIT} \times \text{MIE} \times \text{MEC (Pittsburgh coal dust)}}{\text{MIT} \times \text{MIE} \times \text{MEC (sample dust)}} \quad (14.1)$$

$$\text{Explosion severity} = \frac{\text{MEP} \times \text{MRPR (sample dust)}}{\text{MEP} \times \text{MRPR (Pittsburgh coal dust)}} \quad (14.2)$$

The index of explosibility is then the

$$\text{Ignition sensitivity} \times \text{Explosion severity} \quad (14.3)$$

Table 14.2 shows the relationship between hazard rating and index of explosibility [3]. It is the relative explosion hazard rating as assessed on the basis of the value or the range of values of the three parameters defined in the above equations.

Table 14.2 — Relation between hazard rating and index of explosibility [3]

Relative explosion hazard rating	Ignition sensitivity	Explosion severity	Index of explosibility
Weak	<0.2	<0.5	<0.1
Moderate	0.2–1.0	0.5–1.0	0.1–1.0
Strong	1.0–5.0	1.0–2.0	1.0–10
Severe	>5.0	>2.0	>10

Sources of ignition may include: naked flames, which appear to be the most dangerous source, smouldering matter, hot surfaces, especially horizontal ones, welding and cutting operations, friction and impact — bucket elevators, and also low energy sparks from hand tools, electrical sparks — sources such as electric mains, batteries, static electricity . . . earth, and spontaneous heating (Arrhenious effect).

To reach an acceptable level of protection against dust explosions, certain design and operational procedures include: minimizing dust cloud formation, where necessary providing equipment containment, for example, comminution equipment, separating the plant units to confine the explosion, providing venting according to publicly accepted standards — here, the venting duct is critical, but must not exhaust into occupied or public access area; implementing inerting where necessary (flue gas, N_2), and employing automatic suppression. The incipient problems of dust explosions and their causes are many and varied. Table 14.3 shows some case histories of dust explosions [4].

Table 14.3 — Case histories: dust explosions

Industry	Source	Ignition
furniture factory	extraction ducting	match, fluff
chemical	mixer	mill
chemical	fluid bed drier	static, dust and ethanol
foodstuff	spray drier	spontaneous heating
sugar	bucket elevator	friction
milk	spray drier	spontaneous heat and lub oil
animal feedstuff	silo	electric lamp
chemical	pneumatic system	static

14.4 SOME HEALTH HAZARDS

Lung diseases such as silicosis and various skin ailments have been known to be a product of particulate irritation. There is an increasing technology of particulates which is concerned with physiologically active sprays and the like. These products include chemical and germ warfare agents, insecticides, therapeutic agents, radioactive fallout, and many others.

There are three important categories of effects of fine particles: fibrosis-producing dusts, toxic dusts, and irritants.

Fibrosis is caused by a number of particulates, including quartz (SiO_2), coal, and asbestos, and it produces increased collagen in lymph glands adjacent to the alveoli. SiO_2 produces silicosis, and the finer particles are the most active ones. Silicosis may be 'cured' by Al powder inhalation. Asbestosis is similar to silicosis except that the hydrated Mg silicate is fibrous and that malignant growths may occur more often than in the case of silicosis. Coal dust is another cause of lung disease. Not only does the silica produce silicosis, but the combination of coal dust and SiO_2 is said to produce anthrosilicosis — modified silicosis.

Beryllium poisoning is a serious danger to those persons in contact with Be, its alloys, or its compounds. Friedlander [5] found that good ventilation reduces the plant hazard to tolerable proportions $< 2\,\mu g/m^3$.

Asbestos fibres that may be inhaled are carefully controlled because of their carcinogenic potential. Recent Japanese data on the carcinogenic effects of glass fibres may cause some reconsideration of the shape carcinogen hypothesis [6–10].

Irritants are substances which are normally damaging to lung tissue, and an important aerosol in this connection is a suspension of sulphuric acid droplets in air. Normally, innocuous substances of foreign protein, may produce allergic reactions in some people. For example, pollens and household dust are relatively of large diameter (10–100 μm); hay fever and asthma are therefore largely diseases of the upper respiratory system. Cotton mill workers have an allergic reaction (emphysema, etc.) called byssinosis. In general, it has been found that although small particles rapidly produce irritation, large ones produce effects which wear off more slowly. This is explained by the high rate of solution of the smaller particles — a more rapid process by which its effect quickly wears off [11–13].

Toxic dusts will more likely poison by inhalation than by swallowing because although the coarser particles are removed by cilia to the digestive tract, the finer, or more soluble particles enter the bloodstream. The classic poison is Pb poisoning. It is curable once the victim is removed from contact with the metal. Cd oxide poisoning is often fatal, but none or few chronic cases are said to occur in the United States of America [10]. More common than any of the above is poisoning from the particles of insecticides. These are the three categories:

(1) Organic pyrophosphates $(CH_3O)_4P_2O_3$ (2) Alkyl thiophosphates (e.g. parathion). People have been killed by this substance. It is banned in many countries.	NERVE GAS TYPES
(3) Phosphoramides (e.g. ethyl bis (dimethylamino) phosphate)	TISSUE ACTING TYPE

Industrial air pollution refers to single source pollution. It is easy to track down and can therefore be rectified. The other main type of pollution is smog, of which there are two kinds: one is called photochemical smog produced in automobile-using communities, and the other is that produced in coal-burning communities. Strictly speaking, smog is a combination of smoke and fog. Coal smogs contain large quantities of aromatic, aliphatic and cyclic compounds, and it is quite likely that they contain carcinogens. A major industrial air pollutant is fluorine from the fertilizer and the Al industries. In the manufacture of Al the Hall-Heroult process dissolves Al_2O_3 in molten cryolite (Na_3AlF_6) and the F containing particles get into the atmosphere. The Bessemer steelmaking process and the open hearth process are large producers of very fine particulates, as shown in Table 14.4. The electric process also produces very fine particles. In all of the cases cited the major component is iron oxide [1–15].

Table 14.4 — Open hearth furnace fumes

μm	w/o	cum w/o
1.0–3.0	7	7
0.5–1.0	29	36
0.15–0.5	50	86
<0.15	14	100

The initiation of a coal dust explosion is an easy task to accomplish. Its likelihood is compounded by methane which is always present in mines. Methane is dangerous in concentrations of 2–5 per cent and very dangerous above 5 per cent. Some idea of the seriousness of the problem has been documented by Cybulski [16] in Table 14.5.

Table 14.5 — The number of explosions resulting in fatal accidents including those in which a coal dust factor was ascertained [in British mines from 1936–1969 (deaths in brackets)]

Years	Number of explosions	Number of explosions excluding minor accidents	Number of explosions with the participation of coal dust
1936–1940	43 (306)	29 (287)	3 (172)
1941–1945	51 (224)	32 (199)	4 (79)
1946–1950	31 (211)	16 (172)	4 (152)
1951–1955	15 (129)	9 (116)	1 (81)
1956–1960	14 (123)	13 (122)	4 (76)
Total for years 1936–1960	154 (993)	99 (896)	16 (560)
1961–1965	8 (68)	8 (68)	3 (25)
1966–1969	1 (2)	1 (2)	0

The dust originates in various ways: from mining of the coal, 50%, mechanical preparation of the coal, 37%, and transportation of the coal, 13%. Approximately 5 per cent of the dust is retained on the walis, 5 per cent on pillars and 5 per cent on the road [16]. A conventional method of reducing the hazard is to spread stone dust upon the coal dust layers on the mine surfaces.

A primary variable of importance in coal dust explosions appears to be the percentage of volatile matter in the coal particles. In general, heat causes pyrolysis of the coal, constituting a chemical-physical change in the nature of the coal. A bituminous coal is termed thermolabile. The percentage of volatile matter in coal is determined by heating the coal to 850°C under specified conditions. The processes of coal decomposition are complicated. A rapid heating of a typical Upper Silesian coal produces a product that has the following composition [16–20]: 18.4 per cent H_2, 73.4 per cent CO, 2.8 per cent CO_2, and 4.4 per cent CH_4.

Measures that are routinely taken to help prevent dust explosions in coal mines include limiting dust formation, removal of dust, prevention of cloud formation, prevention of ignition, prevention of propagation, and so on. Stone dust, which may be slate, diatomite, limestone, or gypsum, can be used in the form of stone dust barriers to kill an explosion [21,22]. The dusts usually have about 85 per cent of particles sized less than 75 μm, and their effectiveness varies depending upon their material substance. For example, in the case of coal dust with a volatile matter of 33.3 per cent the limit percentages of the named dusts required to suppress the explosion are as listed by Cybulski in Table 14.6. Tables [21–25] 14.7, 14.8, and 14.9 show typical dust content in working zone during manufacture of sintered articles from iron-base powder, during manufacture of non-ferrous magnets (of Alnico and Magnico types), and particle size analysis [26,27] of dust at operator's positions during manufacture of sintered articles (%).

Table 14.6 — The percentage of dust required to suppress explosion of coal dust with 3.3 per cent V.M.

Slate	67.5
Diatomite	62.5
Anydrite	60.0
Limestone	57.5
Dolomite	57.5
Gypsum	40.0

REFERENCES

[1] Palmer, K., Dust explosions seminar, *International Conference on Powder and Bulk Solids Handling, Chicago* (May, 1976).
[2] Dorsett, H., Jacobson, M., Nagy, W., & Williams, R. P., *U.S. Bureau of Mines, RI 5624*, (1960).
[3] Palmer, K., *Dust explosions and fires*. London: Chapman and Hall (1973).

Table 14.9 — Particle size analysis of dust at operator's positions during manufacture of sintered articles, percent

Operation	Size of dust particles, μm						
	<1	1–2	2–3	3–4	4–5	5	6–7
Unloading of ferronickel master alloy from mill	34.8	32.4	13.8	11.2	6.2	1.6	—
Screening of ferronickel master alloy on vibrating screen (middle of shop)	41	23	4.8	7	21.2	2.4	0.6
Pressing of aluminium-nickel-cobalt magnets	50.8	14.8	11.2	12.2	1.2	9.2	0.6
Pressing of articles (middle of pressing department)	58	16	4.4	5.8	8.2	5.4	2.2
Loading of ferrobarium charge into mixer	37.6	26.8	18.4	8.2	7.6	1.4	—
Pressing of ferrobrium magnets	29	34.4	15.8	10	6.6	3.4	0.8
Pressing of tungsten–copper articles	18.4	15.6	20.6	7.8	16.6	16.6	4.2
Unloading of copper-graphite mixture from mixers	30.4	30.8	14.2	8	10.8	4	1.8
Milling of ferroaluminium master alloy in miller	24.6	20	10.6	11.8	30	3	—
Proportioning of iron-graphite mixture	16.6	20.6	20.8	29.4	8.2	4.4	
Pressing of ferrobarium magnets	33.4	27.4	22.2	8.2	4.6	2.8	1.4
Screening of copper-graphite mixture of vibrating screens	58	27.8	8.2	4	1.4	0.6	—
Milling of ferroaluminium master alloy in muller	18	24.2	18.6	14.4	15.2	6.4	3.2
Pressing of copper-graphite articles (PV-474 press)	57.8	16.6	8.4	4.8	5.6	4.8	2
Pressing of magnesium-nickel-cobalt magnets (Magnico)	50.8	14.8	11.2	12.2	7.2	3.2	0.6
Pressing of tungsten-copper contacts	34.6	32.6	13.8	11.2	6.2	1.6	—
Pressing of bronze-graphite articles (PV-474 press)	30.4	31.0	14.2	9	10.0	4	1.4
Loading and unloading of ferrobarium charge	18	36	26	12	5	1	2
Loading of nonferrous magnets into boats	15	25	38	10	12	—	—
Unloading of nonferrous magnets after sintering	10	24	36	8	15	2	5
Loading of iron-graphite articles into boats with packing material	13	19	42	7	14	3	2
Unloading bronze-graphite articles from boats	14	28	32	15	9	2	—
Separating from packing material	18	30	24	18	10	—	—

[26] Prindle, L., *J. Air Pol. Control Assoc.*, **9**, 12 (1959).
[27] Cadle, R., *Particle size.* New York: Van Nostrand Reinhold (1965).
[28] Palmer, K. N., *Dust explosion and fires.* Chapman and Hall (1973).

15

Glossary of frequently used terms [1] in powder characterization

Anisotropy The condition of a solid when its properties vary with the direction in which they are measured. In bulk solids there are two types of anisotropy, shear and inherent.

In inherent anisotropy, the anisotropy present in a bulk solid is by reason of the anisotropic shape of the individual particles.

In shear anisotropy, the anisotropy induced in a bulk solid is that by physical shear strain. This type of anisotropy would be present in an assembly of perfect spheres when subjected to shear strain.

Bulk density The mass of a quantity of powder inclusive of moisture divided by its total volume.

Specific volume The reciprocal of bulk density.

Voidage The volume of the voids within a quantity of powder divided by its total volume (i.e. volume of voids plus volume of solid).

Void ratio The volume of the voids of a quantity of powder divided by the volume occupied by the solid.

Bulk solid An assembly of particles having a sufficient number of particles for the statistical mean of any property to be independent of the number of particles. Other synonyms are powder and particulate solid.

Compaction The process of permanent volume reduction by means of stress or vibration.

Isostatic compaction Compaction where the internal compaction stresses are equal in all directions. Caution is necessary here before a process is called isostatic compaction. Though hydrostatic compaction may have been employed to produce a powder compact, the powder compaction process is not necessarily isostatic compaction especially if a fixed boundary former is used.

Uniaxial compaction Compaction in which the volume reduction is achieved by permanent deformation in one direction only. In practice, because of wall friction, slight differences occur between single ended uniaxial compaction and double ended uniaxial compaction.

Core flow A type of powder flow pattern which occurs in hoppers and which is characterized by a central core of flowing powder while the powder adjacent to the hopper walls remains stationary.

Critical state line The line describing the relationship in three-dimensional space which relates shear stress, normal stress, and void ratio in steady state or equilibrium conditions of failure in simple shear.

Critical void ratio The void ratio of a powder which would be finally reached if that powder were sheared under a specified constant normal stress (i.e. the void ratio corresponding to a point on the critical state line specified by a normal stress).

Deformation

Elastic deformation This is an aspect of powders and of other materials which has received little experimental attention beyond the simplest stress systems, and our knowledge of elastic behaviour is at best meagre. A simple but narrow definition is that an elastic deformation is one which is entirely recoverable when the stress causing the original deformation is removed. It is well known that if a normal stress is applied then there follows a volume decrease, but not so well known is that an applied shear stress will cause a volume increase [1]. If both shear stress and normal stress are applied simultaneously, the outcome as yet cannot be predicted.

Plastic deformation A plastic deformation is the permanent deformation which remains after the elastic deformation has been removed from a total deformation.

Pure shear Imagine a square contained within a mass of powder about to be deformed and in which the diagonals are aligned with the directions of the principal stresses. Pure shear obtains if the applied principal stresses deform the square into a rhombus while maintaining the same diagonal direction. An example of pure stress is the deformation of a cylinder which becomes a cylinder of different dimensions; note that no restriction has been placed upon the final volume of the cylinder.

Simple shear Simple shear is the deformation of a body in which there is no change in length in the direction of the applied shear stress. An example of simple shear is the deformation of a square into a parallelogram in which one pair of opposite sides maintain their original length.

Dilatancy The capacity to change volume.

Dilation This term was originally used by Reynolds in 1885 to describe the increase in volume of a stable arrangement of spheres when a shear stress was applied. When much later it was realized that contraction as well as expansion was an integral part of powder behaviour, dilation was used by some authors to refer to all volume changes.

Failure The term failure is used in many different disciplines, all of which have had their peculiar definition. Each discipline has often only special interest in failures which have resulted in a confusion of definitions, all of which are special and not sufficiently general for all situations. If failure is defined as exceeding the elastic limit, this definition is no different from permanent deformation. To make a useful definition of failure we must differentiate between the different regimes of failure, i.e. over-consolidated failure, critical failure and under-consolidated failure. Therefore we shall define failure as a permanent change in the 'state of a powder and further refine the definition in the different regimes' of failure.

Critical failure A permanent deformation with no change in the 'state of a powder', i.e. no change in volume or stresses.

Over-consolidated failure A permanent change in the 'state of a powder' which is characterized by an overall increase in volume when the initial 'state of the powder' is changed to a lower or weaker state. During the change, at least one principal stress will pass through a maximum. This maximum is used to define 'the point of failure'.

Unconfined failure A special case of over-consolidated failure where the sample is confined except for the unidirectional stress causing failure.

Under-consolidated failure A permanent change in the 'state of a powder' which is characterized by a decrease in volume, when the initial state of the powder is changed to a higher or stronger state. In contrast to over-consolidated failure there is no identifiable point at which the failure can be said to have taken place.

Failure loci and failure K-lines All the loci and K-lines encountered here are simply a series of stress conditions at failure having some common factor. In powder technology the stresses used for plotting are generally the shear stress and normal stress, while in soil mechanics they are often principal stress and functions of shear and normal stress. Because the normal stress, which is invariably the abscissa, may be used in either a compound stress system or an applied stress system, there are two locus or K-line names for each series of failure points having a common factor, as shown in Fig. 15.1. The difference between locus and a K-line is that a locus is a line through a series of points representing failures while K-lines are lines tangential to the Mohr stress circles which pass through the failure points. The terms 'failure locus' and 'failure K-line' have recently been adopted instead of 'yield locus', which generally signifies the locus of a series of elastic limits. Failure points and elastic limits do not necessarily coincide.

Critical failure locus and Internal friction locus The C.F.L and I.F.L are the loci of a series of points representing the stress condition at critical failure on an applied stress and compound stress basis respectively. Other identical definitions are:

- a locus of equilibrium points;
- a locus of the end points of the Jenike failure locus or compound Jenike failure locus according to the stress basis;
- a projection of the critical state line onto the shear stress-normal stress plane.

Effective friction K-line This was formerly the effective yield locus defined by Jenike [2, 3]. While the definition is unchanged, the name has been changed [1] to be consistent with many other loci and K-lines, of which those given in this glossary will represent only a proportion when further practical applications are developed.

Jenike failure failure locus and compound J.F.L. The J.F.L. and C.J.F.L. are the loci of a series of points representing the maximum value stress conditions occurring in samples having the same initial bulk density during over-consolidated failure, on an applied stress, or compound stress basis, respectively.

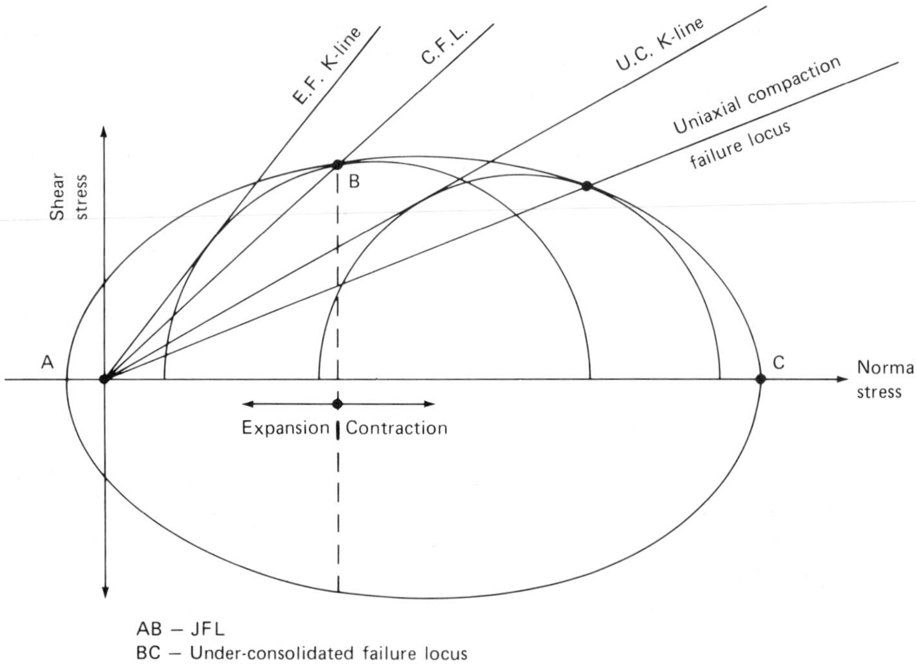

Fig. 15.1 — Failure loci and K-lines on an applied stress basis.

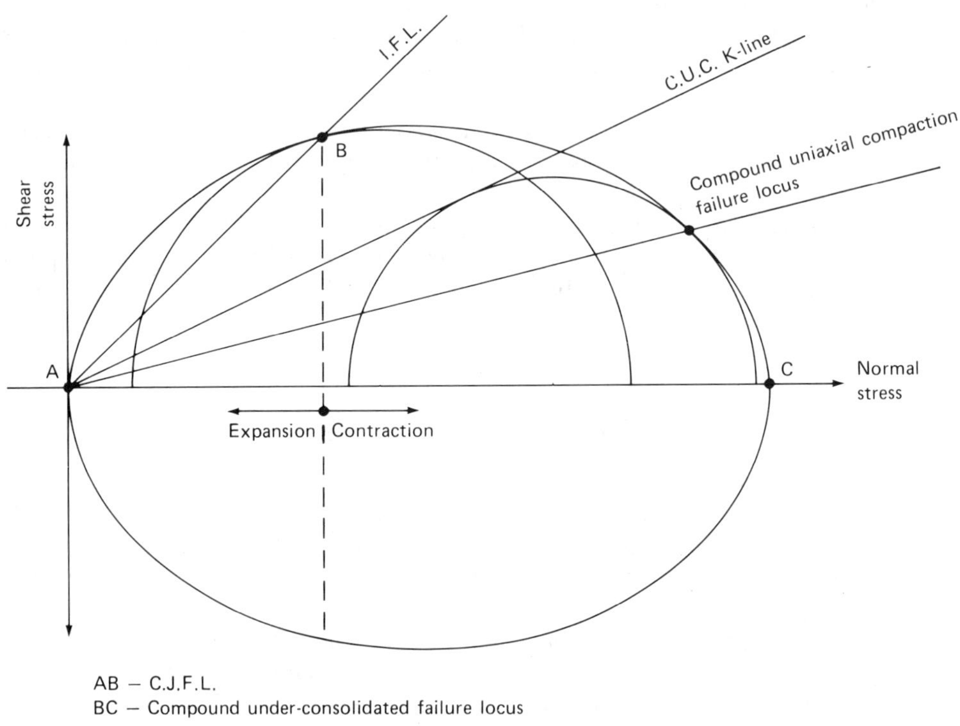

Fig. 15.2 — Failure loci and K-lines on a compound stress basis.

Under-consolidated failure locus and Compound under-consolidated failure locus The under-consolidated failure locus and the compound under consolidated failure locus are the loci of points representing stress conditions occurring in samples having the same initial bulk density during under-consolidated failure, on an applied stress or compound stress basis, respectively, as shown in Fig. 15.2.

Uniaxial compaction failure locus and compound uniaxial compaction failure locus The uniaxial compaction failure locus and the compound uniaxial compaction failure locus are loci of points representing the stress conditions at under-consolidated failure when the sample is undergoing axial compaction on an applied stress and compound stress, respectively.

Uniaxial compaction K-line and compound uniaxial compaction K-line The uniaxial compaction K-line and the compound uniaxial compaction K-line are the lines tangential to the Mohr stress circles (a) which pass through the failure points represented by the uniaxial compaction failure locus or its compound counterpart, and (b) which are tangential to the under-consolidated failure locus or its compound counterpart shown in Fig. 15.1.

Wall failure locus The locus of a series of points representing the component stresses when a powder fails at a wall.

Failure plane During over-consolidated failure, as the stresses increase, the deformation seeks out the weakest point or the longest line of weak points in the powder. Since in over-consolidated failure, deformation makes the powder weaker, these points of weakness become self-propagating and quickly become a failure plane, a few particles thick, bounded on both sides by a mass of stronger powder. In contrast, any points of weakness in a powder undergoing under-consolidated failure becomes stronger, and a failure plane is not allowed to propagate.

Failure properties Strictly speaking, the term should be 'bulk failure properties', but since it is obvious from its context whether it refers to bulk behaviour or individual particle behaviour, the prefix is not necessary. Failure properties are the parameters which control the magnitude and shape of the failure loci and K-lines and of the failure surface.

Angle of internal friction The inclination of the I.F.L. This may be regarded as a measure of the difficulty in maintaining the flow of a powder. Many different definitions of the angle of internal friction have been and are given, but the majority can be immediately discarded on the grounds that they contain unknown quantities of work of expansion in addition to friction.

Angle of shear resistance The slope of the J.Y.L. Sometimes this locus is approximated by a straight line, so that a single numerical value may be used to define it.

Angle of wall friction The inclination of the W.F.L.

Apparent angle of internal friction The inclination of the C.F.L.

Cohesion Coulomb [6] defined the vertical intercept of his failure locus, calling it cohesion, thinking it to be the force between particles. Hence we now have two definitions of cohesion.

Cohesion 1 (bulk failure property) Cohesion is the shear stress at failure under zero applied normal stress, i.e. the intercept of the J.F.L. with the shear stress axis.

Cohesion 2 (particle failure property) The force of attachment between two particles. This is a normal force in contrast to the shear stress involved in cohesion 1.

Effective angle of friction The inclination of the effective friction K-line (E.F. K-line) shown in Fig. 15.1.

End point of the J.Y.L. or equilibrium point Several definitions having identical meaning are possible: the intersection of the J.F.L. with the C.F.L. the intersection of the C.J.F.L. with the I.F.L.; a point on the critical line; a shear stress at failure with no volume change.

Failure function A graph of the unconfined failure stress and the major consolidating stress. This can be regarded as a measure of the difficulty of starting flow in powders.

Flow function Like the above, flow function is the term used by Jenike [2,3,4] for failure function. The term flow implies a rate process which is not present in its determination and is, therefore, a misnomer [1,7,8].

Principal stress ratio A ratio of two principal stresses from the same state of stress at a point, generally the major and minor principal stresses.

Ultimate tensile stress The uniaxial tensile principal stress acting at failure of an unconfined sample. In powders this is often called tensile strength. U.T.S. is the bulk equivalent of cohesion 2 on the particle level. For some solids the word ultimate will possibly be found to refer to a practical limit rather than a theoretical limit.

Unconfined failure stress The major principal stress causing the failure of a sample in unconfined uniaxial compression.

Failure surface The surface which describes the magnitude of the failure stresses in a three-dimensional system representing failure behaviour. Two systems are commonly used to represent failure conditions, one uses principal stress space, i.e. δ_1, δ_2, and δ_3 making up the 3-D coordinates in which density is accounted for by making it a parameter. The other uses the shear and normal stresses acting in the failure plane and the specific volume as the 3-D co-ordinates.

Flow factor of a hopper The ratio of two important stresses in a hopper, the value of which depends upon the type of hopper, the hopper wall slopes, the angle of internal friction, and angle of wall friction of a powder placed in the hopper. The ratio is the vertical major principal stress caused by a stationary powder divided by the horizontal major principal stress of a flowing powder [7,8].

Internal friction The friction when particles slide over each other. The value of this friction is greater than the friction of the massive solid which makes up the individual particles, since the particles are also rising and falling while maintaining an overall constant bulk volume.

Isostatic The condition of equal magnitude in all directions of an external variable acting upon a solid. The word 'hydrostatic' is sometimes used for this meaning, even when liquid pressures are not involved.

Isotropy The condition of a solid when its properties, not an external variable, are independent of the direction in which they are measured.

Mass flow A type of powder flow pattern which occurs in hoppers and which is characterized by the powder flowing at every point including that at the hopper wall.

Plastic Description of a solid capable of suffering permanent deformation.

Sample condition A sample of powder being tested in the laboratory is often referred to as under-consolidated, critically consolidated, or over-consolidated. This implies a relationship between the degree of compaction that the powder has received in preparation and the stresses about to be imposed on it to cause failure such that the sample will be under-consolidated if the imposed stresses were to cause under-consolidated failure, will be critically consolidated if the imposed stresses were to cause critical failure, and will be over-consolidated if the imposed stresses were to cause over-consolidated failure.

Shear cell The part of an apparatus which contains the powder for experimental failure property measurement where the force applied is a shear force.

State of powder or practical definition The condition of the powder as defined by the density of the powder and the applied stress which would produce critical failure. This is analogous to a state in the thermodynamic sense.

Stress/strain Many stress and strain terms have common definitions, therefore it is convenient to deal with the two definitions simultaneously. Where this applies the term will appear simultaneously as stress/strain. Stress is defined as the force per unit area of a given plane where the force is acting in a specified direction with respect to that plane. Strain is defined as the ratio of a displacement to the original length perpendicular to a given plane, where the direction of the displacement is specified with respect to the plane.

Applied stress basis This is the system in which the method of stress representation takes account only of external stress applied to a powder.

Component stresses/strain This term is used to refer simultaneously to shear and normal stress/strain which are the components of a stress/strain vector.

Compound stress basis This is the system in which the method of stress representation sums external and internal stresses, this being the total stress. The internal stress is, of course, the ultimate tensile stress.

Mohr stress circle The Mohr stress circle is a circle drawn on a graph with shear stress and normal stress coordinates with the centre of the circle falling on the normal stress abscissa. This circle then provides a method of relating shear stresses on a specified plane to the principal stresses. Principal stresses are, of course, a special case of normal stress, and they are, therefore, represented by points along the normal stress axis, as can be seen in Fig. 15.1.

Normal stress/strain This is one of the two component stresses/strains. As the name implies, the direction of the stress strain is perpendicular to a given plane.

Principal stresses/strains These are normal stresses/strains which act on the plane in which shear stresses/strains are absent. In any stress system there are three principal stresses which may or may not be of equal magnitude. If of unequal magnitude, the principal stresses/strains are prefixed by the terms major, intermediate, and minor.

Reduced stresses These are principal stresses and component stresses, acting on a powder initially at a given state which have been divided by the normal stress which would produce critical failure in the same state.

Shear stress/strain This is one of the two component stresses/strains. In this case the stress/strain is directed tangentially to the given plane.

State of stress This is a term which allows one to refer collectively to all the stress vectors acting at a point in a powder.

Stress ellipse or ellipsoid The stress ellipse or ellipsoid is the envelope enclosing the magnitude of all the stress vectors acting at a point in a powder in two and three dimensions respectively. This is a graphical representation of the state of stress.

Stress/strain vector This is resultant of the component stresses/strains acting on a given plane.

Stress/strain tensor This term is used to refer to all the nine component stresses/strains, three on each plane of the three planes positioned orthogenally to each other.

Unconfined failure stresses The stresses which cause unconfined failure.

Major consolidating principal stress or major consolidating stress The major principal stress of the Mohr stress circle which passes through the end-point and is tangential to the J.F.L. or I.F.L. according to the normal stress basis used. The term 'consolidating' is in fact incorrectly used; it had come into use historically during the early development of the understanding of powder behaviour, but because it defines a special case of a major principal stress, the term is still useful [1,7,8].

Yield locus This is a locus of elastic limits for different states of stress acting on identical samples. However, the term is still incorrectly used in some quarters to mean the failure locus [1].

REFERENCES

[1] Birks, A. H. & Muzaffar, S. A., *Symposium on Powder Flow and storage, University of Bradford* (1971).

[2] Jenike, A. W. & Shield, R. T., *J. Appl. Mech.*, **B81**, 599 (1959).

[3] Jenike, A. W., Elsey, P. J., & Wooley, R. H., *Proc. Am. Soc. Test. Mater.*, **60** (1960).

[4] Jenike, A. W., Storage and flow of solids, *Bulletin 123*, Utah Engineering Experiment Station, Univ. of Utah, Salt Lake City, Utah, USA (1964).

[5] Wroth, C. P. & Bassett, R. H., *Geotechnique*, **15**, 32 (1965).

[6] Coulomb, C. A., *Mem. Acad. r. Sci., Paris* (1976) See Ref [5].

[7] Williams, J. C., Birks, A. H., & Bhattacharya, D., *Powder Technol.*, **4**, 328 (1970/71).

[8] Williams, J. C. & Birks, A. H., *Rheol. Acta*, **4**, (3), 170 (1965).

Appendix A
Micrographs of spherical bronze powders

Fig. A1 — Spherical bronze powders (×40 magnifications).

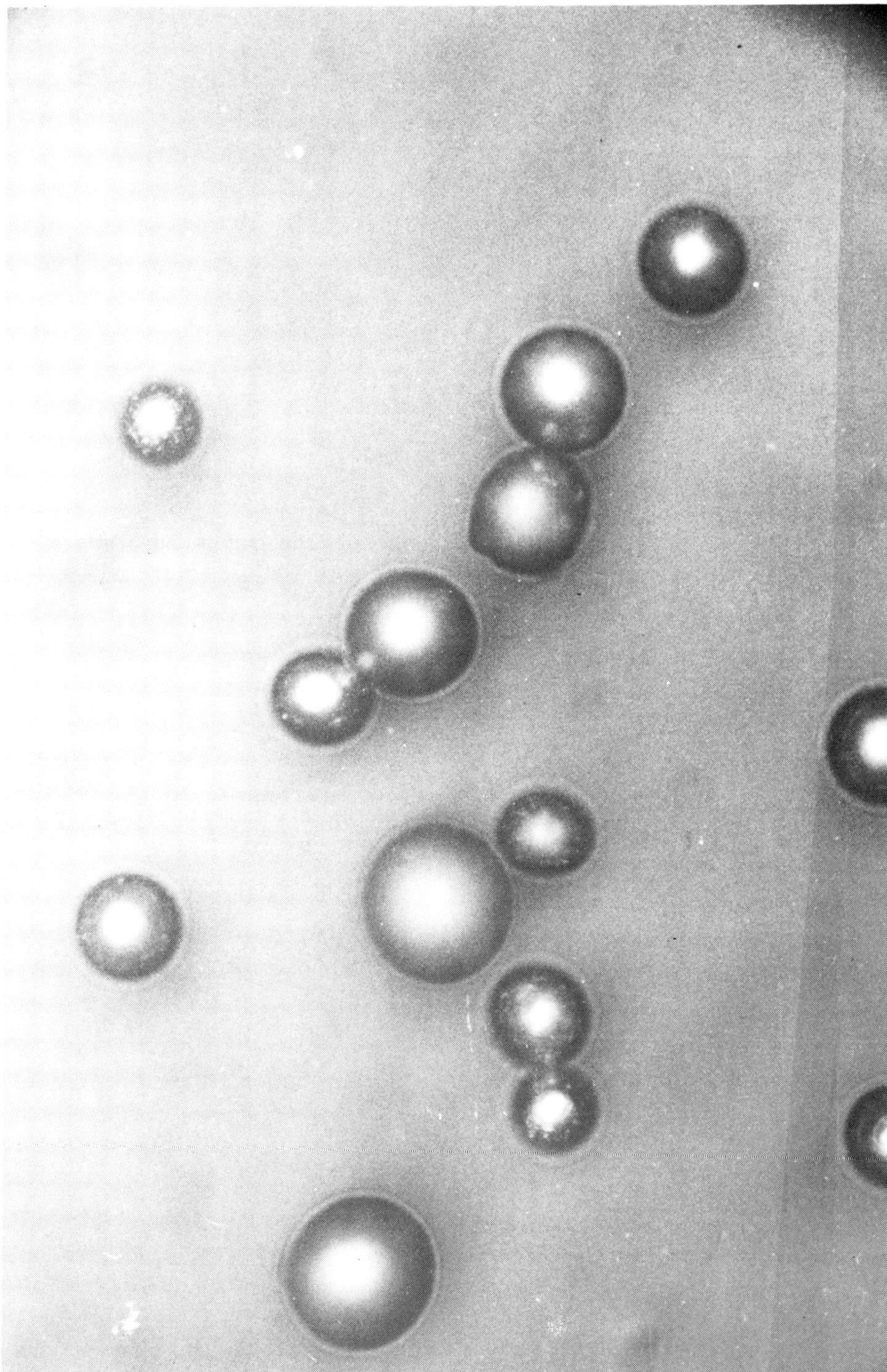

Fig. A2 — Spherical bronze powders (×40 magnifications).

Appendix A1.1
Steady state permeametry

The rate of flow of a gas moving under the influence of a constant pressure difference through a compacted bed of powder is related to the surface area of the powder by the following equation:

$$\frac{dV}{dt} = \frac{A\Delta P}{k\mu L S_V^2}\frac{\varepsilon^3}{(1-\varepsilon)^2} + \frac{8}{3}\frac{\delta k_0}{k}\frac{A\Delta P}{L\bar{P}S_V}\left[\frac{2RT}{\pi M}\right]^{1/2}\frac{\varepsilon^2}{(1-\varepsilon)} \tag{1}$$

Rather than solve this equation directly the calculation can be carried out by evaluating two separate terms and combining them to give S_V.

$$S_K^2 = \frac{A\Delta P}{k\mu L\left(\frac{dV}{dt}\right)}\frac{\varepsilon^3}{(1-\varepsilon)^2} \tag{2}$$

This is the Carman–Kozeny equation and gives the contribution to the surface area of the powder due to streamline flow.

$$S_M = \frac{8}{3}\frac{\delta k_0}{k}\frac{A\Delta P}{L\bar{P}\left(\frac{dV}{dt}\right)}\left[\frac{2RT}{\pi M}\right]^{1/2}\frac{\varepsilon^2}{(1-\varepsilon)} \tag{3}$$

This gives the contribution to the surface area due to molecular slip at the powder surface.

The second term is negligible for coarse powders at atmospheric pressures. The second term becomes important for fine powders at atmospheric pressure and all powders at reduced pressures.

Combining equations (1), (2) and (3) gives the quadratic in S_V:

$$S_V^2 - S_M S_V - S_K^2 = 0 \tag{4}$$

having the following solution:

$$2S_V = S_M \pm \sqrt{S_M^2 + 4S_K^2} \tag{5}$$

CORRECTION TO THE VISCOUS FLOW TERM

For the system under investigation, equation (2) may be written:

$$\frac{1}{\rho}\frac{dm}{dt} = \frac{A\Delta P}{k\mu L S_K^2}\frac{\varepsilon^3}{(1-\varepsilon)^2}$$

For an ideal gas:

$$PV = mRT$$

For steady state flow equation (SSFE), $P_1 = P_2 = \Delta P =$ constant. The inlet pressure $=$ $P_1V_1 = m_1RT$
and

$$\frac{dm}{dt} = \frac{dm_1}{dt} = \frac{dm_2}{dt}$$

Therefore,

$$\frac{dm}{dt} = \frac{P_1}{RT}\frac{dV_1}{dt} + \frac{V_1}{RT}\frac{dP_1}{dt}$$

$$\frac{P_1 dV_1}{dt} + V_1\frac{dP_1}{dt} = RT\frac{dm_1}{dt}$$

Note that

$$\frac{P}{RT} = \rho$$

and

$$\bar{\rho} = \frac{\bar{P}}{RT}$$

$$\frac{1}{\bar{\rho}}\frac{\mathrm{d}m}{\mathrm{d}t} = \frac{P_1}{\bar{P}}\frac{\mathrm{d}V_1}{\mathrm{d}y} + \frac{V_1}{\bar{P}}\frac{\mathrm{d}P_1}{\mathrm{d}t}$$

where

$$P_1 = P_2 + \Delta P$$

$$\bar{P} = \frac{P_1 + P_2}{2} = P_2 + \frac{\Delta P}{2} = P_1 - \frac{\Delta P}{2}$$

$$P_2 = P_1 - \Delta P$$

Now

$$\frac{1}{\bar{\rho}}\frac{\mathrm{d}m}{t} = \frac{2}{2P_2 + \Delta P}\left[P_1\frac{\mathrm{d}V_1}{\mathrm{d}t} + V_1\frac{\mathrm{d}P_1}{\mathrm{d}t}\right] \tag{6}$$

but

$$\frac{1}{\bar{\rho}}\frac{\mathrm{d}m}{\mathrm{d}t} = \frac{A\Delta P}{k\mu L S_K^2}\frac{\varepsilon^2}{(1-\varepsilon)^2}$$

Equating the RHS of equation (6)

$$\frac{2}{2P_2 + \Delta P}\left[\frac{P_1 \mathrm{d}V_1}{\mathrm{d}t} + \frac{V_1 \mathrm{d}P_1}{\mathrm{d}t}\right] = \frac{A\Delta P}{k\mu L S_K^2}\frac{\varepsilon^3}{(1-\varepsilon)^2} \tag{7}$$

Since by the SSFE $P_1 = P_2 = \Delta P =$ constant, equation (7) reduces to

$$\frac{2}{2P_2+\Delta P}\left[P_1\frac{\mathrm{d}V_1}{\mathrm{d}t}\right]=\frac{A\Delta P}{k\mu LS_K^2}\frac{\varepsilon^3}{(1-\varepsilon)^2}$$

but

$$P_2=P_1-\Delta P$$

$$\frac{2}{2(P_1-\Delta P)+\Delta P}P_1\frac{\mathrm{d}V_1}{\mathrm{d}t}=\frac{A\Delta P}{k\mu LS_K^2}\frac{\varepsilon^3}{(1-\varepsilon)^2}$$

$$\frac{P_1\mathrm{d}V_1}{P_1-\dfrac{\Delta P}{2}\mathrm{d}t}=\frac{A\Delta P}{k\mu LS_K^2}\frac{\varepsilon^3}{(1-\varepsilon)^2}$$

Therefore,

$$S_K^2=\frac{\dfrac{A\Delta P}{k\mu L}\dfrac{\varepsilon^3}{(1-\varepsilon)^2}}{\dfrac{P_1\mathrm{d}V/\mathrm{d}t}{P_1\dfrac{\Delta P}{2}}}$$

Let

$$\frac{P_1\dfrac{\mathrm{d}V_1}{\mathrm{d}t}}{P_1-\dfrac{\Delta P}{2}}=I_1$$

Hence

$$S_K^2=\frac{A\Delta P\varepsilon^3}{k\mu LI_1(1-\varepsilon)^2}$$

or

$$= \frac{A\Delta P\bar{P}\varepsilon^3}{k\mu LP_1\left(\frac{dV}{dt}\right)(1-\varepsilon)^2} \tag{8}$$

CORRECTION TO THE SLIP FLOW TERM

$$\frac{1}{\rho}\frac{dm}{dt} = \frac{8}{3}\frac{\delta k_0}{k}\frac{A\Delta P}{L\bar{P}S_M}\left[\frac{2RT}{\pi M}\right]^{1/2}\frac{\varepsilon^2}{(1-\varepsilon)} \tag{9}$$

Equating (6) and (9)

$$\frac{2}{2P_2+\Delta P}\left[P_1\frac{dV_1}{dt}+V_1\frac{dP_1}{dt}\right] = \frac{8}{3}\frac{\delta k_0}{k}\frac{A\Delta P}{L\bar{P}S_M}\left[\frac{2RT}{\pi M}\right]^{1/2}\frac{\varepsilon^2}{(1-\varepsilon)}$$

Again, $P_1 = P_2 = \Delta P =$ constant

$$\frac{2}{2P_2+\Delta P}P_1\frac{dV_1}{dt} = \frac{8}{3}\frac{\delta k_0}{k}\frac{A\Delta P}{L\bar{P}S_M}\left[\frac{2RT}{\pi M}\right]^{1/2}\frac{\varepsilon^2}{(1-\varepsilon)}$$

but

$$P_2 = P_1 - \Delta P$$

$$\bar{P} = P_1 - \frac{\Delta P}{2}$$

Hence

$$\frac{2P_1}{2(P_1-\Delta P)+\Delta P}\frac{dV_1}{dt} = \frac{8}{3}\frac{\delta k_0}{k}\frac{A\Delta P}{LS_M\left(P_1-\frac{\Delta P}{2}\right)}\left[\frac{2RT}{\pi M}\right]^{1/2}\frac{\varepsilon^2}{(1-\varepsilon)}$$

$$\frac{P_1}{P_1-\frac{\Delta P}{2}}\frac{dV_1}{dt} = \frac{8}{3}\frac{\delta k_0}{k}\frac{A\Delta P}{LS_M\left(P_1-\frac{\Delta P}{2}\right)}\left[\frac{2RT}{\pi M}\right]^{1/2}\frac{\varepsilon^2}{(1-\varepsilon)}$$

$$\frac{P_1}{\Delta P}\times\frac{\mathrm{d}V}{\mathrm{d}t}=\frac{8}{3}\frac{\delta k_0}{k}\frac{A\Delta P}{LS_\mathrm{M}}\left[\frac{2RT}{\pi M}\right]^{1/2}\frac{\varepsilon^2}{(1-\varepsilon)}$$

Therefore,

$$S_\mathrm{M}=\frac{8}{3}\frac{\delta k_0}{k}\frac{A\Delta P}{LP_1\left(\dfrac{\mathrm{d}V}{\mathrm{d}t}\right)}\left[\frac{2RT}{\pi M}\right]^{1/2}\frac{\varepsilon^2}{(1-\varepsilon)} \tag{10}$$

THE COMBINED CORRECTION TERM

Equation (4) with values of S_K and S_M corrected for gas expansion may be written fully as:

$$S_V^2=S_\mathrm{M}S_\mathrm{V}+S_\mathrm{K}^2$$

$$S_\mathrm{V}^2=\frac{8}{3}\frac{\delta k_0}{k}\frac{A\Delta P}{LP_1\left(\dfrac{\mathrm{d}V}{\mathrm{d}t}\right)}\left[\frac{2RT}{\pi M}\right]^{1/2}\frac{\varepsilon^2}{(1-\varepsilon)}S_\mathrm{V}+\frac{\dfrac{A\Delta P}{k\mu L}\dfrac{\varepsilon^3}{(1-\varepsilon)^2}}{\dfrac{P_1(\mathrm{d}V/\mathrm{d}T)}{P_1-\Delta P/2}}$$

$$\frac{P_1\left(\dfrac{\mathrm{d}V}{\mathrm{d}t}\right)}{P_1-\dfrac{\Delta P}{2}}=\frac{8}{3}\frac{\delta k_0}{k}\frac{A\Delta P}{LP_1\left(\dfrac{\mathrm{d}V}{\mathrm{d}t}\right)}\frac{P_1\dfrac{\mathrm{d}V}{\mathrm{d}t}}{P_1-\dfrac{\Delta P}{2}}\frac{\varepsilon^2}{\Delta P(1-\varepsilon)}\frac{1}{S_\mathrm{V}}\left[\frac{2RT}{\pi M}\right]^{1/2}+\frac{A\Delta P}{k\mu L}\frac{\varepsilon^3}{(1-\varepsilon)^2}\frac{1}{S_\mathrm{V}^2}$$

$$\frac{P_1\left(\dfrac{\mathrm{d}V}{\mathrm{d}t}\right)}{P_1-\dfrac{\Delta P}{2}}=\frac{8}{3}\frac{\delta k_0}{k}\frac{A\Delta P}{LP_1-\dfrac{\Delta P}{2}}\frac{\varepsilon^2}{\Delta P(1-\varepsilon)}\frac{1}{S_\mathrm{V}}\left[\frac{2RT}{\pi M}\right]^{1/2}+\frac{A\Delta P}{k\mu L}\frac{\varepsilon^3}{(1-\varepsilon)^2}\frac{1}{S_V^2}$$

$$\frac{P_1}{\Delta P}\times\frac{\mathrm{d}V}{\mathrm{d}t}=\frac{8}{3}\frac{\delta k_0}{k}\frac{A}{LS_\mathrm{V}}\left[\frac{2RT}{\pi M}\right]^{1/2}\frac{\varepsilon^2}{(1-\varepsilon)}+\frac{A}{k\mu L}\frac{\varepsilon^3}{(1-\varepsilon)^2}\frac{1}{S_\mathrm{V}^2}\left(P_1-\frac{\Delta P}{2}\right)$$

$$= \frac{8}{3}\frac{\delta k_0}{k}\frac{A}{LS_V}\left[\frac{2RT}{\pi M}\right]^{1/2}\frac{\varepsilon^2}{(1-\varepsilon)} + \frac{A}{k\mu L}\frac{\varepsilon^3}{(1-\varepsilon)^2}\frac{1}{S_V^2}\overline{P}$$

$$\frac{P_1}{\Delta P} \times \frac{dV}{dt} = \frac{A}{k\mu L}\frac{\varepsilon^3}{(1-\varepsilon)^2}\frac{1}{S_V^2}\overline{P} + \frac{8}{3}\frac{\delta k_0}{k}\frac{A}{LS_V}\left[\frac{2RT}{\pi M}\right]^{1/2}\frac{\varepsilon^2}{(1-\varepsilon)}$$

From the slope of the graph of the left-hand function against $\overline{P}$ a value of S_V can be found, and the intercept yields a value of $(\delta k_0/k)$.

Appendix A2.1
The relationship between the hydraulic diameter (r_H) and the mean free path (λ) of the gas molecules

The universal equation relating the flow of gases through a packed bed under non-turbulent flow conditions, to the surface area of the powder, may be written as before, from equation (1.1) as:

$$\frac{P_1}{\Delta P} \times \frac{dV}{dt} = K_1 \frac{\overline{P}}{S_V^2} + \left(\frac{\delta k_0}{k}\right) \frac{K_2}{S_V}$$

where

$$K_1 = \frac{A}{k \mu L} \frac{\varepsilon}{(1-\varepsilon)^2}$$

and

$$K_2 = \frac{8A}{3L} \frac{\varepsilon^2}{(1-\varepsilon)} \left[\frac{2RT}{\pi M}\right]^{1/2}$$

The following relationships are derivable from the gas laws:

(i) $$\overline{u} = \left[\frac{2RT}{\pi M}\right]$$

(ii) $$\rho_g = \frac{M}{RT} \overline{P}$$

(iii) $$\mu = \frac{1}{3} \rho_g \overline{u} \lambda$$

Note also that:

$$d_{sv} = \frac{6}{S_V}$$

and

$$r = \frac{3}{S_V}$$

Therefore,

$$\lambda = \frac{2\mu RT}{\overline{P}\ M}\left[\frac{\pi M}{8RT}\right]^{1/2}$$

$$= \frac{\mu}{\overline{P}}\frac{\pi}{2}\left[\frac{2RT}{\pi M}\right]^{1/2}$$

Now,

$$\varepsilon = \frac{V_v}{V_s + V_v}$$

$$\varepsilon(V_s + V_v) = V_v$$

$$V_v = \left(\frac{\varepsilon}{1-\varepsilon}\right)V_s$$

The hydraulic diameter,

$$r_H = \frac{V_v}{S}$$

$$= \frac{\varepsilon}{(1-\varepsilon)}\frac{V_s}{S}$$

$$= \frac{\varepsilon}{(1-\varepsilon)}\frac{1}{S_V}$$

From the above considerations,

$$\frac{r_H}{\lambda} = \left(\frac{\varepsilon}{1-\varepsilon}\right)\frac{2\overline{P}}{\pi\mu S_V}\Bigg/\left[\frac{2RT}{\pi M}\right]^{1/2}$$

and

$$\frac{K_1}{K_2} = \frac{A}{k\mu L}\frac{\varepsilon}{(1-\varepsilon)^2}\frac{3L(1-\varepsilon)}{8A\varepsilon^2}\left[\frac{\pi M}{2RT}\right]^{1/2}$$

$$= \frac{3}{8}\left(\frac{\varepsilon}{1-\varepsilon}\right)\frac{1}{k\mu}\Bigg/\left[\frac{2RT}{\pi M}\right]^{1/2}$$

$$\left[\frac{2RT}{\pi M}\right]^{1/2} = \frac{3}{8}\left(\frac{\varepsilon}{1-\varepsilon}\right)\frac{1}{k\mu}\frac{K_2}{K_1} = \left(\frac{\varepsilon}{1-\varepsilon}\right)\frac{2\overline{P}}{\pi\mu S_V}\frac{\lambda}{r_H}$$

This yields

$$\frac{r_H}{\lambda} = \frac{2}{\pi}\frac{8}{3}k\frac{\overline{P}}{S_V}\left(\frac{K_1}{K_2}\right)$$

$$= \left[\frac{16k}{3\pi}\frac{\overline{P}}{S_V}\right]\frac{K_1}{K_2}$$

Typical values of the ratio r_H/λ calculated at various mean pressures for dicalcium phosphate powder (DCP) at various porosities, and for spherical bronze powder at various mean particle sizes, are shown in Tables A2.1–A2.11.

Worked example

$\overline{P} = 9.2977 + 10^{-4}\ \mathrm{Nm}^{-2}$

$S_V = 1.8800 \times 10^6\ \mathrm{m^2/m^3}$

$K_1 = 44.3601\ \mathrm{m^2\ s\ kg^{-1}}$

$K_2 = 1.3373\ \mathrm{m^2/s}$

$$\frac{r_H}{\lambda} = \frac{16 \times 5}{3\pi} \times \frac{9.2977 \times 10^{-4}}{1.8800 \times 10^6} \times \frac{44.3601}{1.3373}$$

$$= \underline{0.1393 \times 10^{-6}}$$

Table A2.1 — Dicalcium phosphate powder (DCP)
$\varepsilon = 0.6559$
$S_V = 1.8800 \times 10^6\ \mathrm{m^2/m^3}$
$K_1 = 44.3601\ \mathrm{m^2\ s/kg}$
$K_2 = 1.3373\ \mathrm{m^2/s}$

$\overline{P}$ ($\mathrm{Nm^{-2}} \times 10^{-4}$)	r_H/λ ($\times 10^{-6}$)
9.2977	0.1393
8.3472	0.1251
7.3348	0.1099
6.2612	0.0938
5.2319	0.0784
4.2233	0.0633
3.2140	0.0482
2.1595	0.0324
1.1253	0.0169
0.1988	0.0030

Table A2.2

DCP

$\varepsilon = 0.6243$

$S_V = 2.2995 \times 10^6$ m^2/m^3

$K_1 = 35.0101$ m^2 s/kg

$K_2 = 1.2116$ m^2/s

$\overline{P}$ ($Nm^{-2} \times 10^{-4}$)	r_H/λ ($\times 10^{-6}$)
9.4652	0.1010
8.4312	0.0900
7.3899	0.0789
6.3419	0.0677
5.3279	0.0569
4.2279	0.0451
3.2293	0.0345
2.1646	0.0231
1.1861	0.0127
0.3429	0.0037

Table A2.3

DCP

$\varepsilon = 0.6055$

$S_V = 2.3172 \times 10^6$ m^2/m^3

$K_1 = 30.4382$ m^2 s/kg

$K_2 = 1.1394$ m^2/s

$\overline{P}$ ($Nm^{-2} \times 10^{-4}$)	r_H/λ ($\times 10^{-6}$)
9.6573	0.0946
8.6087	0.0843
7.5993	0.0744
6.4920	0.0636
5.4093	0.0530
4.2886	0.0420
3.3007	0.0323
2.2607	0.0221
1.2775	0.0125
0.6284	0.0062

Table A2.4

DCP

$\varepsilon = 0.5603$

$S_V = 2.3279 \times 10^6$ m^2/m^3

$K_1 = 21.6397$ m^2 s/kg

$K_2 = 0.9757$ m^2/s

$\overline{P}$ (Nm$^{-2} \times 10^{-4}$)	r_H/λ ($\times 10^{-6}$)
9.5926	0.0776
8.5273	0.0690
7.5353	0.0610
6.4840	0.0525
5.3886	0.0436
4.2079	0.0341
3.2046	0.0259
2.2407	0.0181
1.3636	0.0110
0.6437	0.0052

Table A2.5 — Spherical bronze powder

Spherical bronze (13.375 μm)

$\varepsilon = 0.3732$

$S_V = 0.4528 \times 10^6$ m^2/m^3

$K_1 = 2.2658$ m^2 s/kg

$K_2 = 0.2187$ m^2/s

$\overline{P}$ (Nm$^{-2} \times 10^{-4}$)	r_H/l ($\times 10^{-6}$)
9.4678	0.1840
8.2992	0.1613
7.2845	0.1416
6.2693	0.1218
5.2506	0.1020
4.2359	0.0823
3.2207	0.0626
2.1987	0.0427
1.1794	0.0229
0.3082	0.0060

Table A2.6

Spherical bronze (14.664 μm)

$\varepsilon = 0.3811$

$S_V = 0.4138 \times 10^6$ m^2/m^3

$K_1 = 2.5029$ m^2 s/kg

$K_2 = 0.2334$ m^2/s

$\overline{P}$ (Nm$^{-2} \times 10^{-4}$)	r_H/λ ($\times 10^{-6}$)
9.2684	0.2040
8.2545	0.1817
7.2405	0.1594
6.2266	0.1370
5.2126	0.1147
4.1986	0.0924
3.1847	0.0701
2.1700	0.0478
1.1561	0.0254
0.2695	0.0059

Table A2.7

Spherical bronze (15.240 μm)

$\varepsilon = 0.3872$

$S_V = 0.4050 \times 10^6$ m^2/m^3

$K_1 = 2.6978$ m^2 s/kg

$K_2 = 0.2452$ m^2/s

$\overline{P}$ (Nm$^{-2} \times 10^{-4}$)	r_H/λ ($\times 10^{-6}$)
9.2524	0.2135
8.2385	0.1901
7.2245	0.1667
6.2106	0.1433
4.1966	0.1199
3.1687	0.0731
2.1547	0.0497
1.1408	0.0263
0.2415	0.0056

Table A2.8

Spherical bronze (17.153 μm)

$\varepsilon = 0.3783$

$S_V = 0.3596 \times 10^6\ m^2/m^3$

$K_1 = 2.3501\ m^2\ s/kg$

$K_2 = 0.2221\ m^2/s$

$\bar{P}$ ($Nm^{-2} \times 10^{-4}$)	r_H/λ ($\times 10^{-6}$)
9.2423	0.2310
8.2284	0.2056
7.2145	0.1803
6.2006	0.1550
5.1865	0.1296
4.1726	0.1043
3.1587	0.0789
2.1446	0.0536
1.1308	0.0283
0.2482	0.0062

Table A2.9

Spherical bronze (19.576 μm)

$\varepsilon = 0.3841$

$S_V = 0.3101 \times 10^6\ m^2/m^3$

$K_1 = 2.5123\ m^2\ s/kg$

$K_2 = 0.2313\ m^2/s$

$\bar{P}$ ($Nm^{-2} \times 10^{-4}$)	r_H/λ ($\times 10^{-6}$)
9.2057	0.2738
8.1918	0.2437
7.1777	0.2135
6.1639	0.1834
5.1498	0.1532
4.1359	0.1230
3.1219	0.0929
2.1080	0.0627
1.0941	0.0326
0.2215	0.0066

Table A2.10

Spherical bronze (20.841 μm)

$\varepsilon = 0.3855$

$S_V = 0.3023 \times 10^6$ m²/m³

$K_1 = 2.6025$ m² s/kg

$K_2 = 0.2384$ m²/s

$\overline{P}$ ($Nm^{-2} \times 10^{-4}$)	r_H/λ ($\times 10^{-6}$)
9.1990	0.2821
8.1851	0.2510
7.1711	0.2199
6.1572	0.1888
5.1432	0.1577
4.1292	0.1266
3.1153	0.0955
2.1013	0.0644
1.0874	0.0333
0.2082	0.0064

Table A2.11

Spherical bronze (21.708 μm)

$\varepsilon = 0.3990$

$S_V = 0.3010 \times 10^6$ m²/m³

$K_1 = 2.9936$ m² s/kg

$K_2 = 0.2592$ m²/s

$\overline{P}$ ($Nm^{-2} \times 10^{-4}$)	r_H/λ ($\times 10^{-6}$)
9.1889	0.2994
8.1751	0.2664
7.1610	0.2333
6.1472	0.2003
5.1332	0.1672
4.1191	0.1342
3.1053	0.1012
3.0912	0.0681
1.0774	0.0351
0.2208	0.0072

Appendix A3.1
Some common conversion tables and constants

Table A3.1 — Factors for unit conversions

Quantity	Equivalent values
Mass	1 kg = 1000 g = 0.001 metric ton = 2.20462 lb_m = 35.27392 oz
	1 lb_m = 16 oz = 5×10^{-4} ton = 453.593 g = 0.453593 kg
Length	1 m = 100 cm = 1000 mm = 10^6 microns (μm) = 10^{10} angstroms (Å)
	= 39.37 in = 3.2808 ft = 1.0936 yd = 0.0006214 mile
	1 ft = 12 in = 1/3 yd = 0.3048 m = 30.48 cm
Volume	1 m^3 = 1000 litres = 10^6 cm^3 = 10^6 ml
	= 35.3145 ft^3 = 220.83 imperial gallons = 264.17 gal
	= 1056.68 qt
	1 ft^3 = 1728 in^3 = 7.4805 gal = 0.028317 m^3 = 28.317 litres
	= 28 317 cm^3
Force	1 N = 1 kg·m/s^2 = 10^5 dynes = 10^5 g·cm/s^2 = 0.22481 lb_f
	1 lb_f = 32.174 lb_m·ft/s^2 = 4.4482 N = 4.4482×10^5 dynes
Pressure	1 atm = 1.01325×10^5 N/m^2 (Pa) = 101.325 kPa = 1.01325 bars
	= 1.01325×10^6 dynes/cm^2
	= 760 mm Hg at 0°C (torr) = 10.333 n H_2O at 4°C
	= 14.696 lg_f/in2 (psi) = 33.9 ft H_2O at 4°C
	= 29.921 in Hg at 0°C
Energy	1 J = 1 N·m = 10^7 ergs = 10^7dyne·cm
	= 2.778×10^{-7} kW·h = 0.23901 cal
	= 0.7376 ft·lb_f = 9.486×10^{-4} Btu
Power	1 W = 1 J/s = 0.23901 cal/s = 0.7376 ft·lb_f/s = 9.486×10^{-4} Btu/s
	= 1.341×10^{-3} hp

Example: The factor to convert grams to lb_m is $\left(\frac{2.20462\ lb_m}{1000\ g}\right)$.

Table A3.2 — SI units of weights and measures

Base units		Multiples and prefixes	
length	metre (m)	exa $=10^{18}$	deci $=10^{-1}$
mass	kilogram (kg)	peta $=10^{15}$	centi $=10^{-2}$
time	second (s)	tera $=10^{12}$	milli $=10^{-3}$
electric current	ampere (A)	giga $=10^{9}$	micro $=10^{-6}$
temperature	kelvin (K)	mega $=10^{6}$	nano $=10^{-9}$
amount of substance	mole (mol)	kilo $=10^{3}$	pico $=10^{-12}$
luminous intensity	candela (cd)	hecto $=10^{2}$	femto $=10^{-15}$
		deka $=10^{1}$	atto $=10^{-18}$

Table A3.3 — Selected conversions between customary and SI units

Units of length — to SI
1 inch = 2.54 centimetres
1 foot = 0.3048 metre
1 mile = 1.6093 kilometres
Units of length — from SI
1 centimetre = 0.3937 inch
1 metre = 3.2808 feet
1 kilometre = 0.62137 mile
Units of area — to SI
1 square inch = 6.4516 square centimetres
1 square foot = 0.0929 square metre
1 acre = 0.4047 hectare
1 square mile = 2.5900 square kilometres
Units of area — from SI
1 square centimetre = 0.1550003 square inch
1 square metre = 10.764 square feet
1 hectare = 2.471 acres
1 square kilometre = 0.3861 square mile
Units of volume — to SI
1 cubic inch = 16.387 cubic centimetres
1 cubic foot = 0.028317 cubic metre
Units of volume — from SI
1 cubic centimetre = 0.061024 cubic inch
1 cubic metre = 35.31467 cubic feet
Capacity (liquid) — to SI
1 quart = 0.9463 litres
1 gallon = 3.785 litres
Capacity (liquid) — from SI
1 litre = 1.0567 quarts
1 litre = 0.2642 gallon
Capacity (dry) — to SI
1 quart = 1.101191 litres
1 bushel = 35.2381 litres
Capacity (dry) — from SI
1 litre = 0.908 quart
1 litre = 0.028378 bushel
Units of mass — to SI
1 avoirdupois ounce = 28.3495 grams
1 apothecaries' or troy ounce = 31.1035 grams
1 apothecaries' or troy pound = 373.242 grams
= 0.37324 kilogram
1 avoirdupois pound = 0.45359 kilogram
1 short ton = 0.9072 metric ton
1 long ton = 1.016 metric tons
Units of mass — from SI
1 gram = 0.035274 avoirdupois ounce
1 gram = 0.03215 apothecaries' or troy ounce
1 kilogram = 2.679 apothecaries' or troy pounds
1 kilogram = 2.2046 avoirdupois pounds
1 metric ton = 1.102 short tons
1 metric ton = 0.9842 long ton

Table A3.4 — Customary weights and measures

Length
1000 mils = 1 inch
12 inches = 1 foot
3 feet = 1 yard
$5\frac{1}{2}$ yards = 1 rod
40 rods = 1 furlong
8 furlongs (5280 feet) = 1 mile
3 miles = 1 league
Note: The English and American mile is 1760 yards ('statute mile'). Other nations measure differently.

Area
144 square inches = 1 square foot
9 square feet = 1 square yard
$30\frac{1}{4}$ square yards = 1 square rod
100 square rods = 1 acre
640 acres = 1 square mile
36 square miles = 1 township

Mariners' measure
6 feet = 1 fathom
120 fathoms = 1 cable length
8.430 cable lengths = 1 nautical mile
3 nautical miles = 1 marine league
1.15 United States miles = 1 nautical mile
Note: A nautical or geographical mile is one minute of the arc on the earth's surface at the equator. A knot is a speed sufficient to travel one nautical mile in one hour.

Surveyors' measure
7.92 inches = 1 link
25 links = 1 rod
4 rods (66 feet) = 1 chain
10 chains = 1 furlong
8 furlongs = 1 mile
1 square mile = 1 section

Capacity
1728 cubic inches = 1 cubic foot
27 cubic feet = 1 cubic yard
128 cubic feet = 1 cord (wood or stone)
231 cubic inches = 1 United States gallon
277.42 cubic inches = 1 British imperial gallon
2150.4 cubic inches = 1 United States bushel

United states capacity (liquid)
4 fluid ounces = 1 gill
4 gills = 1 pint
2 pints = 1 quart
4 quarts = 1 gallon
9 gallons = 1 firkin
$32\frac{1}{2}$ gallons = 1 barrel
63 gallons = 1 hogshead
Note: A gallon in most states occupies 231 cubic inches.

United States capacity (dry)
2 pints = 1 quart
8 quarts = 1 peck
4 pecks = 1 bushel
2150.42 cubic inches = 1 bushel

Troy weight
3.086 grains = 1 carat
24 grains = 1 pennyweight
20 pennyweights = 1 ounce
12 ounces = 1 pound
Note: 1 troy pound is 0.8229 of an avoirdupois pound.

Avoirdupois weight
$437\frac{1}{2}$ grains = 1 ounce
16 ounces = 1 pound
100 pounds = 1 short hundredweight
2000 pounds = 1 short ton
2240 pounds = 1 long ton
Note: A stone of 14 pounds is a common English measure of weight, as is the long hundredweight of 112 pounds.

Apothecaries' weight
20 grains = 1 scruple
3 scruples = 1 dram
8 drams = 1 ounce
12 ounces = 1 pound

Counting
12 items = 1 dozen
13 items = baker's dozen
12 dozen = 1 gross
12 gross = 1 great gross
20 things = 1 score

Angular measure
60 seconds = 1 minute
60 minutes = 1 degree
360 degrees = 1 circle or circumference

Paper measure
25 sheets = 1 quire
500 sheets (formerly 480) = 1 ream
2 reams = 1 bundle
4 bundles = 1 case

Table A3.5 — Miscellaneous units of weights and measures

1 atmosphere (pressure) = 14.7 pounds per square inch
1 atmosphere = 34 feet head of water
1 atmosphere = 700 millimetres of mercury
absolute zero = −273.16°C = −459.69°F
1 degree C = ($\frac{5}{9}$ F) − 32)
1 degree F = ($\frac{9}{5}$ C) + 32
1 horsepower = 550 foot-pounds per second
1 horsepower = 746 watts = 0.746 kilowatt
1 kilowatt = 1.34 horsepower
1 carat = 200 milligrams (gem weight)
1 stone = 14 pounds
1 load = 1.9 cubic yards

United States official weights of barrel
wheat, barley, rye flour = 196 pounds
resin, tar, pitch (gross) = 500 pounds
pickled fish (net) = 200 pounds
Lime (net) = 200 pounds
cement (4 bags) (net) = 376 pounds

United States official weights of bushel
wheat, beans, peas, potatoes = 60 pounds
rye, corn, linseed, maslin = 56 pounds
barley, buckwheat = 48 pounds
dry onions = 50 pounds
rough rice = 45 pounds
oats = 32 pounds
green peanuts in shell = 22 pounds

1 point (type size) = $\frac{1}{72}$ inch = 0.01384 inch
1 light year = 5.9×10^{12} miles
1 nail = $2\frac{1}{2}$ inches (cloth measure)
1 finger = $4\frac{1}{2}$ inches (cloth measure)
1 span = 9 inches (cloth measure)
1 ell = 45 inches
1 hand = 4 inches (measuring horses)
1 cubit (Biblical) = 18 inches
1 board foot = 144 cubic inches
1 cord foot = 16 cubic feet
1 cord = 8 cord feet
1 pace = $2\frac{1}{2}$ feet

United States common household measures
3 teaspoons = 1 tablespoon
1 tablespoon = $\frac{1}{2}$ fluid ounce
16 tablespoons = 1 cup
1 cup = 8 fluid ounces
8 gills = 1 cup
1 cup = $\frac{1}{2}$ liquid pint
1 cup granulated sugar = $\frac{1}{2}$ pound
1 cup flour = $\frac{1}{4}$ pound
1 cup chopped meat = $\frac{1}{2}$ pound
1 cup lard = $\frac{1}{2}$ pound
1 cup rice = $\frac{1}{2}$ pound
1 cup raisins = 6 ounces
1 cup dry crumbs = 2 ounces

Table A3.6 — Conversion factors for the US Customary System, metric system, and International System

A. UNITS OF LENGTH

units	cm	m	in	ft	yd	mile
1 cm	=1	0.01*	0.3937008	0.03280840	0.01093613	6.213712×10^{-6}
1 m	=100	1	39.37008	3.280840	1.093613	6.213712×10^{-4}
1 in	=2.54*	0.0254	1	0.08333333...	0.02777777...	1.578283×10^{-5}
1 ft	=30.48	0.3048	12*	1	0.3333333...	$1.893939\ldots\times10^{-4}$
1 yd	=91.44	0.9144	36	3*	1	$5.681818\ldots\times10^{-4}$
1 mile	$=1.609344\times10^{5}$	1.609344×10^{3}	6.336×10^{4}	5280*	1760	1

B. UNITS OF AREA

units	cm^2	m^2	in^2	ft^2	yd^2	$mile^2$
1 cm^2	=1	10^{-4}*	0.1550003	1.076391×10^{-3}	1.195990×10^{-4}	3.861022×10^{-11}
1 m^2	$=10^{4}$	1	1550.003	10.76391	1.195990	3.861022×10^{-7}
1 in^2	=6.4516*	6.4516×10^{-4}	1	6.944444×10^{-3}...	7.716049×10^{-4}	2.490977×10^{-10}
1 ft^2	=929.0304	0.0929034	144*	1	0.1111111...	3.587007×10^{-8}
1 yd^2	=8361.273	0.8361273	1296	9*	1	3.228306×10^{-7}
1 $mile^2$	$=2.589988\times10^{10}$	2.589988×10^{6}	4.014490×10^{9}	2.78784×10^{7}*	3.0976×10^{6}	1

continued next page

Table A3.6 — (*Contd.*)

C. UNITS OF VOLUME

units	m^3	cm^3	litre	in^3	ft^3	qt	gal
1 m^3	=1	10^6	10^3	6.102374×10^4	35.31467	1.056688×10^3	264.1721
1 cm^3	$=10^{-6}$	1	10^{-3}	0.06102374	3.531467×10^{-5}	1.056688×10^{-3}	2.641721×10^{-4}
1 litre	$=10^{-3}$	1000*	1	61.02374	0.03531467	1.056688	0.2641721
1 in^3	$=1.638706\times10^{-5}$	16.38706*	0.01638706	1	5.787037×10^{-4}	0.01731602	4.329004×10^{-3}
1 ft^3	$=2.831685\times10^{-2}$	28316.85	28.31685	1728*	1	2.992208	7.480520
1 qt	$=9.46353\times10^{-4}$	946.353	0.946353	57.75	0.0342014	1	0.25
1 gal (US)	$=3.785412\times10^{-3}$	3785.412	3.785412	231*	0.1336806	4*	1

D. UNITS OF WEIGHT

units	g	kg	oz	lb	metric ton	ton
1 g	=1	10^{-3}	0.03527396	2.204623×10^{-3}	10^{-6}	1.102311×10^{-6}
1 kg	=1000	1	35.27396	2.204623	10^{-3}	1.102311×10^{-3}
1 oz (avdp)	=28.34952	0.02834952	1	0.0625	2.834952×10^{-5}	3.125×10^{-5}
1 lb (avdp)	=453.5924	0.04535924	16*	1	4.535924×10^{-4}	0.0005
1 metric ton	$=10^6$	1000*	35273.96	2204.623	1	1.102311
1 ton	=907184.7	907.1847	3200*	200*	0.9071847	1

E. UNITS OF DENSITY

units	$g\cdot cm^{-3}$	$g\cdot L^{-1}$, $kg\cdot m^{-3}$	$oz\cdot in^{-3}$	$lb\cdot in^{-3}$	$lb\cdot ft^{-3}$	$lb\cdot gal^{-1}$
1 $g\cdot cm^{-3}$	=1	1000	0.5780365	0.03612728	62.42795	8.345403
1 $g\cdot L^{-1}$, $kg\cdot m^{-3}$	$=10^{-3}$	1	5.780365×10^{-4}	3.612728×10^{-5}	0.06242795	8.345403×10^{-3}
1 $oz\cdot in^{-3}$	=1.729994	1729.994	1	0.0625	108	14.4375
1 $lb\cdot in^{-3}$	=27.67991	27679.91	16	1	1728	231
1 $lb\cdot ft^{-3}$	=0.01601847	16.01847	9.259259×10^{-3}	5.7870370×10^{-4}	1	0.1336806
1 $lb\cdot gal^{-1}$	=0.1198264	119.8264	4.749536×10^{-3}	4.3290043×10^{-3}	7.480519	1

F. UNITS OF PRESSURE

units	Pa, $N\cdot m^{-2}$	$dyn\cdot cm^{-2}$	bar	atm	$kgf\cdot cm^{-2}$	mmHg (torr)	in Hg	$lbf\cdot in^{-2}$
1 Pa, 1 $N\cdot m^{-2}$	=1	10	10^{-5}	9.869233×10^{-6}	1.019716×10^{-5}	7.500617×10^{-3}	2.952999×10^{-4}	1.450377×10^{-4}
1 $dyn\cdot cm^{-2}$	=0.1	1	10^{-6}	9.869233×10^{-7}	1.019716×10^{-6}	7.500617×10^{-4}	2.952999×10^{-5}	1.450377×10^{-5}
1 bar	$=10^5$*	10^6	1	0.9869233	1.019716	750.0617	29.52999	14.50377
1 atm	=101325.0*	101325.0	1.013250	1	1.033227	760	29.92126	14.69595
1 $kgf\cdot cm^{-2}$	=98066.5	980665	0.980665	0.9678411	1	735.5592	28.95903	14.22334
1 mm Hg (torr)	=133.3224	1333.224	1.333224×10^{-3}	1.3157895×10^{-3}	1.3595099×10^{-3}	1	0.03937008	0.01933678
1 in Hg	=3386.388	33863.88	0.03386388	0.03342105	0.03453155	25.4	1	0.4911541
1 $lbf\cdot in^{-2}$	=6894.757	68947.57	0.06894757	0.06804596	0.07030696	51.71493	2.036021	1

G. UNITS OF ENERGY

units	g mass	J	int J	cal	cal_{IT}	Btu_{IT}	kWh	hp h	ft·lbf	cu ft·lbf in^{-2}	litre·atm
1 g mass	=1	8.987552×10^{13}	8.986069×10^{13}	2.148076×10^{13}	2.146640×10^{13}	8.518555×10^{10}	2.496542×10^7	3.347918×10^7	6.628878×10^{13}	4.603388×10^{11}	8.870024×10^{11}
1 J	$=1.112650\times10^{-14}$	1	0.999835	0.2390057	0.2388459	9.478172×10^{-4}	$2.777777\ldots\times10^{-7}$	3.725062	0.7375622	5.121960×10^{-3}	9.869233×10^{-3}
1 int J	$=1.112834\times10^{-14}$	1.000165	1	0.2390452	0.2388853	9.479735×10^{-4}	2.778236×10^{-7}	3.725676×10^{-7}	0.7376839	5.122805×10^{-3}	9.870862×10^{-3}
1 cal	$=4.655328\times10^{-14}$	4.184*	4.183310	1	0.9993312	3.965667×10^{-3}	$1.1622222\ldots\times10^{-6}$	1.558562×10^{-6}	3.085960	2.143028×10^{-2}	0.04129287

*Numbers followed by an asterisk are definitions of the relation between the two units.

Table A3.7 — Some common SI derived units

Quantity	Unit	SI unit symbol ('abbreviation'). Use roman type	Formula. Use roman type
absorbed dose	gray	Gy	J/kg
acceleration	metre per second squared	...	m/s^2
activity (of radionuclides)	becquerel	Bq	1/s
angular acceleration	radian per second squared	...	rad/s^2
angular velocity	radian per second	...	rad/s
area	square metre	...	m^2
Celsius temperature	degree Celsius	°C	K
density	kilogram per cubic metre	...	kg/m^3
dose equivalent	sievert	Sv	J/kg
electric capacitance	farad	F	A·s/V (=C/V)
electric charge	coulomb	C	A·s
electrical conductance	siemens	S	A/V
electric field strength	volt per metre	...	V/m
electric inductance	henry	H	V·s/A (=Wb/A)
electric potential	volt	V	W/A
electric resistance	ohm	Ω	V/A
electromotive force	volt	V	W/A
energy	joule	J	N·m
entropy	joule per kelvin	...	J/K
force	newton	N	$kg{\cdot}m/s^2$
frequency	hertz	Hz	1/s
illuminance	lux	lx	lm/m^2
luminance	candela par square metre	...	cd/m^2
luminous flux	lumen	lm	cd·sr
magnetic field strength	ampere per metre	...	A/m
magnetic flux	weber	Wb	V·s
magnetic flux density	tesla	T	Wb/m^2
potential difference	volt	V	W/A
power	watt	W	J/s
pressure	pascal	Pa	N/m^2
quantity of electricity	coulomb	C	A·s
quantity of heat	joule	J	N·m
radiant flux	watt	W	J/s
radiant intensity	watt per steradian	...	W/sr
specific heat	joule per kilogram kelvin	...	J(kg·K)
stress	pascal	Pa	N/m^2
thermal conductivity	watt per metre kelvin	...	W/(m·K)
velocity	metre per second	...	m/s
viscosity, dynamic	pascal second	...	Pa·s
viscosity, kinematic	square metre per second	...	m^2/s
voltage	volt	V	W/A
volume*	cubic metre	...	m^3
wave number	1 per metre	...	1/m
work	joule	J	N·m

*In 1964, the General Conference on Weights and Measures adopted litre as a special name for the cubic decimetre but discouraged the use of litre for volume measurement of extreme precision.

Table A3.8 — SI unit prefixes

Multiplication factor	SI prefix	SI prefix symbol. Use roman type	Pronounciation (US)*	Meaning (US)	Meaning in other countries
1 000 000 000 000 000 000 $= 10^{18}$	exa**	E	ex' a (*a* as in *a* bout)	one quintillion times†	trillion
1 000 000 000 000 000 $= 10^{15}$	peta**	P	as in *peta* l	one quadrillion times†	thousand billion
1 000 000 000 000 $= 10^{12}$	tera	T	as in *terra* ce	one trillion times†	billion
1 000 000 000 $= 10^{9}$	giga	G	jig' a (*a* as in *a* bout)	one billion times†	milliard
1 000 000 $= 10^{4}$	mega	M	as in *mega* phone	one million times	
1 000 $= 10^{3}$	kilo	k	as in *kilo* watt	one thousand times	
100 $= 10^{2}$	hecto‡	h	heck' toe	one hundred times	
10 $=$ 10	deka‡	da	deck' a (*a* as in *a* bout)	ten times	
0.1 $= 10^{-1}$	deci‡	d	as in *deci* mal	one tenth of	
0.01 $= 10^{-2}$	centi‡	c	as in *senti* ment	one hundredth of	
0.001 $= 10^{-3}$	milli	m	as in *mili* tary	one thousandth of	
0.000 001 $= 10^{-6}$	micro	μ	as in *micro* phone	one millionth of	
0.000 000 001 $= 10^{-9}$	nano	n	nan' oh (*an* as in *an* t)	one billionth of†	milliardth
0.000 000 000 001 $= 10^{-12}$	pico	p	peek' oh	one trillionth of†	billionth
0.000 000 000 000 001 $= 10^{-15}$	femto	f	fem' toe(*fem* as in *fem* inine)	one quadrillionth of†	thousand billionth
0.000 000 000 000 000 001 $= 10^{-18}$	atto	a	as in an*ato* my	one quintillionth of†	trillionth

*The first syllable of every prefix is accented to assure that the prefix will retain its identity. Therefore, the preferrred pronunciation of kilometre places the accent on the first syllable, not the second.

**Approved by the 15th General Conference of Weights and Measures (CGPM), May–June 1975.

†These terms should be avoided in technical writing because the denominations above 1 million are different in most other countries, as indicated in the last column.

‡While hecto, deka, deci, and centi are SI prefixes, their use generally should be avoided except for the SI unit multiples for area, volume, moment, and non-technical use of centimetre, as for body and clothing measurement.

Table A3.9 — Length equivalents

metres	centi-metres	inches	feet	yards	chains	kilo-metres	miles
1	100 2.00000	39.37 $\bar{1}.59517$	3.281 0.51598	1.0936 0.03886	0.04971 $\bar{2}.69644$	0.001 $\bar{3}.00000$	$0.0_{5}6214$ $\bar{6}.79335$
0.01 $\bar{2}.00000$	**1**	**0.3937** $\bar{1}.59517$	**0.03281** $\bar{2}.51598$	**0.01094** $\bar{2}.03886$	**$0.0_{3}4971$** $\bar{4}.69644$	**10^{-5}** $\bar{5}.00000$	**$0.0_{3}6214$** $\bar{4}.79335$
0.0254 $\bar{2}.40483$	**2.504** 0.40483	**1**	**0.08333** $\bar{2}.92082$	**0.02778** $\bar{2}.44370$	**0.001263** $\bar{3}.10127$	**$0.0_{4}254$** $\bar{5}.40483$	**$0.0_{4}1578$** $\bar{5}.19818$
0.3048 $\bar{1}.48402$	**30.48** 1.48402	**12** 1.07918	**1**	**0.3333** $\bar{1}.52288$	**0.01515** $\bar{2}.18046$	**$0.0_{3}3048$** $\bar{4}48,402$	**$0.0_{3}1894$** $\bar{4}27736$
0.9144 $\bar{1}.96114$	**91.44** 1.96114	**36** 1.55630	**3** 0.47712	**1**	**0.04545** 2.65758	**$0.0_{3}9144$** $\bar{4}.96114$	**$0.0_{3}5682$** $\bar{4}.75449$
20.12 1.30356	**2012** 3.30356	**792** 2.89873	**55** 1.81954	**22** 1.34242	**1**	**0.02012** $\bar{2}.30356$	**0.0125** $\bar{2}.09691$
1,000 3.00000	**100,000** 5.00000	**39,370** 4.59517	**3,281** 3.51598	**1,093.6** 3.03886	**49.71** 1.69644	**1**	**0.6214** $\bar{1}.79335$
1,609 3.20665	**160,935** 5.20665	**63,360** 4.80182	**5,280** 3.72263	**1,760** 3.24551	**80** 1.90309	**1.609** 0.20665	**1**

The equivalents are given in the bold type. Logarithms of the equivalents are given immediately below with the characteristric of the logarithm (i.e. $\bar{1}.59517 = 0.59517$). In some cases, the equivalents have been rounded off, while the logarithm corresponds to the equivalent carried to a greater number of decimal places. Subscripts after any figure — 0_3, 9_4 — mean that that figure is to be repeated the indicated number of times.

Table A3.10 — Conversion of lengths*

	inches to milli-meires	milli-metres to inches	feet to metres	metres to feet	yards to metres	metres to yards	miles to kilo-metres	kilo-metres to miles
1	25.40	0.03937	0.3048	3.281	0.9144	1.094	1.609	0.6214
2	50.80	0.07874	0.6096	6.562	1.829	2.187	3.219	1.243
3	76.20	0.1181	0.9144	9.842	2.743	3.281	4.828	1.864
4	101.60	0.1576	1.219	13.12	3.658	4.374	6.437	2.485
5	127.00	0.1968	1.524	16.40	4.572	5.468	8.047	3.107
6	152.40	0.2362	1.829	19.68	5.486	6.562	9.656	3.728
7	177.80	0.2756	2.134	22.97	6.401	7.655	11.27	4.350
8	203.20	0.3150	2.438	26.25	7.315	8.749	12.87	4.971
9	228.60	0.3543	2.743	29.53	8.230	9.842	14.48	5.592

*For example, 1 in = 25.40 mm.

Table A3.11 — Area equivalents*

square metres	square inches	square feet	square yards	square rods	square chains	roods	acres	square miles or sections
1	**1,550** 3.19033	**10.76** 1.03197	**1.196** 0.07773	**0.0395** 2.59699	**0.002471** 3.39288	**0.0_39884** 4.99494	**0.0_32471** 4.39288	**0.0_63861** 7.58670
0.0_36452 4.80967	**1**	**0.006944** $\bar{3}.84164$	**0.0_37716** $\bar{4}.88740$	**0.0_42551** $\bar{5}.40667$	**0.0_51594** $\bar{6}.20255$	**0.0_66377** 7.80461	**0.0_61594** 7.20255	**0.0_92491** $\bar{1}\bar{0}.39637$
0.09290 $\bar{2}.96803$	**144** 2.15836	**1**	**0.1111** $\bar{1}.04576$	**0.003673** $\bar{3}.56503$	**0.0_32296** $\bar{4}.36091$	**0.0_49184** $\bar{5}.96297$	**0.0_42296** $\bar{5}.36091$	**0.0_73587** $\bar{8}.55473$
0.8361 $\bar{1}.92227$	**1,296** 3.11260	**9** 0.95424	**1**	**0.03306** $\bar{2}.51927$	**0.002066** $\bar{3}.31515$	**0.0_38264** $\bar{4}.91721$	**0.0002066** $\bar{4}.31515$	**0.0_63228** 7.50898
25.29 1.40300	**39,204** 4.59333	**272.25** 2.43497	**30.25** 1.48072	**1**	**0.0625** $\bar{2}.79588$	**0.02500** $\bar{2}.39794$	**0.00625** $\bar{3}.79588$	**0.0_59766** $\bar{6}.98970$
404.7 2.60712	**627,264** 5.79745	**4,356** 3.63909	**484** 2.68484	**16** 1.20412	**1**	**0.4** $\bar{1}.60206$	**0.1** $\bar{1}.00000$	**0.0001562** $\bar{4}.19382$
1,012 3.00506	**1,568,160** 6.19539	**10,890** 4.03703	**1,210** 3.08278	**40** 1.60206	**2.5** 0.39794	**1**	**0.25** $\bar{1}.39794$	**0.0_33906** $\bar{4}.59176$
4,047 3.60712	**6,272,640** 6.79745	**43,560** 4.63909	**4,840** 3.68484	**160** 2.20412	**10** 1.00000	**4** 0.60206	**1**	**0.001562** $\bar{3}.19281$
2,589,998 6.41330		**27,878,400** 7.44527	**3,097,500** 6.49102	**102,400** 5.01030	**6,400** 3.80618	**2,560** 3.40824	**640** 2.80618	**1**

The equivalents are given in the bold type. Logarithms of the equivalents are given immediately below with the characteristic of the logarithm (i.e. $\bar{2}.59699 = 0.059699$). In some cases, the equivalents have been rounded off, while the logarithm corresponds to the equivalent carried to a greater number of decimal places. Subscripts after any figure — 0_3, 9_4, etc. — mean that the figure is to be repeated the indicated number of times.

* 1 hectare = 100 acres = 10 000 m^2.

Table A3.12 — Volume and capacity equivalents

litres	cubic inches	cubic feet	cubic yards	US apothecary fluid ounces	US quarts liquid	US quarts dry	US gallons	US bushels
1	**61.03**	**0.03532**	**0.001308**	**33.81**	**1.057**	**0.9081**	**0.2642**	**0.02838**
	1.78551	2.54796	$\bar{3}$.11659	1.52909	0.02394	$\bar{1}$.95812	$\bar{1}$.42188	$\bar{2}$.45297
0.01639	**1**	**$0.0_{3}5787$**	**$0.0_{4}2143$**	**0.5541**	**0.01732**	**0.01488**	**$0.0_{2}4329$**	**$0.0_{3}4650$**
$\bar{2}$.21450		$\bar{4}$.76246	5.33109	$\bar{1}$.74360	$\bar{2}$.23845	$\bar{2}$.17263	3.63639	4.66748
28.32	**1728**	**1**	**0.03704**	**957.5**	**29.92**	**25.71**	**7.481**	**0.8036**
1.45205	3.23754		2.56864	2.98114	1.47599	1.41017	0.87393	1.90502
764.5	**46,656**	**27**	**1**	**25,853**	**807.9**	**694.3**	**202.2**	**21.70**
2.88340	4.66891	1.43136		4.41251	2.90736	2.84153	2.30530	1.33638
0.02957	**1.805**	**0.001044**	**$0.0_{4}3868$**	**1**	**0.03125**	**0.02686**	**0.007812**	**$0.0_{3}8392$**
$\bar{2}$.47091	0.25640	3.01886	$\bar{5}$.58749		$\bar{2}$.49485	$\bar{2}$.42903	3.89279	4.92388
0.9463	**57.75**	**0.03342**	**0.001238**	**32**	**1**	**0.8594**	**0.25**	**0.02686**
$\bar{1}$.97604	1.76155	$\bar{2}$.52401	$\bar{3}$.09264	1.50515		1.93418	1.39794	2.42903
1.101	**67.20**	**0.03889**	**0.001440**	**37.24**	**1.164**	**1**	**0.2909**	**0.03125**
0.04188	1.82737	2.58983	$\bar{3}$.15847	1.57097	0.06582		$\bar{1}$.46376	$\bar{2}$.49485
3.785	**231**	**0.1337**	**0.004951**	**128**	**4**	**3.437**	**1**	**0.1074**
0.57812	2.36361	$\bar{1}$.12607	$\bar{3}$.69470	2.10721	0.60206	0.53624		1.03109
35.24	**2,150**	**1.244**	**0.04609**	**1,192**	**37.24**	**32**	**9.309**	**1**
1.54703	3.33252	0.09498	$\bar{2}$.66362	3.07612	1.57097	1.50515	0.96891	

The equivalents are given in the bold type. Logarithms of the equivalents are given immediately below with the characteristic of the logarithm (i.e. $\bar{2}.54796 = 0.054796$). In some cases, the equivalents have been rounded off, while the logarithm corresponds to the equivalent carried to a greater number of decimal places. Subscripts after any figure — 0_3, 9_4, etc. — mean that the figure is to be repeated the indicated number of times.

Table A3.13 — Conversion of areas*

	sq in to cm^2	cm^2 to sq in	sq ft to m^2	m^2 to sq ft	sq yd to m^2	m^2 to sq yd	acres to hectares	hectares to acres	sq miles to km^2	km^2 to sq miles
1	6.452	1.1550	0.0929	10.76	0.8361	1.196	0.4047	2.471	2.590	0.3861
2	12.90	0.3100	0.1858	21.53	1.672	2.392	0.8094	4.942	5.180	0.7722
3	19.35	0.4650	0.2787	32.29	2.508	3.588	1.214	7.413	7.770	1.158
4	25.81	0.6200	0.3716	43.06	3.345	4.784	1.619	9.884	10.360	1.544
5	32.26	0.7750	0.4645	53.82	4.181	5.980	2.023	12.355	12.950	1.931
6	38.71	0.9300	0.5574	64.58	5.017	7.176	2.428	14.826	15.540	2.317
7	45.16	1.085	0.6503	75.35	5.853	8.372	2.833	17.297	18.130	2.703
8	51.61	1.240	0.7432	86.11	6.689	9.568	3.237	19.768	20.720	3.089
9	58.06	1.395	0.8361	96.87	7.525	10.764	3.642	22.239	23.310	3.475

* For example, 1 sq in = 6.452 cm^2.

Table A3.14 — Conversion of volumes or cubic measures

	cu in to cm^3	cm^3 to cu in	cu ft to m^3	m^2 to cu ft	cu yd to m^3	m^3 to cu yd	gal to cu ft	cu ft to gal
1	16.39	0.06102	0.02832	35.31	0.7646	1.308	0.1337	7.481
2	32.77	0.1220	0.05663	70.63	1.529	2.616	0.2674	14.96
3	49.16	0.1831	0.08495	105.9	2.294	3.924	0.4010	22.44
4	65.55	0.2441	0.1133	141.3	3.058	5.232	0.5347	29.92
5	81.94	0.3051	0.1416	176.6	3.823	6.540	0.6684	37.40
6	98.32	0.3661	0.1699	211.9	4.587	7.848	0.8021	44.88
7	114.7	0.4272	0.1982	247.2	5.352	9.156	0.9358	52.36
8	131.1	0.4882	0.2265	282.5	6.116	10.46	1.069	59.84
9	147.5	0.5492	0.2549	317.8	6.881	11.77	1.203	67.32

Table A3.15 — Conversion of volumes or capacities

	fluid oz to cm^3	cm^3 to fluid oz	liquid pints to litres	litres to liquid pints	liquid quarts to litres	litres to liquid quarts	gal to litres	litres to gal	bushels to hecto-litres	hecto-litres to bushels
1	29.57	0.03381	0.4732	2.113	0.9463	1.057	3.785	0.2642	0.3524	2.838
2	59.15	0.06763	0.9463	4.227	1.893	2.113	7.571	0.5284	0.7048	5.676
3	88.72	0.1014	1.420	6.340	2.839	3.170	11.36	0.7925	1.057	8.513
4	118.3	0.1353	1.893	8.454	3.785	4.227	15.14	1.057	1.410	11.35
5	147.9	0.1691	2.366	10.57	4.732	5.284	18.93	1.321	1.762	14.19
6	177.4	0.2092	2.839	12.68	5.678	6.340	22.71	1.585	1.114	17.03
7	207.0	0.2367	3.312	14.79	6.624	7.397	26.50	1.849	2.467	19.86
8	236.6	0.2705	3.785	16.91	7.571	8.454	30.28	2.113	2.819	22.70
9	266.2	0.3043	4.259	19.02	8.517	9.510	34.07	2.378	3.171	25.54

Table A3.16 — Mass equivalents

kg	grains	oz		lbm		tons		
		troy and apothecary	avoir-dupois	troy and apothecary	avoir-dupois	short	long	metric
1	**15,432**	**32.15**	**35.27**	**2.6792**	**2.205**	**$0.0_{2}1102$**	**$0.0_{3}9842$**	**0.001**
	4.18843	1.50719	1.54745	0.42801	0.34333	3.04230	$\bar{4}.99309$	3.00000
$0.0_{4}6480$	**1**	**$0.0_{2}2083$**	**$0.0_{2}2286$**	**$0.0_{3}1736$**	**$0.0_{3}1429$**	**$0.0_{7}7143$**	**$0.0_{7}6378$**	**$0.0_{7}6480$**
$\bar{5}.81157$		$\bar{3}.31876$	$\bar{3}.35902$	$\bar{4}.23958$	$\bar{4}.15490$	$\bar{8}.85387$	$\bar{8}.80465$	8.81157
0.03110	**480**	**1**	**1.09714**	**0.08333**	**0.06857**	**$0.0_{4}3429$**	**$0.0_{4}3061$**	**$0.0_{4}3110$**
$\bar{2}.49281$	2.68124		0.04026	$\bar{2}.92082$	$\bar{2}83614$	$\bar{5}.53511$	$\bar{5}.48590$	$\bar{5}.49281$
0.02835	**437.5**	**0.9115**	**1**	**0.07595**	**0.0625**	**$0.0_{4}3125$**	**$0.0_{4}2790$**	**$0.0_{4}2835$**
$\bar{2}.45255$	2.64098	$\bar{1}.95974$		$\bar{2}.88056$	$\bar{2}.79588$	$\bar{5}.49485$	$\bar{5}.44563$	$\bar{5}.45255$
0.3732	**5,760**	**12**	**13.17**	**1**	**0.8229**	**$0.0_{3}4114$**	**$0.0_{2}3673$**	**$0.0_{3}3732$**
$\bar{1}.57199$	3.76042	1.07918	1.11944		$\bar{1}.91532$	$\bar{4}.61429$	$\bar{4}.56508$	$\bar{4}.57199$
0.4536	**7,000**	**14.58**	**16**	**1.215**	**1**	**0.0005**	**$0.0_{3}4464$**	**$0.0_{3}4536$**
$\bar{1}.65667$	3.84510	1.16386	1.20412	0.08468		$\bar{4}.69897$	$\bar{4}.64975$	$\bar{4}.65667$
907.2	**140_{6}**	**29,167**	**320_{3}**	**2,431**	**2,000**	**1**	**0.8929**	**0.9072**
2.95770	7.14613	4.46489	4.50515	3.38571	3.30103		$\bar{1}.95078$	$\bar{1}.95770$
1,016	**$15,680_{4}$**	**32,667**	**35,840**	**2,722**	**2,240**	**1.12**	**1**	**1.016**
3.00691	7.19535	4.51411	4.55437	3.43492	3.35025	0.04922		0.00691
1,000	**15,432,356**	**32,151**	**35,274**	**2,679**	**2,205**	**1.102**	**0.9842**	**1**
3.00000	7.18843	4.50719	4.54745	3.42801	3.34333	0.04230	1.99309	

The equivalents are given in the bold type. Logarithms of the equivalents are given immediately below with the characteristic of the logarithm (i.e. $\bar{3}.04230 = 0.0004230$). In some cases, the equivalents have been rounded off, while the logarithm corresponds to the equivalent carried to a greater number of decimal places. Subscripts after any figure — 0_3, 9_4, etc. — mean that the figure is to be repeated the indicated number of times.

Table A3.17 — Conversion of masses

	lbm (avoir-dupois) to kg	kg to lbm (avoir-dupois)	grains to grams	grams to grains	oz (avoir-dupois) to grams	grams to oz (avoir-dupois)	short tons (2,000 lbm) to metric tons	metric tons (1,000 kg) to short tons	long tons (2,240 lbm) to metric tons	metric tons to long tons
1	0.4536	2.205	0.06480	15.43	28.35	0.03527	0.907	1.102	1.016	0.984
2	0.9072	4.409	0.1296	30.86	56.70	0.07055	1.814	2.205	2.032	1.968
3	1.361	6.614	0.1944	46.30	85.05	0.1058	2.722	3.307	3.084	2.953
4	1.814	8.818	0.2592	61.73	113.40	0.1411	3.629	4.409	4.064	3.937
5	2.268	11.02	0.3240	77.16	141.75	0.1764	4.536	5.512	5.080	4.921
6	2.722	13.23	0.3888	92.59	170.10	0.2116	5.443	6.614	6.096	5.905
7	3.175	15.43	0.4536	108.03	198.45	0.2469	6.350	7.716	7.112	6.889
8	3.629	17.64	0.5184	123.46	226.80	0.2822	7.257	8.818	8.128	7.874
9	4.082	19.84	0.5832	138.89	255.15	0.3175	8.165	9.921	9.144	8.858

Table A3.18 — Velocity equivalents

m/s	m/min	km/h	cm/s	ft/s	ft/min	mph	knots
1	**60**	**3.6**	**100**	**3.281**	**196.85**	**2.237**	**1.943**
	1.77815	0.55630	2.00000	0.51598	2.29414	0.34965	0.28836
0.01667	**1**	**0.06**	**1.667**	**0.05488**	**3.281**	**0.03728**	**0.03238**
$\bar{2}$.22184		$\bar{2}$.77815	0.22184	2.73783	0.51598	$\bar{2}$.57105	$\bar{2}$.51022
0.2778	**16.67**	**1**	**27.78**	**0.9113**	**54.68**	**0.6214**	**0.53959**
$\bar{1}$.44370	1.22184		1.44370	$\bar{1}$.95908	1.73783	$\bar{1}$.79335	1.73205
0.01	**0.6**	**0.036**	**1**	**0.03281**	**1.9685**	**0.02237**	**0.01943**
	1.77815	2.55630		$\bar{2}$.51598	0.29414	$\bar{2}$.34965	$\bar{2}$.28836
0.3048	**18.29**	**1.097**	**30.48**	**1**	**60**	**0.6818**	**0.59209**
$\bar{1}$.48402	1.26217	0.04032	1.48402		1.77815	$\bar{1}$.83367	$\bar{1}$.77238
0.005080	**0.3048**	**0.01829**	**0.5080**	**0.01667**	**1**	**0.01136**	**0.00987**
3.70586	1.48402	2.26217	1.70586	2.22185		$\bar{2}$.05553	$\bar{3}$.99423
0.4470	**26.82**	**1.609**	**44.70**	**1.467**	**88**	**1**	**0.86839**
1.65035	1.42850	0.20670	1.65035	0.16633	1.94448		1.93871
0.51479	**30.887**	**1.8532**	**51.479**	**1.68894**	**101.337**	**1.15155**	**1**
$\bar{1}$.71163	1.48978	0.26793	1.71163	0.22761	2.00577	0.06128	

The equivalents are given in the bold type. Logarithms of the equivalents are given immediately below with the characteristic of the logarithm (i.e. $\bar{2}$.59699 = 0.059699). In some cases, the equivalents have been rounded off, while the logarithm corresponds to the equivalent carried to a greater number of decimal places.

Table A3.19 — Conversion of linear and angular velocities

	rev/min to rads/s	rads/s to rev/min	cm/s to ft/min	ft/min to cm/s	cm/s to mph	mph to cm/s	ft/s to mph	mph to ft/s
1	0.1047	9.55	1.97	0.508	0.0224	44.70	0.682	1.47
2	0.2094	19.10	3.94	1.016	0.0447	89.41	1.364	2.93
3	0.3142	28.65	5.91	1.524	0.0671	134.1	2.045	4.40
4	0.4189	38.20	7.87	2.032	0.0895	178.8	2.727	5.87
5	0.5236	47.75	9.84	2.540	0.1118	223.5	3.409	7.33
6	0.6283	57.30	11.81	3.048	0.1342	288.2	4.091	8.80
7	0.7330	66.84	13.78	3.556	0.1566	312.9	4.773	10.27
8	0.8378	76.39	15.75	4.064	0.1790	357.6	5.455	11.73
9	0.9425	85.94	17.72	4.572	0.2013	402.3	6.136	13.20

Table A3.20 — Conversion of pressures*

	lbm per sq in to kilopascals	kilopascals to lbm per sq in	atmospheres to lbm per sq in	lbm per sq in to atmospheres	atmospheres to kilopascals	kilopascals to atmospheres
1	6.8948	0.1450	14.70	0.0680	101.3	0.00987
2	13.789	0.2901	29.39	0.1361	202.6	0.01974
3	20.684	0.4351	44.09	0.2041	304.0	0.02961
4	27.579	0.5802	58.78	0.2722	405.3	0.03947
5	34.474	0.7252	73.48	0.3402	506.6	0.04935
6	41.369	0.8702	88.18	0.4083	608.0	0.05922
7	48.263	1.0153	102.9	0.4763	709.3	0.06908
8	55.158	1.1603	117.6	0.5444	810.6	0.07895
9	62.053	1.3053	132.3	0.6124	911.9	0.08882

*For example, 1 lbm/sq in = 6.8948 kilopascals.

Table A3.21 — Pressure equivalents

kilo-pascals	kg/cm^2 (metric atmos-pheres)	lbm/sq in	short tons per sq ft	atmos-spheres	columns of mercury at temperature 0°C and g=9.80665 m/s^2		columns of water at temperature 15°C and g=9.80665 m/s^2		
					(m)	(in)	(m)	(in)	(ft)
1	**0.010197**	**0.14504**	**0.010443**	**0.009869**	**0.007501**	**0.2953**	**0.1021**	**4.018**	**0.3349**
	$\bar{2}$.00848	1.16148	2.01882	$\bar{3}$.87510	$\bar{3}$.99427	$\bar{1}$.47025	$\bar{1}$.00886	0.60402	$\bar{1}$.52485
98.066	**1**	**14.22**	**1.024**	**0.9678**	**0.7356**	**28.96**	**10.01**	**394.1**	**32.84**
1.99152		1.15300	0.01034	$\bar{1}$.98579	1.86662	1.46177	1.00038	2.59556	1.51636
6.8948	**0.07031**	**1**	**0.072**	**0.06805**	**0.05171**	**2.036**	**0.7037**	**27.70**	**2.309**
0.83852	2.84700		2.85733	2.83280	2.71360	0.30876	$\bar{1}$.84738	1.44254	0.36336
95.760	**0.97765**	**13.89**	**1**	**0.9451**	**0.7183**	**28.28**	**9.774**	**384.8**	**32.07**
1.98119	1.98966	1.14267		1.97547	1.85628	1.45143	0.99006	2.58521	1.50604
101.325	**1.0332**	**14.70**	**1.058**	**1**	**0.76**	**29.92**	**10.34**	**407.1**	**33.93**
2.00573	0.01420	1.16722	0.02453		$\bar{1}$.88081	1.47598	1.01459	2.60975	1.53058
133.322	**1.3595**	**19.34**	**1.392**	**1.316**	**1**	**39.37**	**13.61**	**535.7**	**44.64**
2.12490	0.13338	1.28640	0.14373	0.11919		1.59517	1.13378	2.72894	1.64976
3.386	**0.03453**	**0.4912**	**0.03536**	**0.03342**	**0.02540**	**1**	**0.3456**	**13.61**	**1.134**
0.52975	$\bar{2}$.53823	$\bar{1}$.69124	$\bar{2}$.51857	$\bar{2}$.52402	$\bar{2}$.40484		$\bar{1}$.53861	1.13378	0.05460
9.798	**0.09991**	**1.421**	**0.1023**	**0.09670**	**0.7349**	**2.893**	**1**	**39.37**	**3.281**
0.99114	$\bar{2}$.99962	0.15262	1.00996	$\bar{2}$.98541	$\bar{2}$.86622	0.46139		1.59517	0.51598
0.2489	**0.002538**	**0.03609**	**0.002599**	**0.002456**	**0.001867**	**0.07349**	**0.02540**	**1**	**0.08333**
$\bar{1}$.39598	3.40446	$\bar{2}$.55745	$\bar{3}$.41479	$\bar{3}$.39024	$\bar{3}$.27106	$\bar{2}$.86622	$\bar{2}$.40484		$\bar{2}$.92082
2.926	**0.03045**	**0.4331**	**0.03119**	**0.02947**	**0.02240**	**0.8819**	**0.3048**	**12**	**1**
0.47516	2.48364	1.63663	2.49397	2.46942	2.35024	1.94540	1.48402	1.07918	

The equivalents are given in bold type. Logarithms of the equivalents are given immediately below with the characteristic of the logarithm (i.e. $\bar{1}$.00848 = 0.00848). In some cases, the equivalents have been rounded off, although the logarithm corresponds to the equivalent carried to a greater number of decimal places. Subscripts after any figure — 0_3, 9_4, etc. — mean that the figure is to be repeated the indicated number of times.

Table A3.22 — Energy or work equivalents

J	kg-m	ft-bf	kW hours	metric hp hours	hp hours	L-atm	kilo-calories	Btus
1	**0.10197**	**0.7376**	**$0.0_{6}2778$**	**$0.0_{6}3777$**	**$0.0_{6}3725$**	**0.009869**	**$0.0_{3}2388$**	**$0.0_{3}9478$**
	1.00848	1.86780	7.44370	7.57711	7.57113	3.99427	4.37809	4.97670
9.80665	**1**	**7.233**	**$0.0_{5}2724$**	**$0.0_{5}37037$**	**$0.0_{5}3653$**	**0.09678**	**0.002342**	**0.009295**
0.9915207		0.85932	6.43521	6.56863	6.56265	2.98579	3.36961	3.96825
1.356	**0.1383**	**1**	**$0.0_{6}3766$**	**$0.0_{6}51206$**	**$0.0_{6}50505$**	**0.01338**	**$0.0_{3}3238$**	**0.001285**
0.13220	1.14068		7.57590	7.70932	7.70333	2.12647	4.51029	3.10890
3.600×10^{5}	**3.671×10^{5}**	**2.6555×10^{6}**	**1**	**1.3596**	**1.341**	**35.528**	**859.9**	**3,412**
6.55630	5.56478	6.42410		0.13342	0.12743	4.55057	2.93443	3.53303
2.648×10^{6}	**270,000**	**1.9529×10^{6}**	**0.7355**	**1**	**0.9863**	**26.131**	**632.4**	**2,510**
6.42288	5.43136	6.29068	1.86658		1.99401	4.41715	2.80098	3.39961
2.6845×10^{8}	**2.7375×10^{5}**	**1.98×10^{6}**	**0.7457**	**1.0139**	**1**	**26.493**	**641.2**	**2,544**
6.42887	5.43735	6.29667	1.87356	0.00598		4.42314	2.80699	3.40557
101.33	**10.333**	**74.74**	**$0.0_{4}2815$**	**$0.0_{4}3827$**	**$0.0_{4}3775$**	**1**	**0.02420**	**0.09604**
2.00573	1.01421	1.87353	5.44952	5.58284	5.57686		2.38282	2.98246
4,187	**426.9**	**3,088**	**0.001163**	**0.001581**	**0.001560**	**41.32**	**1**	**3.968**
3.62191	2.63036	3.48971	3.06558	3.19902	3.19304	1.61618		0.59861
1,055	**107.6**	**778.2**	**$0.0_{3}2931$**	**$0.0_{3}3985$**	**$0.0_{3}3930$**	**10.41**	**0.25200**	**1**
3.02360	2.03178	2.89110	4.46697	4.60042	4.59444	1.01757	1.40139	

The equivalents are given in bold type. Logarithms of the equivalents are given immediately below with the characteristic of the logarithm (i.e. $\bar{1}.00848 = 0.00848$). In some cases, the equivalents have been rounded off, although the logarithm corresponds to the equivalent carried to a greater number of decimal places. Subscripts after any figure — 0_3, 9_4, etc. — mean that the figure is to be repeated the indicated number of times.

Table A3.23 — Conversion of energy, work, heat

	calories to J	J to calories	ft-bf to kg-m	kg-m to ft-bf	ft-bf to Btu	Btu to ft-bf	kg-m to kilocalories	kilo-calories to kg-m
1	4.187	0.2388	0.1383	7.233	0.001285	778.2	0.002342	426.9
2	8.374	0.4777	0.2765	14.47	0.002570	1,556	0.004685	853.9
3	12.56	0.7165	0.4148	21.70	0.003855	2,334	0.007027	1,281
4	16.75	0.9554	0.5530	28.93	0.005140	3,113	0.009369	1,708
5	20.93	1.194	0.6913	36.16	0.006425	3,891	0.01172	2,135
6	25.12	1.433	0.8295	43.40	0.007710	4,669	0.01405	2,562
7	29.31	1.672	0.9678	50.63	0.008995	5,447	0.01640	2,989
8	33.49	1.911	1.106	57.86	0.01028	6,225	0.01874	3,415
9	37.68	2.150	1.244	65.10	0.01156	7,003	0.02108	3,842

Table A3.24 — Power equivalents*

kW	hp	metric hp	poncelets	kg-m/s	ft-bf/s	kilocalories/s	Btu/s
1	**1.341** 0.12743	**1.360** 0.13343	**1.020** 0.00848	**102.0** 2.00848	**737.6** 2.86780	**0.2388** 1.37813	**0.9478** $\bar{1}$.97673
0.7457 1.87256	**1**	**1.014** 0.00599	**0.7604** 1.88105	**76.04** 1.88105	**550** 2.74036	**0.1781** 1.25066	**0.7068** 1.84936
0.7355 1.86658	**0.9863** 1.99402	**1**	**0.75** 1.87506	**75** 1.87506	**542.5** 2.73438	**0.1757** $\bar{1}$.24467	**0.6971** $\bar{1}$.84328
0.9807 1.99152	**1.315** 0.11896	**1.333** 0.12493	**1**	**100** 2.00000	**723.3** 2.85932	**0.2342** 1.36961	**0.9295** 1.96825
0.009807 3.99152	**0.01315** 2.11896	**0.01333** 2.12493	**0.01** 2.00000	**1**	**7.233** 0.85932	**0.002342** $\bar{3}$.36961	**0.009295** 3.96825
0.001356 3.13220	**0.00182** $\bar{3}$.25946	**0.00184** 3.26562	**0.00138** $\bar{3}$.14067	**0.1383** $\bar{1}$.14067	**1**	**$0.0_3 3238$** 4.51029	**0.001285** $\bar{3}$.10890
4.187 0.62187	**5.615** 0.74934	**5.692** 0.75530	**4.269** 0.63036	**426.9** 2.63036	**3,088** 3.48971	**1**	**3.968** 0.59861
1.055 0.02320	**1.415** 0.15074	**1.434** 0.15668	**1.076** 0.03178	**107.6** 2.03178	**778.2** 2.89110	**0.2520** $\bar{1}$.40138	**1**

*For example, 1 hp = 0.7457 kW.
The equivalents are given in bold type. Logarithms of the equivalents are given immediately below with the characteristic of the logarithm (i.e. $\bar{1}$.00848 = 0.00848). In some cases, the equivalents have been rounded off, although the logarithm corresponds to the equivalent carried to a greater number of decimal places. Subscripts after any figure — 0_3, 9_4, etc. — mean that the figure is to be repeated the indicated number of times.

Table A3.25 — Conversion of power

	hp to kW	kW to hp	metric hp to kW	kW to metric hp	hp to metric hp	metric hp to hp
1	0.7457	1.341	0.7355	1.360	1.014	0.9863
2	1.491	2.682	1.471	2.719	2.028	1.973
3	2.237	4.023	2.206	4.079	3.042	2.959
4	2.983	5.364	2.942	5.438	4.055	3.945
5	3.729	6.705	3.677	6.798	5.069	4.932
6	4.474	8.046	4.412	8.158	6.083	5.918
7	5.220	9.387	5.147	9.520	7.097	6.904
8	5.966	10.73	5.883	10.88	8.111	7.891
9	6.711	12.07	6.618	12.24	9.125	8.877

Table A3.26 — Density equivalents and conversion factors*

Equivalents						Conversion factors			
kg/m^3	g/cm^3	lbm/cu in	lbm/cu ft	lbm/US gal		lbm/cu ft to kg/m^3	kg/m^3 to lbm/cu ft	g/cm^3 to kg/m^3	kg/m^3 to g/cm^3
1	**0.001**	**$0.0_4 3613$**	**0.0624**	**0.00835**	1	**16.019**	**0.0624**	**1,000**	**0.001**
		$\bar{5}.55787$	$\bar{2}.79539$	$\bar{3}.92143$	2	32.037	0.1249	2,000	0.002
1000	**1**	**0.03613**	**62.43**	**8.345**	3	**48.055**	**0.1873**	**3,000**	**0.003**
		$\bar{2}.55787$	1.79539	0.92143	4	64.074	0.2497	4,000	0.004
27,680	**27.68**	**1**	**1,728**	**231**	5	**80.092**	**0.3121**	**5,000**	**0.005**
4.44217	1.44217		3.23754	2.36361	6	96.111	0.3746	6,000	0.006
16.019	**0.01602**	**$0.0_3 5787$**	**1**	**0.1337**	7	**112.129**	**0.4370**	**7,000**	**0.007**
1.20466	$\bar{2}.20466$	$\bar{4}.76245$		$\bar{1}.12613$	8	128.148	0.4994	8,000	0.008
119.83	**0.1198**	**0.004329**	**7.481**	**1**	9	**144.166**	**0.5619**	**9,000**	**0.009**
2.07855	$\bar{1}.07855$	$\bar{3}.63639$	0.87396		10	160.185	0.6243	10,000	0.010

* For example, 1 $kg/m^3 = 0.0624$ lbm/cu ft.
The equivalents are given in bold type. Logarithms of the equivalents are given immediately below with the characteristic of the logarithm (i.e. $\bar{1}.00848 = 0.00848$). In some cases, the equivalents have been rounded off, although the logarithm corresponds to the equivalent carried to a greater number of decimal places. Subscripts after any figure — 0_3, 9_4, etc. — mean that the figure is to be repeated the indicated number of times.

Table A3.27 — The gas constant

8.314 m^3·Pa/mol·K
0.08314 litre·bar/mol·K
0.08206 litre·atm/mol·K
62.36 litre·mm Hg/mol·K
0.7302 ft^3·atm/lb·mole·°R
10.73 ft^3·psia/lb·mole·°R
8.314 Jmol·K
1.987 cal/mol·K
1.987 Btu/lb-mole·°R

Index